INTELLIGENT FAULT DIAGNOSIS AND REMAINING USEFUL LIFE PREDICTION OF ROTATING MACHINERY

INTELLIGENT FAULT DIAGNOSIS AND REMAINING USEFUL LIFE PREDICTION OF ROTATING MACHINERY

YAGUO LEI

西安交通大学出版社
XI'AN JIAOTONG UNIVERSITY PRESS

ELSEVIER

AMSTERDAM • BOSTON • HEIDELBERG • LONDON • NEW YORK • OXFORD
PARIS • SAN DIEGO • SAN FRANCISCO • SINGAPORE • SYDNEY • TOKYO
Butterworth-Heinemann is an imprint of Elsevier

BH

Butterworth-Heinemann is an imprint of Elsevier
The Boulevard, Langford Lane, Kidlington, Oxford OX5 1GB, United Kingdom
50 Hampshire Street, 5th Floor, Cambridge, MA 02139, United States

Notices

Knowledge and best practice in this field are constantly changing. As new research and experience broaden our understanding, changes in research methods, professional practices, or medical treatment may become necessary.

Practitioners and researchers must always rely on their own experience and knowledge in evaluating and using any information, methods, compounds, or experiments described herein. In using such information or methods they should be mindful of their own safety and the safety of others, including parties for whom they have a professional responsibility.

To the fullest extent of the law, neither the Publisher nor the authors, contributors, or editors, assume any liability for any injury and/or damage to persons or property as a matter of products liability, negligence or otherwise, or from any use or operation of any methods, products, instructions, or ideas contained in the material herein.

Library of Congress Cataloging-in-Publication Data
A catalog record for this book is available from the Library of Congress

British Library Cataloguing-in-Publication Data
A catalogue record for this book is available from the British Library

ISBN: 978-0-12-811534-3

For information on all Butterworth-Heinemann publications visit our website at https://www.elsevier.com/

Publisher: Joe Hayton
Acquisition Editor: Simon Tian
Editorial Project Manager: Vivi Li
Production Project Manager: Julie-Ann Stansfield
Designer: Maria Inês Cruz

Typeset by Thomson Digital

CONTENTS

PREFACE

Rotating machinery is commonly used and plays an important role in industrial applications. With rapid development of science and technology, rotating machinery in modern industry is growing larger and more precise. It has become a challenging task to enhance the availability, the reliability, and the security of rotating machinery. Prognostics and Health Management (PHM) is an effective tool for dealing with this task. Therefore, it has attracted considerable attention during the last few decades.

PHM of rotating machinery is composed of several modules, like signal acquisition, signal processing, diagnostics, prognostics, and maintenance decision. Some books have introduced many fault diagnostics and prognostics methods for rotating machinery. However, the earliest research achievements in the area of fault diagnostics and prognostics are not addressed, which is the motivation for writing this book. This book will provide an introduction of intelligent fault diagnostics and RUL prediction based on the current achievements appeared in academic journals, conference proceedings, and technical reports.

The book involves the fundamental theories and the advanced methods of intelligent fault diagnostics and RUL prediction for rotating machinery. These methods are paralleled by experimental investigations and real applications for rotors, rolling element bearings, and gears. It is able to provide a guide for the readers from the area of PHM to know the basic concepts, the fundamental theories, and the cutting edge research. It can be used as a text for master courses at both a fundamental and more advanced level. It also benefits engineers as well as researchers in the area of PHM of rotating machinery.

Most of the fault diagnostic and RUL prediction methods in this book were developed at Xi'an Jiaotong University (XJTU), China. So, I am grateful to my colleagues, Professor Jing Lin, Professor Yanyang Zi, Professor Xuefeng Chen, Professor Zhousuo Zhang, Professor Bing Li, etc., at XJTU, who have assisted me to conduct many theoretical and practical research projects about the subject of this book. I especially appreciate the guidance and the support of my PhD supervisor, Professor Zhengjia He, who died 3 years ago. I cherish the memory of the time when I was still his PhD student working with him together. The longer the time he leaves me, the more I miss him. This book is the best souvenir I can present to Professor He in the third year of his passing away.

Prior to joining XJTU, I worked at the University of Alberta, Canada, as a postdoctoral research fellow. In addition, I also worked at the University of Duisburg-Essen, Germany, as an Alexander von Humboldt fellow. Therefore, I have enjoyed a considerable amount of interaction and collaboration with overseas universities, which include the University of Alberta, University of Duisburg-Essen, Warsaw University of Technology, Carleton University, etc. Without the support and help from these overseas universities, I think that I could not finish writing this book.

Over many years I have received research support from the National Natural Science Foundation of China, Ministry of Education of China, Central Organization Department of China, and Provincial Natural Science Foundation research project of Shaanxi. The research supported by such grants has enriched the materials in this book. I have also received support from industrial partners, in particular Huadian Electric Power Research Institute, CRRC Zhuzhou Institute Co., LTD, Sinopec Jinan Company of China, etc. I acknowledge with thanks the permission given by these industrial partners to reuse the materials in this book. I also thank the Case Western Reserve University and the Institute of FEMTO-ST for sharing their bearing data generously.

Moreover, my PhD students at XJTU, those who worked with intelligent fault diagnosis and RUL prediction topics, supported and helped me to prepare materials and complete this book. I would like to acknowledge specially the help of Mr. Naipeng Li, Mr. Feng Jia, and Mr. Zijian Qiao.

Finally, I would like to thank the editors from Elsevier for their excellent cooperation.

Yaguo Lei

Xi'an, January, 2016

ABOUT THE AUTHOR

Yaguo Lei is a Full Professor in the School of Mechanical Engineering at Xi'an Jiaotong University (XJTU), China, which he joined as an Associate Professor in 2010. Prior to that, he worked at the University of Alberta, Canada, as a postdoctoral research fellow. He also worked at the University of Duisburg-Essen, Germany, as an Alexander von Humboldt fellow in 2012. He was promoted to Full Professor in 2013. He received the BS and PhD degree, both in Mechanical Engineering from XJTU, in 2002 and 2007, respectively. He is an associate editor or a member of the editorial boards of more than 10 journals, including Mechanical Systems and Signal Processing, Measurement Science & Technology, and Neural Computing & Applications. He is also a fellow of the Institution of Engineering and Technology (IET), a senior member of IEEE and a member of ASME, respectively. He has pioneered many signal processing techniques, intelligent fault diagnosis methods, and remaining useful life prediction models for machines. He has published one monograph and more than 100 peer-reviewed papers on signal processing, fault diagnosis, and remaining useful life prediction.

1

INTRODUCTION AND BACKGROUND

1.1 Introduction

Rotating machinery is commonly used in mechanical systems and plays an important role in industrial applications (Lei et al., 2013). It generally operates under tough working environment and is therefore frequently subject to faults. Any fault of the rotating machinery possibly causes a breakdown of the entire mechanical system, which may reduce the reliability, security, and availability of the machinery. With the rapid development of science and technology, rotating machinery in modern industry is growing larger and more precise. The structure of rotating machinery is becoming more complex. As a result, its potential faults become more difficult to be detected. Therefore, how to maintain the normal operation of rotating machinery has attracted considerable attention in recent years.

There are various kinds of rotating machinery in various industry fields, such as the aeroengine in the field of aerospace, the gas turbine and wind turbine in the field of energy, and the automobile transmission in the field of traffic. Even though the rotating machinery is diversified, it generally includes some common essential rotating parts, such as rotors, rolling element bearings, and gears.

A rotor is defined as a rotating part of a machine that is generally supported by bearings. They are the indispensable components in rotating machinery. With the increasing requirement of reliability and precision of rotating machinery, rotors become more flexible and operate under tight clearances and harsh environment. Under such circumstances, one incipient fault possibly causes severe damages in other components and results in failures of the entire machine. For example, a little unbalance of a rotor might cause rub or serious impact between the rotating parts and the stationary parts under a high-speed condition. Severe thermal and mechanical stresses might lead to a fatigue crack in the rotors. The common fault types of rotors include mass unbalance, bent, misalignment, rub, resonance, and so on.

A rolling element bearing is a component that carries loads by placing rolling elements (such as balls or rollers) between two races. The relative motion of the rings causes the rolling elements to roll with little rolling resistance and little sliding. The rolling element bearing is developed from an ancient transportation strategy where sets of logs are laid on the ground with a large stone block on the top. As the stone is pulled, the logs roll along the ground with little sliding friction. A rolling element bearing generally includes three components: an outer race, an inner race, and several rolling elements. Rolling elements, such as balls or rollers, are able to reduce the friction forces between the contacting elements. A rolling element bearing is generally used to connect a shaft and a much larger hole with rollers tightly filling the space between the shaft and hole. As the shaft turns, each roller acts as the logs in the above example. Since the special position and function, rolling element bearings generally suffer from various attacks, such as improper mounting, mishandling, poor lubrication, entry of foreign matter, and abnormal heat generation. All of these may cause different types of faults on rolling element bearings. The common fault types include flaking, spalling, peeling, abrasion, scoring, corrosion, pitting, crack, material failure, and so on.

A gear is a rotating machine part having cut teeth, which meshes with another toothed part to transmit torque. Two or more gears working in a sequence (train) are called a gear train or, in many cases, a transmission. Such gear arrangements are able to produce a mechanical advantage through a gear ratio. Geared devices can change the speed, torque, and direction of a power source. Although, the most common situation is that a gear meshes with another gear. A gear meshes with a nonrotating toothed part, called a rack, thereby producing translation instead of rotation. The gears in a transmission are similar to the wheels in a crossed belt pulley system. An advantage of gears is that the teeth of a gear can prevent slippage. When two gears mesh, and one gear is bigger than the other, a mechanical advantage is produced, with the rotational speeds and the torques of the two gears differing in an inverse relationship. On account of the characteristics of the gears, they are widely used to transmit torque and rotation in mechanical systems.

Rotating machinery plays an important role in the industry applications because of its specific functions for mechanical systems. However, due to the specific function requirement, rotating machinery generally operates under tough working environment. Consequently, it always has a higher fault rate compared with other components, and most maintenance costs are directly or indirectly caused by the fault of rotating machinery. Here, we take the wind turbine system, for example, and give some reports about its fault rates and maintenance costs.

The worldwide wind markets have been dramatically developed in recent years because of its economic advantages and environmental protection compared with other sources of electricity. According to the half-year report in 2014 published by the World Wind Energy Association (WWEA) (The World Wind Energy Association, 2014), the total installed capability of the worldwide wind turbine presents a stable increasing trend from 2011 to 2014 as shown in Fig. 1.1. The worldwide wind turbine capacity reached 336,327 MW by the end of June 2014. The total worldwide installed wind turbine capacity by mid-2014 has generated around 4% of the world's electricity demand.

A major issue with the wind turbine system is the relatively high costs of operation and maintenance (OM). Wind turbines are hard-to-access structures, and they are often located in remote areas. These factors increase the OM costs for wind turbine systems. In addition, poor reliability directly reduces the availability of wind power due to the turbine downtime. For a turbine with over 20 years of operating life, the OM costs are estimated to be 10–15% of the total income for a wind farm, and the OM costs for offshore wind turbine are estimated to be 20–25% of the total income (Lu et al., 2009). The main fault types of the wind turbine are shown in Fig. 1.2, including imbalance, wear, fatigue, and impending cracks in rotor blades, bearings, shafts, the gearbox, the generator, the yaw, and the pitch angle mechanism. A study result for the causes of failures of wind turbines is shown in Fig. 1.3. It is seen that, the gearbox and generator are responsible for 17% of failures and 30% of the maintenance costs, which are clearly leading candidates for the causes of failures. Even in the wind turbine generators, most

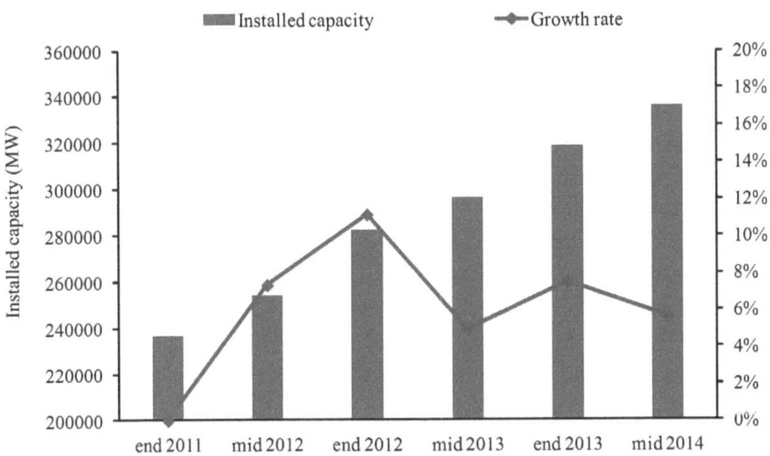

Figure 1.1. Total installed capacity of the worldwide wind turbine.

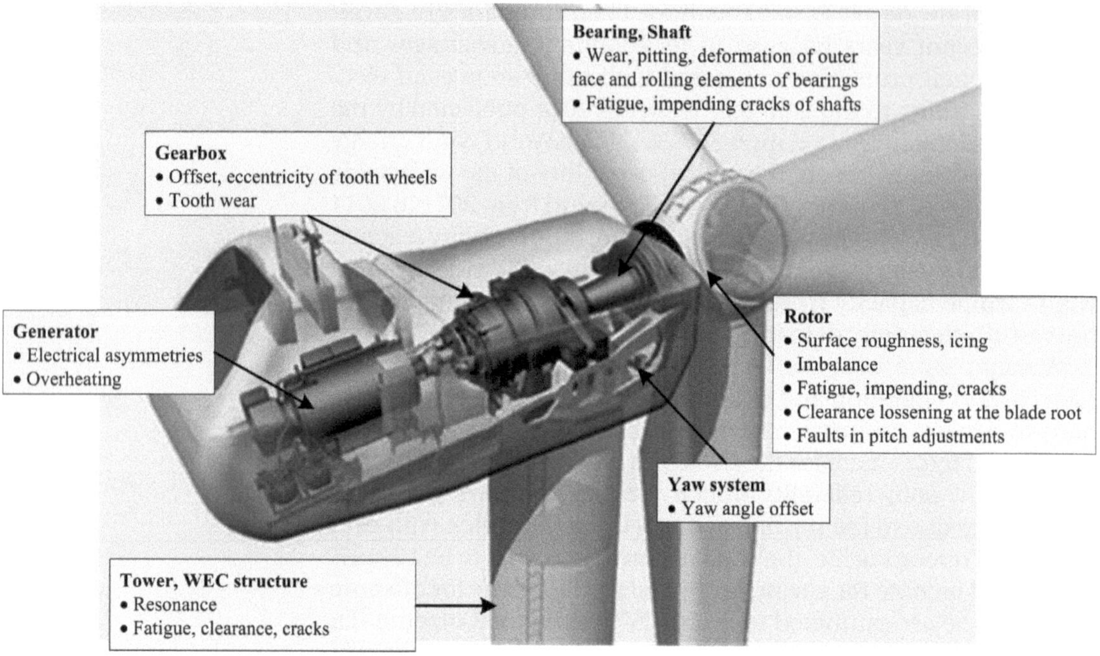

Figure 1.2. Overview of the main faults of wind turbines.

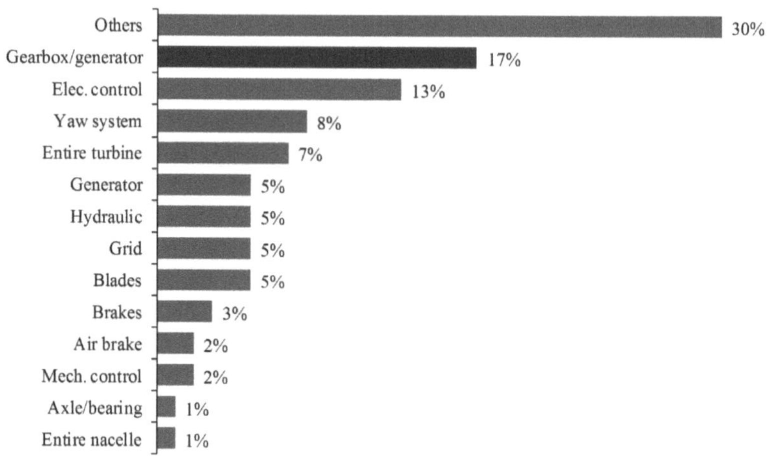

Figure 1.3. Statistic results for the causes of failure of wind turbines.

faults are still caused by the rotating machinery. For induction machines, about 40% failures are related to bearings, 38% to the stator and 10% to the rotor (Hyers et al., 2006).

It is concluded from the above reports that the fault of rotating machinery is the main cause of the failure of wind turbines. This trend is similar in many other mechanical systems. So the health management of rotating machinery is significant for reducing the OM costs of mechanical systems. Analysis of maintenance costs has shown that a repair made after failure always wastes lots of costs compared with the same maintenance when the failure has not occurred. A survey carried out by major organizations has revealed that with an investment of 10,000 – 20,000 dollars in health management, one can save up to 500,000 dollars annually (Saranga and Knezevic, 2001). Thus, it is apparent that an appropriate health management strategy is essential to sustain the inherent reliability of mechanical systems and reduce the OM costs.

The health management strategies of rotating machinery experienced the following three development stages (Randall, 2011) (Fig. 1.4).

1. Reactive maintenance (run-to-failure)
   This is a traditional maintenance strategy where machines keep running until they break down because of the final failure. This strategy in principle gives the longest operation time before failures. However, it only provides a passive reaction after a failure occurs, which may be catastrophic and result in severe damages or accidents. This strategy is acceptable in some cases where large numbers of machines are alternative even one machine has failed and the failure is not catastrophic. However, for machines like wind-turbine generator sets, steam-turbine generator sets, and heavy oil catalytic cracking units, the number of available machines is limited or the failure of a machine possibly leads to a great cost, making this strategy helpless to prevent the shutdown of the machines. Furthermore, if a failure would cause severe damages or catastrophes, this strategy is also invaluable.
2. Preventative maintenance (time-based)
   To prevent the happening of failures, machines are repeatedly checked at regular time intervals in the preventative maintenance, which is more conservative than reactive maintenance.

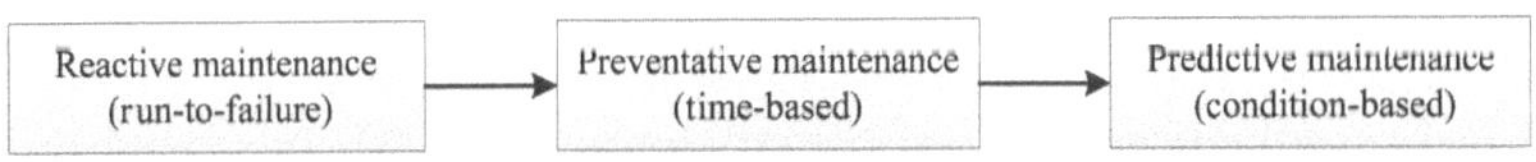

Figure 1.4. Development stages of the health maintenance strategies.

Once an incipient fault is detected, it is common to shut down the machine immediately and replace the fault components to avoid catastrophic consequences. An appropriate time interval for checking plays an important role in this strategy. If the time interval is too long, some unforeseen failures are still possible to occur. In contrast, if the time interval is too short, too much maintenance will be carried out and excessive replacement components will be consumed. This strategy is supposed to recognize the fault occurrence in time and plan the maintenance strategy in advance to reduce the catastrophic failure. However, it needs to shut down the machines frequently, which inevitably reduces the production efficiency and consumes a lot of maintenance costs. The preventative maintenance is appropriate when the failure time is reasonably accurately predicted according to the statistical properties of large numbers of the similar machines. However, for most rotating machines, such as rolling element bearings, there is a large statistical spread around the mean, leading to the estimates given by large numbers of machines severely deviating from the actual failure time of a single machine.

3. Predictive maintenance (condition-based)
   Predictive maintenance, also called condition-based maintenance (CBM), is a maintenance strategy that recommends maintenance actions based on the information collected through condition monitoring. In this strategy, the degradation trend of the rotating machinery is first revealed through the analysis of condition monitoring data. Then the degradation trend in the future is predicted using some prediction models or techniques. With a prespecified failure threshold, the remaining useful life (RUL) of the rotating machinery is predicted. Based on the predicted RUL, an optimal maintenance strategy is scheduled before the real occurrence of the final failure. The predictive maintenance has obvious advantages compared with either reactive maintenance or preventative maintenance. It is able to predict the potential breakdown of a machine through regular condition monitoring thus preventing the happening of catastrophes. In addition, it schedules maintenance at an optimum time according to the predicted RUL. Therefore, it is able to make the machine have a maximum uptime with minimum maintenance costs. Thanks to the superiority of the predictive maintenance, it has attracted substantial attention in recent years.
   Based on the predictive maintenance strategy, a new concept of prognostics and health management (PHM) has been developed in recent years. This book aims to provide an essential

guide to the PHM of rotating machinery, from the basic concepts and the fundamental theories to the latest techniques and their applications.

1.2 Overview of PHM

The flowchart of the PHM of rotating machinery is shown in Fig. 1.5. It is generally composed of five major processes, that is, data acquisition, signal processing, diagnostics, prognostics, and maintenance decision.

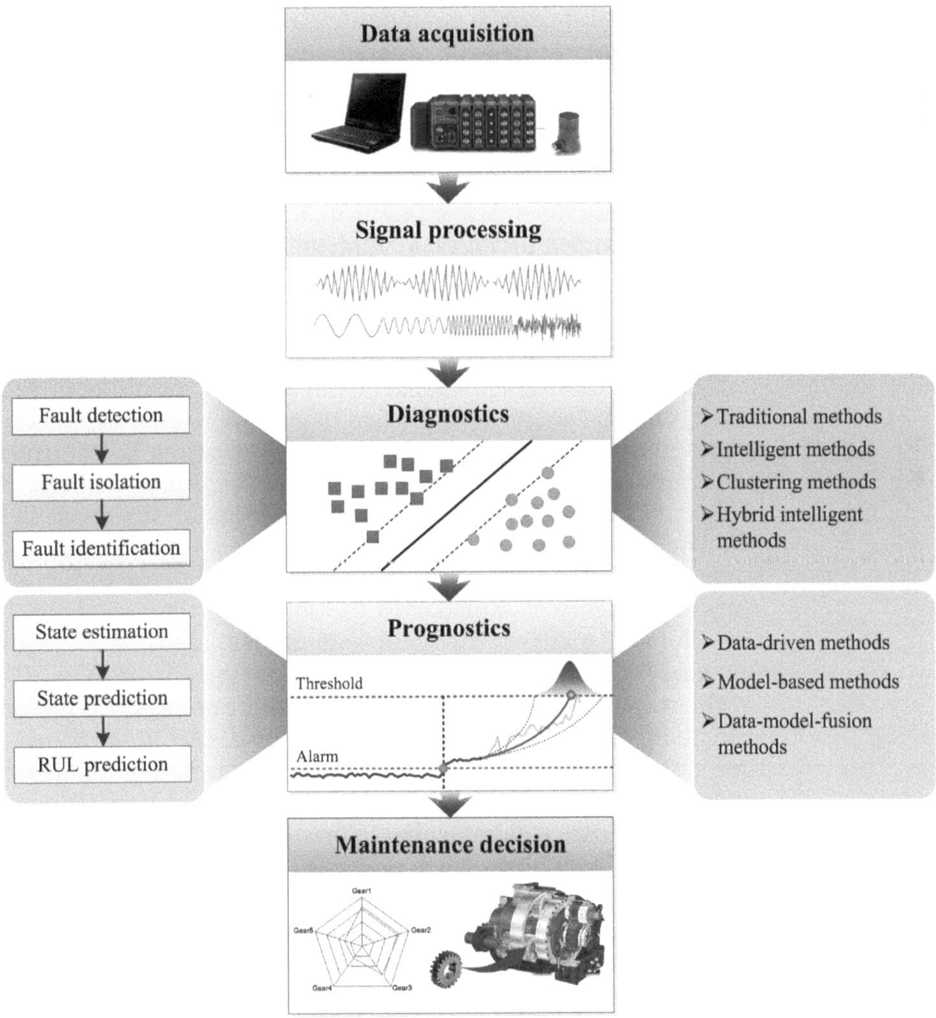

Figure 1.5. Flowchart of the PHM of rotating machinery.

1.2.1 Data Acquisition

Data acquisition is a process of capturing measurement signals using different kinds of sensors from monitored machines and storing the data into a computer. The measurement signals are supposed to be related to the health conditions of the monitored machines. In other words, the measurement signals incorporate some useful information, which reflects the health conditions of the machines. There are many different types of measurement signals, such as vibration signals, acoustic signals, temperatures, and electric currents. Various sensors including accelerometers, acoustic emission sensors, infrared thermometers, ultrasonic sensors, and so on, have been designed to collect different types of signals. The captured signals are transmitted into a PC through a data acquisition (DAQ) equipment and stored into a memory location for further analysis. With the rapid development of advanced computer and sensor technologies, lots of new data acquisition facilities and techniques have been designed and applied in modern industries. These powerful and versatile facilities have made data acquisition for PHM implementation more convenient and feasible.

1.2.2 Signal Processing

Signal processing is to analyze the stored measurement signals in the data acquisition process using signal processing techniques and methods. The task of signal processing is to extract useful information that is able to reveal the health conditions of the machines from the original measurement signals. It has been fully developed till now and numerous signal processing techniques and algorithms have been proposed in the literature. They are roughly classified into the following three categories: time–domain analysis, frequency–domain analysis, and time–frequency–domain analysis.

The original measurement signals that are generally sampled repeatedly between prespecified time intervals are in the form of time domain. Thus, the time-domain analysis is directly based on the original measurement signals. Traditional time-domain analysis calculates statistic characteristics describing the health conditions of machines, such as mean, peak, root mean square (RMS), kurtosis, and skewness. These statistic characteristics are named as time–domain features. Other commonly used time-domain analysis methods include time synchronous average (TSA), the autoregressive (AR) model, the autoregressive moving average (ARMA) model, principal component analysis (PCA), and so on.

Frequency–domain analysis is based on the transformed signals in frequency domain. The advantage of frequency–domain analysis over time–domain analysis is its ability to decompose the original signals into a series of frequency components. The most widely used frequency–domain analysis is the spectrum analysis by means of fast Fourier transform (FFT). The main idea of spectrum analysis is to isolate and locate certain frequency components of interest relating to the fault characteristics of machines. Power spectrum is a commonly used FFT-based method. Cepstrum is also widely used because of its capability to detect harmonics and sideband patterns in power spectrum. Some useful auxiliary tools for spectrum analysis include frequency filters, envelope analysis, side band structure analysis, and so on. Hilbert transform which is a useful tool in envelope analysis has also been used for machine fault detection and diagnostics.

One limitation of frequency–domain analysis is that it is only effective in handling stationary measurement signals and is unable to deal with nonstationary measurement signals. However, the measurement signals of machinery generally present nonstationary characteristics. Thus, time–frequency–domain analysis, which investigates measurement signals in both time and frequency domains, has been applied into the nonstationary measurement signal analysis. The time–frequency analysis describes the characteristics of measurement signals in two-dimensional functions of both time and frequency to better reveal the fault patterns of the machines. One of the commonly used time–frequency–domain analysis tools is short-time Fourier transform (STFT), which divides the measurement signals into different segments with short-time windows and then applies Fourier transform to each segment. Due to the signal segmentation, the frequency resolution is decreased inevitably. In addition, each signal segment is approximately considered to be a stationary process. Therefore, the STFT can only be applied to nonstationary signals with slow change in their rotating speeds. Another commonly used time–frequency–domain analysis tool is Wigner–Ville distribution. It is based on bilinear transform instead of signal segmentation. Therefore, it overcomes the frequency resolution limitation of STFT. However, due to the interference terms produced by the transformation itself, it is difficult to explain the estimated distribution. Another tool for time–frequency analysis is the wavelet transform. Different from the STFT, the wavelet transform can be used for multiscale analysis of a signal through dilation and translation, so it is able to extract time–frequency features of a signal effectively. Due to the multiscale analysis ability, the wavelet transform is able to produce a high frequency resolution at low frequencies and a high time resolution at high frequencies for signals with long duration low frequencies and short duration high frequencies. A predetermined wavelet function is necessary in the implementation of wavelet transform, which restricts its implementation. Therefore, another time–frequency–domain analysis tool, empirical mode decomposition (EMD), is developed. It is based on the local characteristic time scales of measurement signals and could decompose signals into a set of complete and almost orthogonal components that are called intrinsic mode functions (IMFs). Each IMF presents a nature oscillatory mode included in the signals. Thus, the EMD method is able to decompose the measurement signals into different modes according to their local characteristic time scales adaptively. Although the EMD method shows outstanding performance in processing nonlinear and nonstationary signals, it still has some weaknesses, such as lacking a theoretical foundation, end effects, sifting stop criterion, and extreme interpolation.

It should be noticed that, all the signal analysis tools aforementioned have their own advantages and disadvantages. In real cases, case-dependent knowledge and investigations are required to select appropriate ones among a number of available signal analysis methods.

1.2.3 Diagnostics

Diagnostics is summarized as a process of identifying and determining the relationship between the information obtained in the measurement space and the machine fault patterns in the fault space. There are three major steps in diagnostics, that is, fault detection, fault isolation, and fault identification, as shown in Fig. 1.5. Fault detection is a task to indicate whether a fault has already occurred in the monitored machines. Fault isolation is to find the fault component and the position of the fault. Fault identification is the final step in diagnostics, which attempts to determine the pattern and severity of the fault. The three steps correlate with each other. The latter step relies on the output of the former step and therefore is unable to be carried out individually.

Traditional diagnostic methods are to compare the signal processing results of machines under unknown health conditions with those under normal health conditions and to recognize the differences between them. So these manual diagnostic methods require a high degree of expertise in the specific fields of diagnostics. What's more, mechanical equipment in modern industries becomes more automatic, precise, and efficient. To fully inspect its health conditions, condition-monitoring systems are used to collect real-time data from the equipment, and massive data are acquired after the long-time operation, which promotes machinery health monitoring to enter the era of Big Data. Effectively extracting characteristics from such data and accurately identifying the machinery health conditions with advanced theories become new issues in machinery health monitoring. Intelligent fault diagnosis is famous for replacing diagnosis specialists to automatically and efficiently provide accurate fault recognition results. Therefore, it is supposed to be the best tool to handle massive data of machinery fault diagnosis in Big Data Era.

The book provides a comprehensive introduction of the proposed intelligent fault diagnosis methods in recent years as well as the latest contribution, that is, deep learning based diagnosis method, which may be the essential information that the reader needs currently. The intelligent fault diagnosis techniques appeared in literature generally incorporate the following three categories: individual intelligent techniques, clustering algorithms, and multidimensional hybrid intelligent techniques. More details about the fundamental theories and the advanced methods of the intelligent fault diagnosis will be discussed in the following chapters.

1.2.4 Prognostics

Diagnostics is to detect fault presence in machinery as early as possible and identify kinds, positions, and degrees of faults. However, it is only able to be conducted after fault occurrence, which is insufficient to prevent the happening of catastrophes. In order to reduce unscheduled risks and maintenance costs, prognostics should be conducted during the PHM. Prognostics are to forecast the future performance of machinery using prediction methods and to obtain the available time before machinery loses their operation ability, that is, the RUL. An accurate RUL prediction of machinery allows for predictive maintenance or replacement, thus reducing costly unscheduled maintenance. Consequently, RUL prediction appears to be a hot issue attracting more and more attention in recent years.

Since prognostics are able to schedule the maintenance strategy before the occurrence of failures, they are much more efficient than diagnostics to achieve zero-downtime performance. However, prognostics are unable to completely replace diagnostics. The major reasons are as follows: (1) prognostics rely on diagnostic outputs and therefore are unable to be done in isolation; (2) there are always some abrupt failures which are unable to be predicted effectively; and (3) none prognostic method is able to predict the

RUL exactly the same as the real one. Therefore, diagnostics is still necessary for providing maintenance decision support.

As shown in Fig. 1.5, there are three major steps in the prognostic process of machinery, that is, state estimation, state prediction, and RUL prediction. State estimation relies on the output of fault identification in diagnostics, which has determined the pattern and severity of the fault. The task of state estimation is to further quantitatively identify the severity of the fault and to estimate the degradation state of machinery. State prediction is to predict the degradation trend and speed of the state according to the information of the historical degradation curve. The RUL prediction step is carried out by calculating the time length of the degradation curve from current state to the final failure based on a predetermined failure threshold.

Machinery prognostic methods are roughly categorized into three major classes: data-driven methods, model-based methods, and data–model fusion methods. Data-driven methods attempt to derive degradation processes of machinery from measurement signals instead of building models based on comprehensive system physics and human expertise. These methods rely on the assumption that the statistical characteristics of data are relatively consistent unless a fault occurs. They produce RUL prediction results based on historical measurement signals. Therefore, the prediction accuracy of data-driven methods depends on both the quantity and the quality of the historical measurements. However, it is generally difficult to collect enough qualified measurements in real cases.

Model-based methods attempt to set up mathematical or physical models to describe degradation processes of machinery, and update model parameters using measured data. Model-based methods could incorporate both expert knowledge and real-time information from machinery. Consequently, they may work well in RUL prediction of machinery. They also generally require less data than data-driven methods. The commonly used models include the Markov-based models (Rausand and Hoyland, 2004), the Winner process models (Si et al., 2013), the inverse Gaussian process models (Ye, 2013), and so on. Model-based methods, however, may not be suitable to deal with the prognostics of rotating machinery in the following scenarios: (1) machinery fault patterns vary under different operation conditions; (2) the mechanical system is so complex that understanding the principles of system operation is not straightforward and developing an accurate model is prohibitively expensive.

Data-driven methods and model-based methods both have their superiorities and shortcomings for the RUL prediction of rotating machinery. To complement the advantages of them, data–model fusion methods are developed in recent years. They aim to integrate the strengths of both the data-driven and the model-based methods for a more reliable machinery RUL prediction.

The book first explains fundamental theories of prognostics, and then takes some prognostic methods as examples and comprehensively introduces their basic methodologies and application cases in the RUL prediction of rotating machinery. These examples include a representative data-driven method based on relevance vector machine (RVM) and some typical model-based methods. The application cases provide an essential guide for the readers to understand the application process of the prognostic methods.

1.2.5 Maintenance Decision

The aim of diagnostics and prognostics is to provide decision support for maintenance policies. Therefore, maintenance decision is the final process in PHM. Maintenance decision is a process of analyzing outputs from diagnostics and prognostics and making optimal maintenance strategies and accurate replacement decisions for failure components of machinery (Garg and Deshmukh, 2006; Wang, 2002; Tsan, 2002). The optimal maintenance decisions rely on the estimation of the RUL and the effectiveness of maintenance actions. Therefore, it is necessary to extend the diagnostics and prognostics to monitor the changes of the machinery health conditions after a maintenance action and estimate the consequent change of the machinery reliability.

Based on the conception of PHM, plenty of intelligent maintenance systems have been developed recently. An effective PHM system is expected to provide early detection, isolation, and identification of incipient faults; estimate and predict the degradation states of machinery; and make an optimal maintenance schedule and asset management decision. The incipient faults should be detected and monitored from a small fault as it progresses to a larger fault, until the fault severity approaches a prespecified failure threshold or reduces to an acceptable level of reliability. By employing such a system, the health of rotating machinery or the critical components can be monitored at any point of time, and the eventual occurrence of a failure can be predicted and prevented, enabling the achievement of near-zero downtime performance. Meanwhile unnecessary and costly preventative maintenance can be eliminated, maintenance scheduling can be optimized, and lead-time for spare parts and resources can be reduced, all of which result in significant cost savings.

With the development of e-technology, a new term "e-maintenance" has emerged since early 2000s and is very common recently in maintenance-related literature. It is defined by Muller et al. (2008) as a maintenance support, which includes the resources, services, and management necessary to enable proactive decision process execution. This support includes not only e-technologies (ICT, web-based, tether-free, wireless, infotronics technologies, etc.) but also e-maintenance activities (operations or processes), such as e-monitoring, e-diagnosis, and e-prognosis. The e-maintenance emergence is attributed to two main factors. (1) The development of advanced e-technologies makes it possible to increase the maintenance efficiency, velocity, and proactivity, thus optimizing the maintenance-related workflow. (2) The need of integrating business performance increases the requirements of the maintenance area including openness, integration, and collaboration with the other services of the e-enterprise. The e-maintenance has created several potential improvements in maintenance types and strategies. First, with the help of wireless and Internet technologies, any operator, manager, or expert has the capability to remotely link to a factory's equipment through Internet, allowing them to take remote maintenance actions. Second, an e-maintenance system gets the consumers, managers, engineers, and machinery together using an information infrastructure connecting geographically dispersed subsystems, allowing making collaborative maintenances between different human actors, different enterprise areas, and different companies. Third, the real-time remote monitoring and high-rate communications enable the maintenance operators to monitor the health condition of remote equipment, quickly obtain expertise and suggestions, and respond to any situation of the monitoring equipment.

1.3 Preface to Book Chapters

This book will provide a comprehensive introduction of the intelligent fault diagnosis and the RUL prediction based on the current achievements appeared in academic journals, conference proceedings, and technical reports. The main contents include multidomain signal processing and feature extraction, intelligent diagnosis models, clustering algorithms, hybrid intelligent diagnosis strategies, RUL prediction approaches, and so on. The book involves the fundamental theories and the advanced methods of intelligent fault diagnosis and RUL prediction. Beside experimental demonstrations, many application cases are presented and illustrated to test the methods mentioned in the book.

The rest of the book is structured as follows. Chapter 2 first introduces the basic concepts and fundamental theories of machinery vibration signals, and then reviews commonly used signal-processing methods in the time–domain, frequency–domain, and time–frequency–domain, respectively.

Chapters 3–5 discuss the same issue in the PHM of rotating machinery, that is, the diagnostics. However, these three chapters focus on different aspects. In Chapter 3, fault diagnosis methods based on individual intelligent techniques are introduced comprehensively. The fundamental theories of some traditional intelligent techniques, such as artificial neural networks (ANN), statistical learning theories, and deep learning, are explained in front of each section. Then, some advanced techniques developed from these traditional ones are presented, including adaptive neuro-fuzzy inference system (ANFIS), multiclass relevant vector machine, denoising autoencoder, and so on. Finally, several representative application cases of rotating machinery are offered as tutorials of the individual intelligent diagnosis techniques, such as heavy oil catalytic cracking units, planetary gearboxes, and motor bearings.

Chapter 4 focuses on fault diagnosis based on clustering algorithms. Two typical clustering algorithms, that is, K nearest neighbor (KNN) and fuzzy c-means are introduced, and then some improved technologies based on these algorithms are exhibited. In Section 4.4, a hybrid clustering algorithm and its application of rotating machinery diagnostics are provided.

In Chapter 5, some advanced diagnosis methods based on multidimensional hybrid intelligent techniques are discussed. First, a diagnosis method based on multiple weighted K nearest neighbor (WKNN) combination is described and implemented to the fault diagnosis of rolling element bearings. Then, a method based on multiple ANFIS combination with genetic algorithms (GAs) is presented and also applied to the fault diagnosis of rolling element bearings. Finally, a hybrid intelligent method in terms of different classifiers is introduced and a case study in fault diagnosis of gearboxes of this method is investigated.

Chapter 6 addresses key issues regarding the RUL prediction of rotating machinery. At first, the basic concepts and fundamental theories of RUL prediction are introduced. Then, some representative data-driven methods and model-based methods are exhibited. Finally, experimental demonstrations on the RUL prediction of rotating machinery are illustrated.

We hope this book is able to provide an essential guide for the readers from the areas of machinery fault diagnosis and RUL prediction to understand the basic concepts and the fundamental theories, and know the cutting-edge research and the latest methods. Although the research object in this book is rotating machinery, the presented methods, and techniques are not limited to this object but also benefit the readers from other research areas.

References

Garg, A., Deshmukh, S.G., 2006. Maintenance management: literature review and directions. J. Qual. Mainten. Eng. 12, 205–238.

Hyers, R.W., Mcgowan, J.G., Sullivan, K.L., Manwell, J.F., Syrett, B.C., 2006. Condition monitoring and prognosis of utility scale wind turbines. Energy Mater. 1, 187–203.

Lei, Y., Lin, J., He, Z., Zuo, M.J., 2013. A review on empirical mode decomposition in fault diagnosis of rotating machinery. Mech. Syst. Signal Process. 35, 108–126.

Lu, B., Li, Y., Wu, X., Yang, Z.A., 2009. Review of recent advances in wind turbine condition monitoring and fault diagnosis. Power Electronics and Machines in Wind Applications, PEMWA, 24–26 June 2009, Lincoln, NE. IEEE, pp. 1–7.

Muller, A., Crespo Marquez, A., Iung, B., 2008. On the concept of e-maintenance: review and current research. Reliab. Eng. Syst. Saf. 93, 1165–1187.

Randall, R.B., 2011. Vibration-Based Condition Monitoring. John Wiley & Sons Ltd, United Kingdom.

Rausand, M., Hoyland, A., 2004. System Reliability Theory: Models, Statistical Methods and Applications. Wiley-Interscience, Hoboken, NewJersey.

Saranga, H., Knezevic, J., 2001. Reliability prediction for condition-based maintained systems. Reliab. Eng. Syst. Saf. 71, 219–224.

Si, X.S., Wang, W.B., Hu, C.H., Chen, M.Y., Zhou, D.H., 2013. A Wiener-process-based degradation model with a recursive filter algorithm for remaining useful life estimation. Mech. Syst. Signal Process. 35, 219–237.

The World Wind Energy Association, half year report, 2014. Available from: http://www.wwindea.org/webimages/WWEA_half_year_report_2014.pdf

Tsan, A.H.C., 2002. Strategic dimensions of maintenance management. J. Qual. Mainten. Eng. 8, 7–39.

Wang, H.Z., 2002. A survey of maintenance policies of deteriorating systems. Eur. J. Operat. Res. 139, 469–489.

Ye, Z.S., 2013. On the conditional increments of degradation processes. Stat. Probabil. Lett. 83, 2531–2536.

2

SIGNAL PROCESSING AND FEATURE EXTRACTION

2.1 Introduction

A signal is commonly defined as a sign or notice which is perceptible by sight or hearing especially for the purpose of conveying warning, direction or information (Oxford, 2009). For the purpose of description and analysis, a signal can also be defined simply as a mathematical function associated with time and frequency, etc.

According to statistical properties of the signal itself, the signal can be divided into different categories as shown in Fig. 2.1. It is common that signals are divided into the stationary and the nonstationary, where the stationary denotes that the statistical properties of a signal are invariant with time. For deterministic signals, this basically implies that they are composed entirely of some sinusoids with discrete frequencies and thus their frequency spectra consist of discrete lines at the frequencies of these sinusoids. Once the frequencies, amplitudes, and initial phases (i.e., at time zero) of these sinusoids are known, the value of the signal can be determined anytime. Therefore, they are called deterministic signals. Random signals are somewhat more complicated because their values cannot be predicted anytime. For stationary random signals, however, their statistical properties are unchanging with time. Particularly, Gaussian white noise is a typical representation. The nonstationary means that anything which does not satisfy the conditions for the stationary and once again it can be separated into two main categories: the continuous and the transient. There is no hard and fast rule to distinguish them, but the terms "energy" and "power" are generally used to distinguish them. The true power of a signal is related to the square of its amplitude and it is commonly simple to name the squared value as the "power." A transient signal has an instantaneous squared value or power at each point of the time, but is characterized by the integral of this "power" in its whole time, which is named "energy." A stationary random signal using this definition has a constant power and an infinite energy. Cyclostationary signals have the power which changes periodically with time, therefore, the total energy is also infinite. Other nonstationary signals also have a

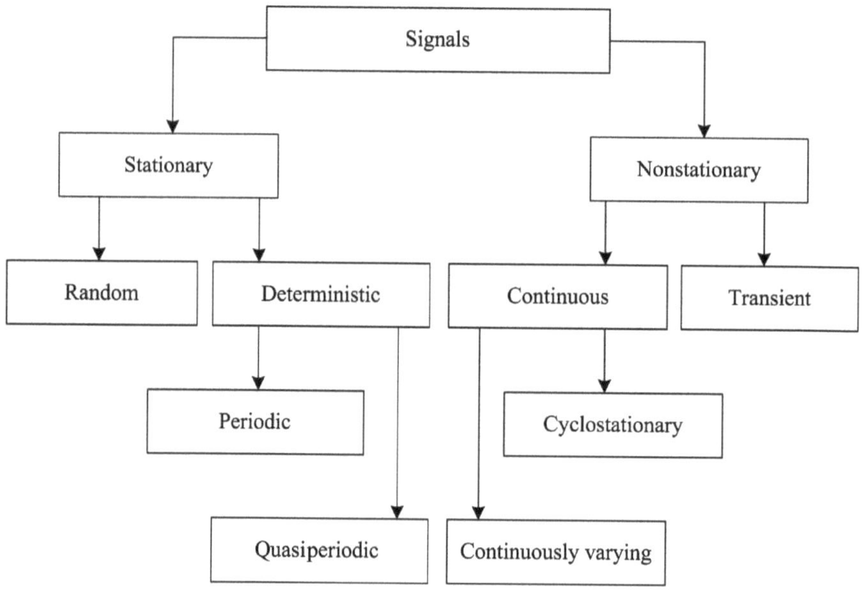

Figure 2.1. Signal categories.

finite length, but are more likely to be considered as continuously varying nonstationary signals rather than the transient since they are typically analyzed to see how their power varies with time by being divided into short quasistationary sections. Generally, the acquired signals from experimental test rigs or real plants contain some noise and useless components. Therefore, signal processing is necessary to be conducted after signal acquisition.

Signal processing is to extract fault-related information of rotating machinery, and further eliminate the bootless imbedded in the signal and increase the signal-to-noise ratio (SNR) (Lei et al., 2013a; Lei et al., 2014). Signal processing is more and more important in scientific research and industrial applications. With the rapid development of science and technology, scientific experiments, signal measurement, and the control of automatic production lines during industrial production are all built on the basis of signal processing. Moreover, the performance tests of new material, research, and development of new equipment and experiments of new process and method are all built on the basis of precise processing and analysis of the acquired signals. In addition, with the applications of automatically electromechanical products, their designs, tests, and operations are all closely associated with signal processing. During the manufacture and operation of modern equipment, signal processing has taken the top spot and it takes 50–70% of the total cost of the equipment.

Signal processing has a long and rich history. Up to now, lots of signal processing techniques have been developed and applied widely. Traditional signal processing methods can be divided into three categories: time–domain ones, frequency–domain ones, and time–frequency–domain ones as shown in Fig. 2.2. Signal preprocessing attempts to cancel noise in the signal and improve the SNR preparing for postprocessing, which is suitable for any signal categories. Signal processing in the time domain mainly includes correlation analysis and time–domain statistical feature extraction. Autocorrelation analysis can capture the periodic characteristics of a signal, while the cross-correlation analysis can measure the relevance degree between two different signals. In addition, the time–domain statistical features can be extracted to reflect the statistical properties of a signal from its time–domain waveform. Signal processing in the frequency domain generally denotes

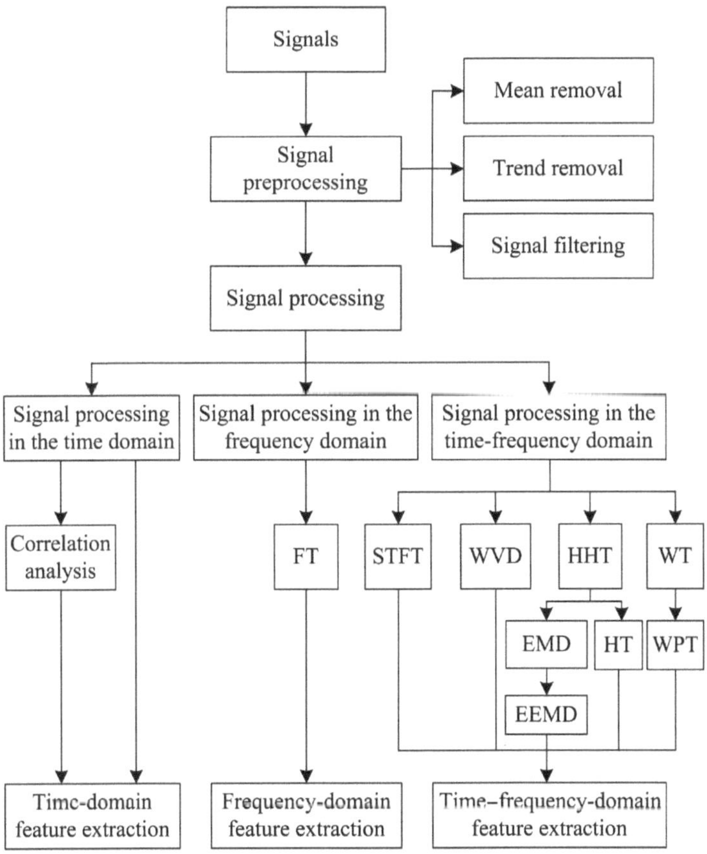

Figure 2.2. Signal processing and feature extraction.

Fourier transform (FT) based on Fourier series and later fast Fourier transform (FFT). Their basis functions are a set of trigonometric functions. Thus, the FT technique is limited to processing stationary signals. By supposing that a signal in short enough time is stationary, some frequency–domain features can be extracted to reflect the frequency–domain statistical properties from the frequency or power spectra of the signal. In practice, a nonstationary signal cannot be completely described in the time domain or frequency domain alone. To reflect the distribution of a nonstationary signal from both time domain and frequency domain, signal processing techniques in time–frequency domain have been developed rapidly. There are lots of representative techniques, such as short-time Fourier transform (STFT), Wigner–Ville distribution (WVD), wavelet transform (WT), wavelet packet transform (WPT), and even Hilbert–Huang transform (HHT). These techniques are widely applied to mechanical fault diagnosis. Since different signal processing techniques characterize distinct performance, it is important for different signals to select a suitable signal processing technique to acquire more useful information.

2.2 Signal Preprocessing

Fault characteristics in vibration signals collected from rotating machinery are very weak since they are embedded in heavy background noise, especially for early faults. Therefore, the vibration signals should be preprocessed to reduce background noise and the errors of measurement systems, to further acquire more useful information and improve the SNR (Jia et al., 2015). In this section, two common techniques for signal preprocessing are introduced, which are trend removal and signal filtering.

2.2.1 Trend Removal

Usually, the direct current component, also called mean value, hardly contains any useful information reflecting mechanical faults because it is a constant. Therefore, the direct current component should be removed before the signal processing. To illustrate this case, a sinuous signal with a direct current component and the corresponding frequency spectrum are plotted in Fig. 2.3A and B. It is observed that the direct current component is able to cause the fact that the time–domain waveform of signal $x(t)$ fluctuates around the direct current component $x = 1$ rather than x-axis ($x = 0$). The signal amplitude is uplifted integrally by the direct current component. However, the corresponding frequency spectrum indicates that the direct current component not only produces an interference frequency at 0 Hz

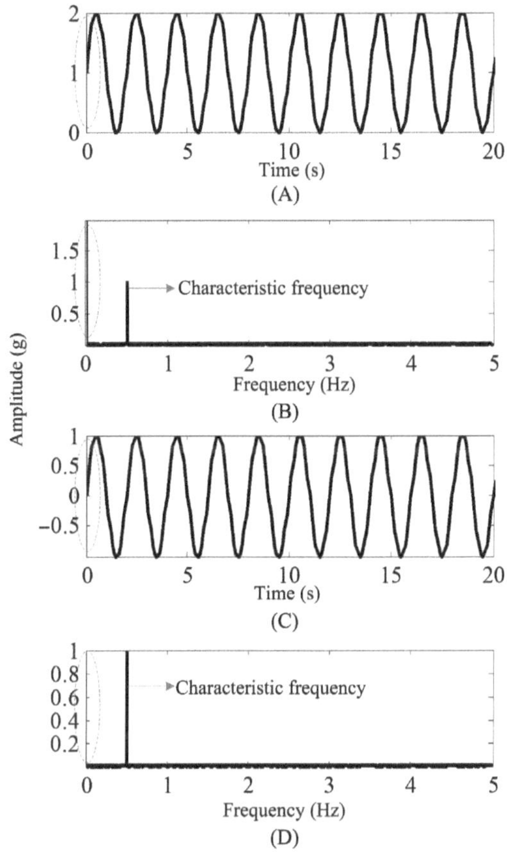

Figure 2.3. Mean removal. (A) Original time–domain waveform with a direct current component, (B) corresponding frequency spectrum, (C) time–domain waveform after removing mean, and (D) the corresponding frequency spectrum.

but also does not carry any useful information of characteristic frequency. The direct current component can be isolated by removing its mean value, and the time–domain waveform and corresponding frequency spectrum are illustrated in Fig. 2.3C and D. Obviously, the influence of direct current component on the characteristic frequency is eliminated and the useful information is remained completely.

Similarly, the trend is a monotonic function associated with time t. Therefore, the trends in the collected vibration signals are unimpressive in some cases. For example, the dark gray line represents the trend included in a signal as shown in Fig. 2.4A and the corresponding frequency spectrum is plotted in Fig. 2.4B. It is clear that the characteristic frequencies f_1 and f_2 are seriously influenced by the trend which is defined as a bootless interference

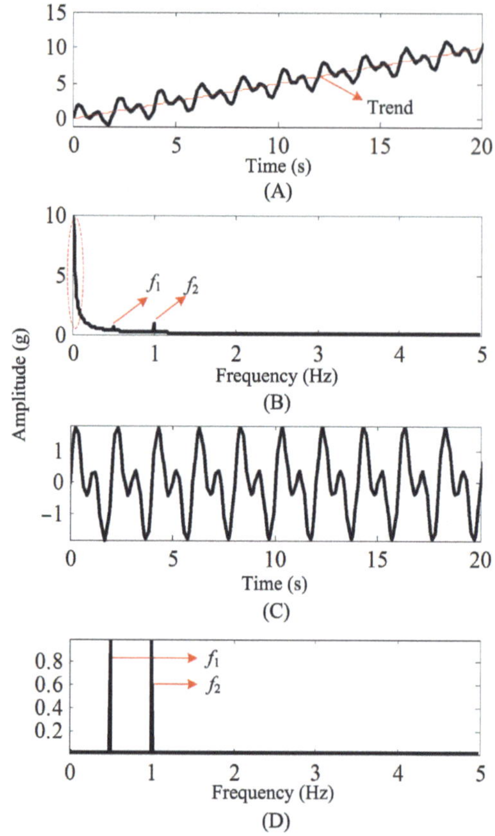

Figure 2.4. Trend removal. (A) Original time–domain waveform with an increasing trend, (B) corresponding frequency spectrum, (C) time–domain waveform after removing the trend, and (D) the corresponding frequency spectrum.

near zero frequency. It is noticed that the frequency interference of the trend is similar to that of the mean. The trend is obtained by computing the least-squares fit of a straight line (or composite line for piecewise linear trends) to the signal, and then the trend is subtracted from the original signal as illustrated in Fig. 2.4C and the corresponding frequency spectrum is shown in Fig. 2.4D. It is noticed that the characteristic frequencies are completely remained and the influence of the trend is canceled. So it is necessary to remove the trend of a signal during signal preprocessing.

2.2.2 Signal Filtering

Signal filtering is to eliminate the interference of noise and remove the unwanted frequency band signals, thereby reserving more useful information. The systems to implement signal filtering are called as filters. According to different amplitude–frequency characteristics, the filters are commonly divided into four categories: low-pass filters, high-pass ones, band-pass ones, and band-stop ones. According to the properties of the signals to be processed, the filters can also be divided into analog filters and digital filters. For the digital filters, they are further classified into infinite impulse response (IIR) filters and finite impulse response (FIR) ones.

The concept of conventional filter is developed on the basis of signal analysis in the frequency domain. Suppose that a signal $x(t)$ contains a useful component $s(t)$ and a noise component $n(t)$, that is, $x(t) = s(t) + n(t)$. Assuming that $y(t)$ is a filtered signal, then the transfer function of the filter is

$$H(\omega)=Y(\omega)/X(\omega) \tag{2.1}$$

where $Y(\omega)$ and $X(\omega)$ represent FT of the signals $y(t)$ and $x(t)$, respectively. $H(\omega)$ denotes the transfer function of a filter. When noise and useful components are in different frequency bands, the noise can be canceled by designing the corresponding filter

$$\begin{cases} H(\omega)=1, & S(\omega)\neq 0 \\ H(\omega)=0, & S(\omega)=0 \end{cases} \tag{2.2}$$

where $S(\omega)$ is the FT of useful component $s(t)$. Therefore the product between $H(\omega)$ and FT of input signal $X(\omega) = S(\omega) + N(\omega)$ is FT of the output $Y(\omega) = S(\omega)$. Obviously, we can obtain

$$y(t)=x(t)*h(t) \tag{2.3}$$

where this notation "*" denotes convolution operator and $h(t) = F^{-1}\{H(\omega)\}$ is inverse FT of $H(\omega)$, namely unit impulse response function of the filter. In other words, the input signal $x(t)$ is equal to an unit impulse signal $\delta(t)$ (delta function), the output signal $y(t)$ equals $h(t)$.

Similarly, suppose that sampling interval is Δt, for discrete time series $x(n\Delta t)$, $y(n\Delta t)$ and $h(n\Delta t)$, and corresponding FT is $X_\Delta(\omega)$, $Y_\Delta(\omega)$ and $H_\Delta(\omega)$, respectively, hence in the frequency band $[-1/(2\Delta t),\ 1/(2\Delta t)]$, we can obtain

$$\begin{cases} y(n\Delta t)=x(n\Delta t)*h(n\Delta t) \\ Y_\Delta(\omega)=H_\Delta(\omega)X_\Delta(\omega) \end{cases} \tag{2.4}$$

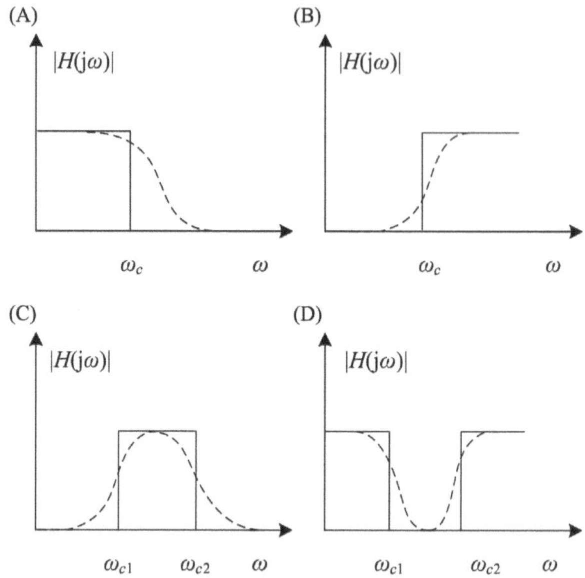

Figure 2.5. Amplitude–frequency characteristic of filters (——ideal analog filters - - - - real analog filters). (A) Low-pass filters, (B) high-pass filters, (C) band-pass filters, and (D) band-stop filters.

According to the signal to be processed, the filters are classified into analog filters and digital filters. Ideal analog filters are ideal models and unable to be implemented in real applications, but they are helpful to design approximate ideal filters by learning their structural characteristics. Their amplitude–frequency characteristics are illustrated in Fig. 2.5. Through a low-pass filter, if the frequency components in a signal are less than ω_c, they will be remained, and conversely removed. Here, the frequency ω_c is called as cutoff frequency of the filter, meanwhile the frequency range $\omega < \omega_c$ is called passband of the low-pass filter and the frequency range $\omega > \omega_c$ is called stopband of the low-pass filter. Similarly, the passband of a high-pass filter is $\omega > \omega_c$ and stopband is $\omega < \omega_c$. The passband of a band-pass filter is $\omega_{c1} < \omega < \omega_{c2}$ and the stop-band is $\omega < \omega_{c1}$ and $\omega_{c2} < \omega$. Note that the ideal low-pass filters are characterized by rectangular amplitude–frequency and linear phase–frequency characteristics as shown in Fig. 2.6. These characteristics can guarantee the filtered signals without distortion. Additionally, ideal high-pass, band-pass, and band-stop filters can be obtained by the cascade of several low-pass filters. Therefore, we take ideal low-pass filters for an example to illustrate the characteristics of the

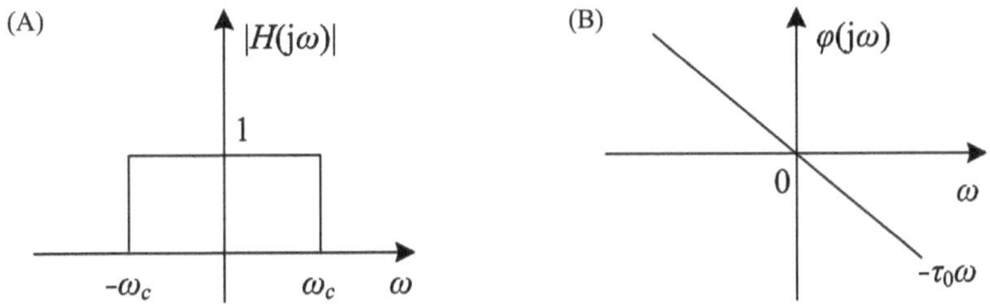

Figure 2.6. Amplitude-frequency and phase–frequency characteristics of ideal analog filters. (A) Amplitude–frequency characteristics and (B) phase–frequency characteristics.

filters, and its amplitude– and phase–frequency characteristics are represented as

$$\begin{cases} |H(j\omega)| = 1, & |\omega| < \omega_c \\ |\varphi(j\omega)| = -\tau_0\omega, & |\omega| < \omega_c \end{cases} \tag{2.5}$$

The corresponding unit impulse response function is

$$h(t) = \frac{\sin\left[\omega_c(t-\tau_0)\right]}{\pi(t-\tau_0)} = \frac{\omega_c}{\pi}\mathrm{sinc}\left[\omega_c(t-\tau_0)\right] \tag{2.6}$$

Therefore, the unit impulse response function of an ideal low-pass filter is a sinc function with time delay τ_0, which is shown in Fig. 2.7. It is noticed that with the increase of cutoff frequency ω_c the unit impulse response function $h(t)$ shrinks at the center of τ_0. Moreover, if $\omega_c \to \infty$, then $h(t) \to \delta(t)$ where $\delta(t)$ is an unit impulse response.

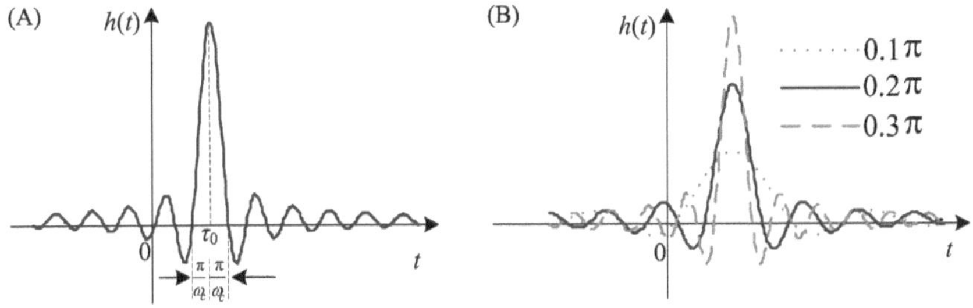

Figure 2.7. Unit impulse response function of an ideal analog filter. (A) A unit impulse response function and (B) Unit impulse response functions with different cutoff frequencies.

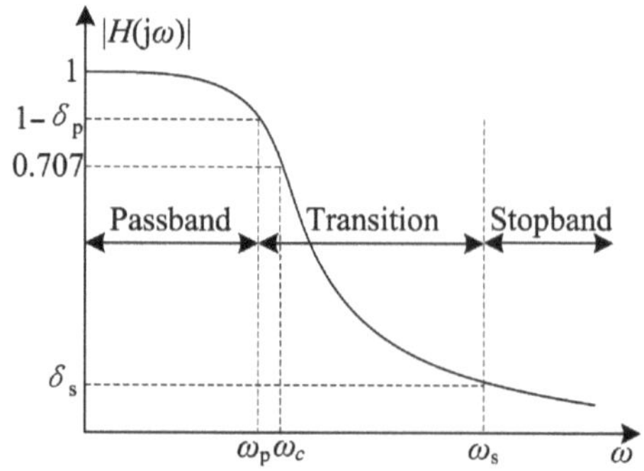

Figure 2.8. A real low-pass analog filter.

The ideal low-pass filters are generally described by their cut-off frequencies, while real low-pass filters have certain transition between passband and stopband in real design. If the amplitude–frequency characteristic $|H(\mathrm{j}\omega)_{\omega=0}|$ is normalized to 1, a real low-pass filter is illustrated in Fig. 2.8, where the cut off frequency ω_c denotes the corresponding frequency at $|H(\mathrm{j}\omega)| = 0.707\ (1/\sqrt{2})$. The frequency range ($\omega < \omega_p$) is called passband, $\omega_p < \omega < \omega_s$ is transition and $\omega > \omega_s$ is stop-band. However, the $|H(\mathrm{j}\omega)|$ is not strictly equal to 1 or 0 at pass-band or stop-band, allowing a certain fluctuation δ_p and δ_s, respectively.

The digital filters can be further divided into two categories: IIR filters and FIR filters. The transfer function of an IIR filter is

$$H(z)=\frac{\sum_{r=0}^{M} b_r z^{-r}}{1+\sum_{k=1}^{N} a_k z^{-k}} \tag{2.7}$$

where $z = \exp(\mathrm{j}\omega\Delta t)$, $a_k (k=1,2,\cdots,N)$ and $b_r (r=0,1,2,\cdots,M)$ are coefficients of denominator polynomial and numerator one, respectively. The transfer function of a FIR digital filter is

$$H(z)=\sum_{n=0}^{N-1} h(n) z^{-n} \tag{2.8}$$

where $h(n)$ $(n=0,1,2,\cdots,N-1)$ is the unit impulse response function of a digital filter. There are large gaps of performance and design between IIR digital filters and FIR digital filters. FIR digital filters can be directly designed by specific amplitude–frequency characteristic. The design methods of FIR digital filters are built on the approximation for ideal filters. The used approximate methods contain the window function methods, frequency-sampling methods, etc. The performance of the filters, therefore, is closely related to their designs. The detailed filter design can refer to the literature (Oppenheim et al., 1989). Moreover, the filtered signal should be further processed to acquire desirable or valuable information, for example, signal processing in the time domain.

2.3 Signal Processing in the Time Domain

Signal processing in the time domain mainly includes correlation analysis and time–domain statistical feature extraction. Through signal preprocessing in Section 2.2, the SNR can be improved to some extent and we can further extract more features reflecting the health conditions of rotating machinery.

2.3.1 Correlation Analysis

Correlation analysis represents mutually dependent relationship between two signals. Generally, the relationship can be written as the inner product between two signals. Assuming that there are two signals $x(t)$ and $y(t)$, then the inner product can be written as

$$\langle x, y \rangle = \int_0^T x(t)y(t)dt \tag{2.9}$$

where T is the observing time of signal $x(t)$ or $y(t)$.

Apparently, if there is a large similarity between two signals, their inner product is also large. In practice, the time delay should be considered. Therefore, the Eq. (2.9) is also written as

$$R(\tau) = \lim_{T \to \infty} \frac{1}{T} \int_0^T x(t)y(t+\tau)dt \tag{2.10}$$

where T is the observing time of signal $x(t)$ or $y(t)$, τ is the time delay of signal $y(t)$ and $R(\tau)$ is a mathematical function associated with time delay τ. The change of $R(\tau)$ can reflect the correlation between $x(t)$ and $y(t+\tau)$.

2.3.1.1 *Autocorrelation Analysis*

In order to reflect the correlation between two same signals, the signal $y(t)$ in Eq. (2.10) is substituted by signal $x(t)$, thus we can obtain $R_x(\tau)$, the autocorrelation function of the signal $x(t)$

$$R_x(\tau) = \lim_{T \to \infty} \frac{1}{T} \int_0^T x(t)x(t+\tau)dt \tag{2.11}$$

where T is the observing time of signal $x(t)$. $R_x(\tau)$ can describe the correlation between $x(t)$ and $x(t+\tau)$. The autocorrelation coefficient in practical applications is often defined as

$$\rho_x(\tau)=\frac{R_x(\tau)}{\sigma_x^2} \tag{2.12}$$

where σ_x is the standard deviation of signal $x(t)$, and $R_x(\tau)$ is the autocorrelation function of signal $x(t)$ and satisfies the following requirements:

1. $R_x(\tau)$ is a real function;
2. $R_x(\tau)$ is an even function, that is, $R_x(\tau) = R_x(-\tau)$;
3. $R_x(0)=\psi_x^2$, where ψ_x^2 is the mean square value of signal $x(t)$;
4. If the signal is ergodic, $|R_x(\tau)| \le R_x(0)$ and $\max R_x(\tau) = R_x(0)$;
5. If μ_x is the mean value of a stochastic signal, $\lim_{\tau\to\infty} R_x(\tau)=\mu_x^2$. In contrast, if μ_x is the mean value of a deterministic signal, $\lim_{\tau\to\infty} R_x(\tau)\neq\mu_x^2$;
6. If the stationary signal $x(t)$ contains a certain periodic component, its autocorrelation function $R_x(\tau)$ also has a certain periodic component and their periods are the same.

Autocorrelation functions of several kinds of common signals are illustrated in Fig. 2.9. It is observed that the periodic components in signal $x(t)$ do not decay in the corresponding autocorrelation function. Moreover, their periods are the same. Therefore, autocorrelation function is able to extract the periodic components from the noisy signals.

2.3.1.2 Cross-Correlation Analysis

The cross-correlation function between signals $x(t)$ and $y(t)$ is defined as

$$R_{xy}(\tau)=\lim_{T\to\infty}\frac{1}{T}\int_0^T x(t)y(t+\tau)dt \tag{2.13}$$

where T is the observing time of signal $x(t)$ or $y(t)$. $R_{xy}(\tau)$ is a function associated with τ and denotes the correlation between two different signals. Similarly, the covariance function can also represent the correlation between them, which is defined as

$$C_{xy}(\tau)=\lim_{T\to\infty}\frac{1}{T}\int_0^T\left\{\left[x(t)-\mu_x\right]\left[y(t+\tau)-\mu_y\right]\right\} \tag{2.14}$$

where μ_x and μ_y are the mean values of signals $x(t)$ and $y(t)$, respectively. Therefore, the cross-correlation coefficient between signals $x(t)$ and $y(t)$ is

$$\rho_{xy}(\tau)=\frac{C_{xy}(\tau)}{\sigma_x\sigma_y} \tag{2.15}$$

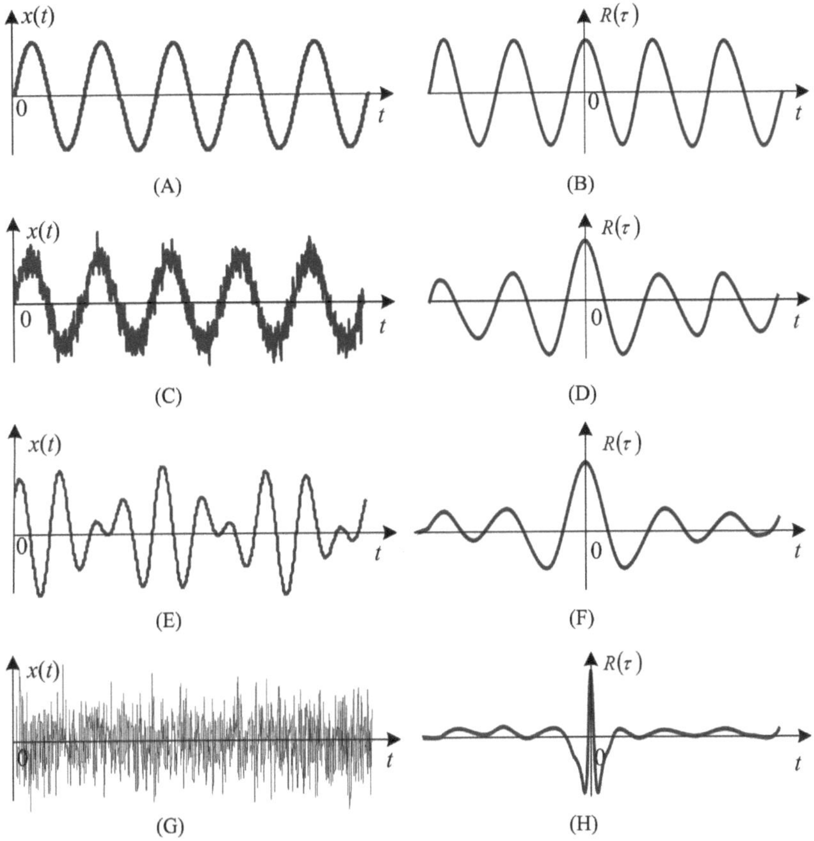

Figure 2.9. Autocorrelation functions of several kinds of common signals. (A) Periodic signal; (B) autocorelation function of (A); (C) periodic signal with noise; (D) autocorelation function of (C); (F) narrow-band noise; (F) autocorelation function of (E); (G) broad-band noise; and (H) autocorelation function of (G).

where $C_{xy}(\tau)$ is the covariance function, σ_x and σ_y are the standard deviations of signals $x(t)$ and $y(t)$, respectively. The cross-correlation function $R_{xy}(\tau)$ satisfies the following conditions.

1. Cross-correlation function $R_{xy}(\tau)$ is a real function but not an even function and max $R_{xy}(\tau)$ is not always at $\tau = 0$. If max $R_{xy}(\tau) = 0$, it is uncorrelated between signals $x(t)$ and $y(t)$. If the max $R_{xy}(\tau)$ occurs at $\tau = \tau_0$, it suggests the maximum correlation between signals $x(t)$ and $y(t + \tau_0)$. If signals $x(t)$ and $y(t)$ meet statistical independence and at least one of μ_x and μ_y is equal to zero, $R_{xy}(\tau) = 0$ for an arbitrary τ. On the contrary, if $R_{xy}(\tau) = 0$, signals $x(t)$ and $y(t)$ are not always independent.
2. For an arbitrary τ, $R_{xy}(\tau)$ satisfies the condition: $[R_{xy}(\tau)]^2 \le R_x(0)R_y(0)$. It can be seen from Fig. 2.9.

3. If the mean value of a signal is zero, when $\tau \to \infty$, $R_{xy}(\pm\infty) = 0$.
4. The cross-correlation function $R_{xy}(\tau)$ has antisymmetry, that is, $R_{xy}(-\tau) = R_{yx}(\tau)$.
5. If the signal $x(t+T) = x(t)$ and $y(t+T) = y(t)$, $R_{xy}(\tau + T) = R_{xy}(\tau)$.

2.3.2 Common Statistical Features in the Time Domain

When a fault occurs, the stiffness of mechanical structures around the fault must change, which can produce an impulse or shock. This impulse may further result in the variation of the collected vibration signals. The amplitudes and distributions of these time–domain signals may be changed. Therefore, time–domain statistical features can be extracted from their time–domain waveforms to reflect the mechanical faults. Common statistical features in the time domain are shown in Table 2.1. Here, mean value (X_{m}) reflects the average of a signal. Root amplitude (X_{root}), root mean square (X_{rms}), and peak (X_{peak}) may reflect the vibration amplitude and energy of a time–domain signal. Generally, a fault can excite mechanical vibration and thereby uplift the mean value, root amplitude, root mean square, and peak. When a fault becomes more and more severe, the above four features can distinctly reflect the severe degree of the fault. However, they are not

Table 2.1 Common Statistical Features in the Time Domain

$X_{\mathrm{m}} = \dfrac{\sum_{n=1}^{N} x(n)}{N}$	$X_{\mathrm{sd}} = \sqrt{\dfrac{\sum_{n=1}^{N}\left(x(n) - X_{\mathrm{m}}\right)^2}{N-1}}$	$X_{\mathrm{root}} = \left(\dfrac{\sum_{n=1}^{N}\sqrt{\lvert x(n)\rvert}}{N}\right)^2$	$X_{\mathrm{rms}} = \sqrt{\dfrac{\sum_{n=1}^{N}(x(n))^2}{N}}$
$X_{\mathrm{peak}} = \max\lvert x(n)\rvert$	$X_{\mathrm{skewness}} = \dfrac{\sum_{n=1}^{N}\left(x(n) - X_{\mathrm{m}}\right)^3}{(N-1)X_{\mathrm{sd}}^3}$	$X_{\mathrm{kurtosis}} = \dfrac{\sum_{n=1}^{N}\left(x(n) - X_{\mathrm{m}}\right)^4}{(N-1)X_{\mathrm{sd}}^4}$	$X_{\mathrm{crest}} = \dfrac{X_{\mathrm{peak}}}{X_{\mathrm{rms}}}$
$X_{\mathrm{clearance}} = \dfrac{X_{\mathrm{peak}}}{X_{\mathrm{root}}}$	$X_{\mathrm{shape}} = \dfrac{X_{\mathrm{rms}}}{\frac{1}{N}\sum_{n=1}^{N}\lvert x(n)\rvert}$	$X_{\mathrm{impulse}} = \dfrac{X_{\mathrm{peak}}}{\frac{1}{N}\sum_{n=1}^{N}\lvert x(n)\rvert}$	

where $x(n)$ is a signal series for $n = 1, 2, \ldots, N$ and N is the number of data points.

sensitive to weak incipient faults. Skewness ($X_{skewness}$), kurtosis value ($X_{kurtosis}$), crest factor (X_{crest}), clearance factor ($X_{clearance}$), shape factor (X_{shape}) and impulse factor ($X_{impulse}$) may be used to represent the time series distribution of the signals in the time domain. The kurtosis value, crest factor, impulse factor and clearance factor can measure the impulse existing in vibration signals. The kurtosis value and crest factor are robust to varying operating conditions and are good indicators for incipient faults. The impulse and clearance factors are also good indicators of spikiness of the sharp impulses generated by the contact of a defect with the bearing mating surfaces. Note that kurtosis value is very sensitive to early faults. With the increase of severe degree the kurtosis value can also increase gradually. However, it unexpectedly decreases for more severe faults. Therefore, the kurtosis value is unsuitable to more severe faults.

It can be noticed from Table 2.1 that different statistical features in the time domain can reflect mechanical health conditions from different aspects. They can compensate each other. In different fault levels, it is different for each indicator to contain fault information. Even many indicators cannot reflect the changes of faults to some extent. Therefore, more features should be extracted to effectively diagnose the faults of rotating machinery and moreover sensitive features should also be further screened out from them.

2.4 Signal Processing in the Frequency Domain

Signal processing in the frequency domain is to compensate the shortcoming of signal processing in the time domain, which just reflects the variation from the waveform of a signal and cannot disclose the inherent nature of a signal. Frequency spectrum, as a most widely used analysis tool, can reflect the frequency components and distribution of a signal. Frequency spectrum analysis is the cornerstone of signal analysis in the frequency domain. Its fundamental theory is FT. First, a periodic signal is divided into a summation of several sinusoidal components by virtues of Fourier series. Then the frequency spectrum of the periodic signal can be obtained by the projection of these sine and cosine signals in the frequency axis. The amplitude of every frequency is the magnitude of corresponding sine or cosine signal. Therefore, the Fourier integral transform is proposed to transform a continuous periodic signal into the frequency spectrum. However, the Fourier integral transform can only process the continuous signals rather than discrete ones. It is unsuitable for us to process the signals by computers. Based on this motivation, subsequently, the discrete Fourier transform (DFT) is presented to obtain frequency spectrum of discrete signals by computers. With the increase of data length, traditional DFT requires more time to compute the DFT of a long enough signal. Thus, FFT is proposed to fast calculate DFT of a signal. The spectrum analysis based on FFT is to estimate the frequency spectrum by FT of sample data. In this section, FT is introduced to pave the way for Section 2.5. In addition, common statistical features in the frequency domain are also listed in this section.

2.4.1 Fourier Transform

Signal processing in the frequency domain is based on FT, which transforms a time–domain signal into a frequency spectrum to analyze signal characteristics in the frequency domain. FT mainly contains four important parts: Fourier series, Fourier integral transform, DFT, and FFT.

2.4.1.1 Fourier Series

The basic concept of FT is to decompose a signal as the summation of several sinusoidal components and almost all signals are decomposed in this way. This Fourier's original analysis, namely Fourier series, is applied to process the signals with finite length. In mechanical vibration analysis, it is used to process periodic signals which are generated by a machine rotating at a constant speed. Thus, for any periodic signal $x(t)$ with the period T it satisfies

$$x(t)=x(t+nT) \tag{2.16}$$

where n is an integer, it can be expanded as

$$x(t)=\frac{a_0}{2}+\sum_{k=1}^{\infty}a_k\cos(k\omega_0 t)+\sum_{k=1}^{\infty}b_k\sin(k\omega_0 t) \tag{2.17}$$

where ω_0 denotes the fundamental angular frequency. The coefficients of the cosine and sine components can be described as

$$a_k=\frac{2}{T}\int_{-T/2}^{T/2}x(t)\cos(k\omega_0 t)dt \tag{2.18}$$

$$b_k=\frac{2}{T}\int_{-T/2}^{T/2}x(t)\sin(k\omega_0 t)dt \tag{2.19}$$

Therefore, a given periodic signal can be decomposed into sine and cosine terms dependent on an arbitrary time assignment of zero time, but all frequency components at frequency $\omega_k(=k\omega_0)$ are given by

$$a_k \cos(\omega_k t) + b_k \sin(\omega_k t) \tag{2.20}$$

which is also written as

$$C_k \cos(\omega_k t + \varphi_k) \tag{2.21}$$

where $C_k = \sqrt{a_k^2 + b_k^2}$ and $\varphi_k = \tan^{-1}(b_k / a_k)$. Eq. (2.21) makes it clearer that the sinusoid component possesses constant amplitude and phase angle at the arbitrarily defined zero time. Here, zero time only affects the initial phase φ_k. Eq. (2.21) can also be expressed in an exponential form

$$\frac{C_k}{2}\{\exp[\mathrm{j}(\omega_k t + \varphi_k)] + \exp[-\mathrm{j}(\omega_k t + \varphi_k)]\} \tag{2.22}$$

which is illustrated as two rotating vectors with length $C_k/2$, one rotating at angular frequency ω_k with initial phase φ_k and the other rotating at angular frequency $-\omega_k$ with initial phase $-\varphi_k$, as illustrated in Fig. 2.10 (Randall, 1987).

According to this illustration of Fourier analysis, $x(t)$ can be expressed as a sum of two rotating vectors and further the Eq. (2.17) is replaced as

$$x(t) = \sum_{k=-\infty}^{\infty} A_k \exp(\mathrm{j}\omega_k t) \tag{2.23}$$

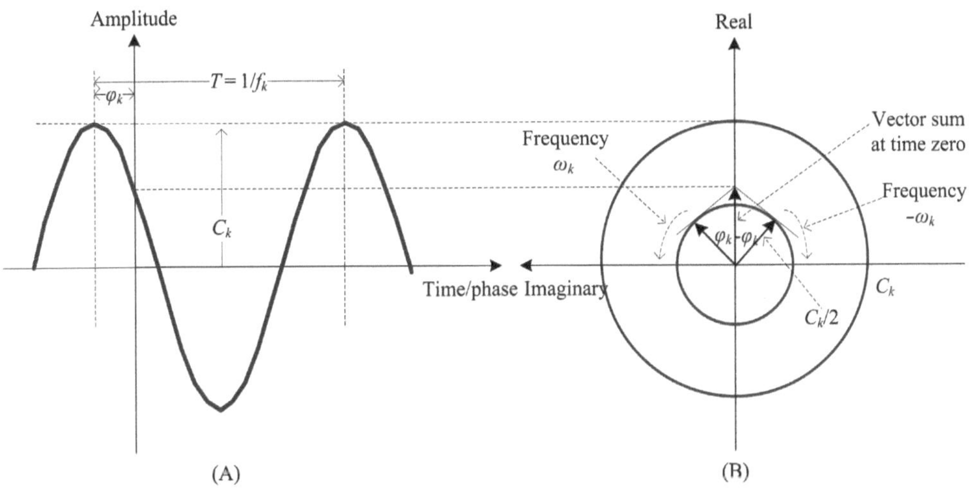

Figure 2.10. Representation of a sinusoid component as a sum of two rotating vectors. (A) A sinusoid component, (B) corresponding two rotating vectors.

where the coefficient A_k including the phase shift is

$$A_k = \frac{C_k}{2}\exp(j\varphi_k) \tag{2.24}$$

In addition, the coefficient A_k can also be calculated by

$$A_k = \frac{1}{T}\int_{-T/2}^{T/2} x(t)\exp(-j\omega_k t)dt \tag{2.25}$$

Its physical implication is that the multiplication by $\exp(-j\omega_k t)$ would subtract angular frequency ω_k from each component. It suggests that the one originally rotating at ω_k would be stopped in the position with zero time, which can be extracted by the integral. However, all other components still rotate at some other positive or negative multiples of ω_k. Each frequency component A_k, denotes the position and value of the corresponding rotating vector at zero time. To obtain its position at an arbitrary time t, we can make it rotate at angular frequency ω_k by multiplying an exponential term $\exp(j\omega_k t)$ and then all frequency components are summed via Eq. (2.23).

Fourier series is interpreted as a sum of rotating vectors with each of amplitude half $C_k/2$, which produces a two-sided spectrum because the component in each positive frequency is accompanied by its complex conjugate at negative frequency. The same result can also be achieved by retaining the positive frequency components only, but doubling their length to C_k and then taking the projection of each vector on real axis as illustrated in Fig. 2.11B (Randall, 2011). A signal with a one-sided spectrum like this is complex, for example, an analytic signal, but later it is interpreted that the projection on the imaginary axis is the Hilbert transform (HT) of the real part.

2.4.1.2 Fourier Integral Transform

An aperiodic signal can also be decomposed as a sum of complex exponentials to implement FT, especially for transients. We can allow the periodic time tend to infinity. Moreover, since transients have finite energy rather than finite power, the division of periodic T is removed. Finally, the FT can be derived from the Fourier series in Eqs. (2.25) and (2.23) as

$$X(f) = \int_{-\infty}^{\infty} x(t)\exp(-j2\pi ft)dt \tag{2.26}$$

$$x(t) = \int_{-\infty}^{\infty} X(f)\exp(j2\pi ft)df \tag{2.27}$$

respectively, where angular frequency ω_k has been substituted by the continuous frequency function f (Hz). Eqs. (2.26) and (2.27)

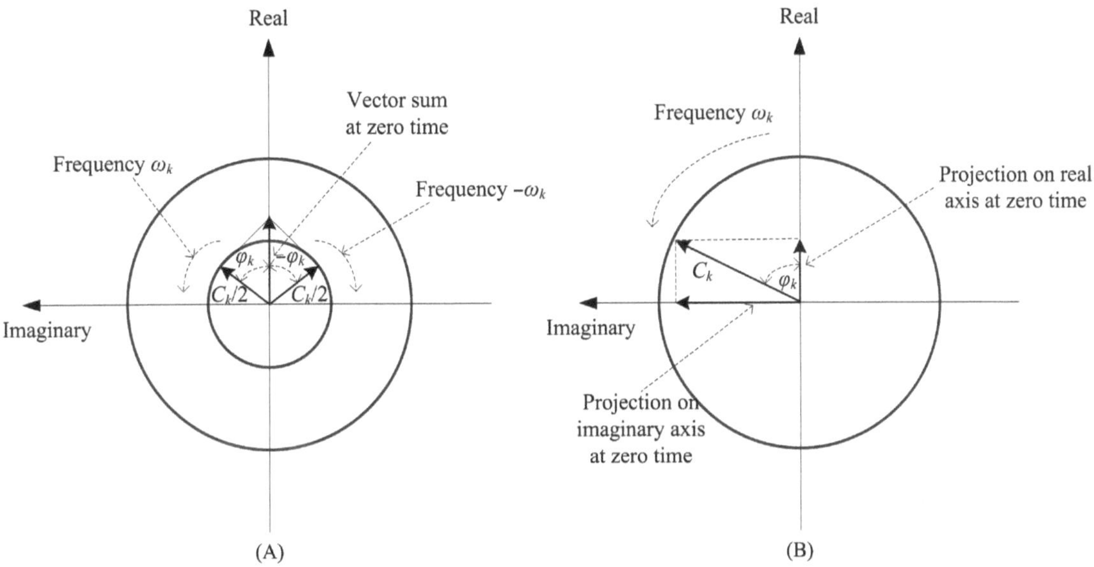

Figure 2.11. (A) Equivalence of the vector sum of positive and negative frequency components and (B) projection on the real axis of a positive frequency component.

are well known as the forward and inverse Fourier integral transforms, respectively.

The forward and inverse transforms in Eqs. (2.26) and (2.27) are almost symmetrical, and the only difference is the sign of the exponential term. This indicates that the equation which is used in the forward transform is also often used in the inverse transform, for example, the convolution theorem in next subsection. This is a special case for real and even functions, for which it makes no difference whether time or frequency runs forwards or backwards.

2.4.1.3 Discrete Fourier Transform

All signals to be processed digitally must be digitized or discretely sampled. The sampled time signal is discrete in the time domain, while it is periodic in the frequency domain due to symmetry of the FT. The corresponding versions of the forward and inverse transforms are

$$X(f)=\sum_{n=-\infty}^{\infty}x(t_n)\exp(-j2\pi ft_n) \tag{2.28}$$

$$x(t_n)=\frac{1}{f_s}\int_{-f_s/2}^{f_s/2}X(f)\exp(j2\pi ft_n)df \tag{2.29}$$

where $t_n = n\Delta t = n/f_s$, f_s is the sampling frequency and Δt is the sample interval. In general, the sampled time signals are infinite length, but the record length is all finite. Thus, this result in the same situation is the same as that in the Fourier series in the aspect that the frequency spectrum is discrete and the time record is implicitly periodic. Therefore, continuous infinite integrals of the FT will be interpreted as finite sums

$$X(k)=\frac{1}{N}\sum_{n=0}^{N-1}x(n)\exp\left(\frac{-\mathrm{j}2\pi kn}{N}\right) \tag{2.30}$$

$$x(n)=\sum_{k=0}^{N-1}X(k)\exp\left(\frac{\mathrm{j}2\pi kn}{N}\right) \tag{2.31}$$

This aforementioned transform pair, known as the DFT, corresponds most closely to that of Fourier series. Here the forward transform is divided by the length of record N to give correctly scaled Fourier series components. If the DFT is applied to other types of signals, such as transients or stationary random signals, the scaling has to be tuned accordingly. The forward DFT expression can also be written as the form of matrix multiplication

$$\mathbf{X}_k=\frac{1}{N}\mathbf{M}_{kn}\mathbf{x}_n \tag{2.32}$$

where $\mathbf{X}_k$ denotes the vector of N frequency components, that is, the $X(k)$ in Eqs. (2.30) and (2.31), while $\mathbf{x}_n$ denotes the N time samples $x(n)$. $\mathbf{M}_{kn}$ denotes a square matrix of unit vectors exp $(-\mathrm{j}2\pi kn/N)$ with angular orientation which is determined by the frequency index k (the rows) and time sample index n (the columns). Finally, the matrix expression is illustrated graphically in Fig. 2.12 (Randall, 2011). The zero frequency value $X(0)$ at $k = 0$ is simply the mean value of the time samples $x(n)$. When $k = 1$ the

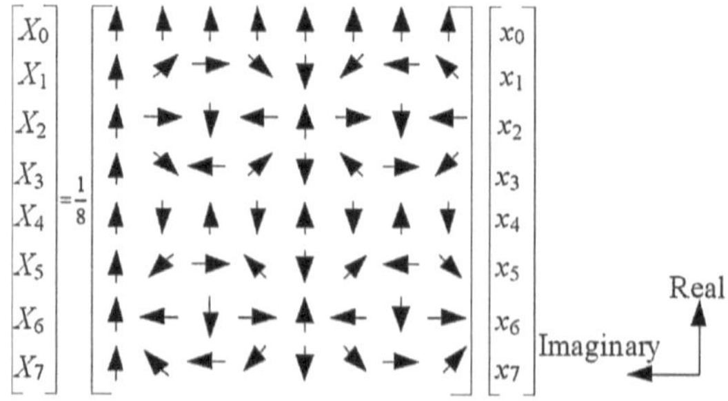

Figure 2.12. Matrix representation of the discrete Fourier transform (DFT).

unit vector rotates $-1/N$th of a revolution for each time sample increment, forming one complete and negative revolution after N samples. The rotation speed is proportionally increased with the increase of k value. For $k = N/2$, that is, the half sampling frequency, namely "Nyquist frequency," the vector turns through $-\pi$ for each time sample, but it is impossible to observe in the direction it has rotated about. For $k > N/2$ the vector rotates more than π in the negative direction but is also explained as rotating less than π in the positive direction and therefore if the signal has been filtered by a low-pass filter at half of the sampling frequency, the second half of $\mathbf{X}_k$ would include the negative frequency components ranging from minus one-half of the Nyquist frequency to just below zero.

2.4.1.4 Fast Fourier Transform

The fast Fourier transform (FFT) (Cooley and Tukey, 1965) is a high-efficient algorithm for computing the DFT in Eqs. (2.30) and (2.31). Beginning with the matrix version in Eq. (2.32), so-called radix 2 algorithm, the FFT is based on N which is a power of 2 and factorizes a modified version of the matrix $\mathbf{M}_{kn}$ into $\log_2 N$ matrices. Each matrix has the property that multiplication by them only needs N complex operations instead of N^2 operations which are required for direct multiplication by $\mathbf{M}_{kn}$. Therefore, the total number of complex operations is decreased from N^2 to $N\log_2 N$. For a typical case where $N = 1024 = 2^{10}$, the algorithm will save effectively a factor of more than 100.

For convenience, the matrix $\mathbf{M}_{kn}$ and its modified version are named as matrices **A** and **B**, respectively. The rows of the matrix **B** are arranged in "bit-reversed order," which are illustrated in Fig. 2.13. This indicates that the most significant bit is indexed rather than the least significant bit, and so the phase increments, for example, go from coarse to fine with the increase of row

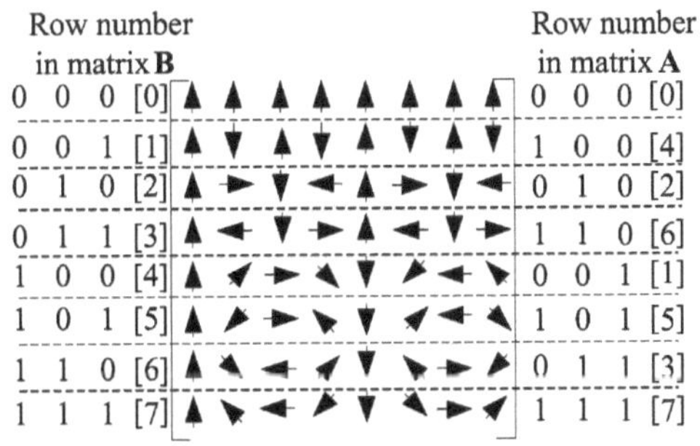

Figure 2.13. Modified matrix B with rows shifted to bit-reversed address.

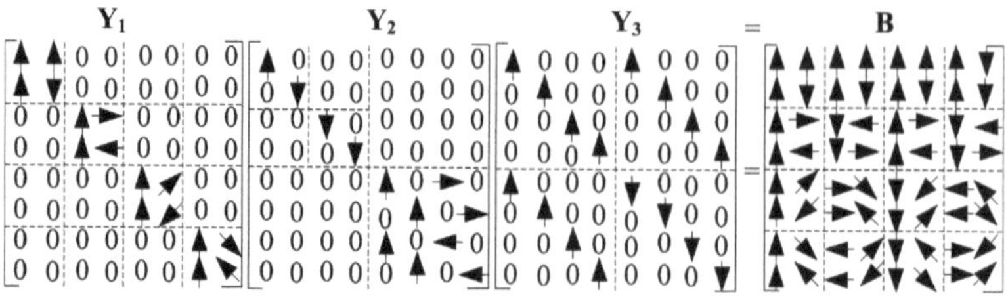

Figure 2.14. Matrix B factorized into three factor matrices Y_1, Y_2, Y_3.

number. Multiplication by matrix **B** indicates that the results are also in bit-reversed order. However, it is a simple operation that the results with bit-reversed order are reshuffled to the correct address pairwise. Moreover, this operation takes negligible time in comparison with the multiplications. We take $N = 8$ (and $\log_2 N = 3$) for an example as shown in Fig. 2.14. It illustrates that how matrix **B** can be factorized into three matrices, $\mathbf{Y}_1$, $\mathbf{Y}_2$, $\mathbf{Y}_3$, in each row of which there are only two nonzero elements, one of which is unity. Therefore, multiplication by each factor matrix needs only N complex multiplications and additions. From the pattern of the factor matrices it is clear that the decomposition can be generalized to higher powers of 2. The factor matrices include progressively finer rotations, for example, in this case 1/2, 1/4, and 1/8 rotations, respectively.

Therefore, more computation time can be further saved in special cases, such as the radix 4 and radix 8 transforms. Similarly, a not very effective saving can also be made for factorization rather than in powers of 2, but the main point is that the properties of the FFT are based on those of the DFT.

2.4.2 Common Statistical Features in the Frequency Domain

When a fault occurs in the machinery, some abnormal frequencies would emerge in the frequency spectra of the vibration signals, which can reflect the health conditions of the machinery. Moreover, frequency spectra are more sensitive to incipient faults since an imperceptible change will produce a spectrum line in corresponding frequency spectrum. Therefore, it is vital for fault diagnosis to extract some indicators in the frequency domain. In addition, these frequency–domain features or indicators may contain some fault-related information that is not present in the time domain. In other words, these features in the frequency domain are an effective compensation for those time–domain features alone.

Table 2.2 Common Statistical Features in the Frequency Domain

$$p_1 = X_{mf} = \frac{\sum_{k=1}^{K} s(k)}{K} \qquad p_2 = \frac{\sum_{k=1}^{K} \left(s(k) - p_1\right)^2}{K-1} \qquad p_3 = \frac{\sum_{k=1}^{K} \left(s(k) - p_1\right)^3}{K\left(\sqrt{p_2}\right)^3} \qquad p_4 = \frac{\sum_{k=1}^{K} \left(\left(s(k) - p_1\right)\right)^4}{Kp_2^2}$$

$$p_5 = X_{fc} = \frac{\sum_{k=1}^{K} f_k s(k)}{\sum_{k=1}^{K} s(k)} \qquad p_6 = \sqrt{\frac{\sum_{k=1}^{K} \left(f_k - p_5\right)^2 s(k)}{K}} \qquad p_7 = X_{rmsf} = \sqrt{\frac{\sum_{k=1}^{K} f_k^2 s(k)}{\sum_{k=1}^{K} s(k)}} \qquad p_8 = \sqrt{\frac{\sum_{k=1}^{K} f_k^4 s(k)}{\sum_{k=1}^{K} f_k^2 s(k)}}$$

$$p_9 = \frac{\sum_{k=1}^{K} f_k^2 s(k)}{\sqrt{\sum_{k=1}^{K} s(k) \sum_{k=1}^{K} f_k^4 s(k)}} \qquad p_{10} = \frac{p_6}{p_5} \qquad p_{11} = \frac{\sum_{k=1}^{K} \left(f_k - p_5\right)^3 s(k)}{Kp_6^3} \qquad p_{12} = \frac{\sum_{k=1}^{K} \left(f_k - p_5\right)^4 s(k)}{Kp_6^4}$$

$$p_{13} = \frac{\sum_{k=1}^{K} \left(f_k - p_5\right)^{1/2} s(k)}{K\sqrt{p_6}} \qquad p_{14} = \sqrt{\frac{\sum_{k=1}^{K} \left(f_k - p_5\right)^2 s(k)}{\sum_{k=1}^{K} s(k)}}$$

where $s(k)$ is a spectrum for $k = 1, 2, \ldots, K$, K is the number of spectrum lines and f_k is the frequency value of the kth spectrum line.

Here, 13 common statistical features in the frequency domain are shown in Table 2.2. They are all based on statistics, so they are called statistical features. Feature p_1, namely mean frequency, may indicate the vibration energy in the frequency domain, which represents the average of the amplitudes of all the frequencies. With the increase of fault degrees, the mean frequency must become larger. Features $p_2 - p_4$, p_6, and $p_{11} - p_{13}$ may describe the convergence of the spectrum power, reflecting the energy of frequency spectrum. Features p_5 and $p_7 - p_9$ may show the position change of main frequencies which are dominant in the frequency spectrum. Therefore, these frequency–domain features can reflect the mechanical health conditions from different aspects. Different fault will result in the variation of the whole frequency spectrum and power spectrum. According to different fault categories, therefore, we should select suitable statistical features to achieve accurate diagnosis results.

2.5 Signal Processing in the Time-Frequency Domain

Since extensively used FT is restricted to stationary signals, it is unsuitable to process the fault signals of machinery which are nonstationary when a fault occurs. Moreover, the frequency spectrum acquired by FT cannot reflect the change of each frequency with time. To overcome these disadvantages, time–frequency analysis techniques are developed to analyze a nonstationary signal from time–frequency distribution. Up to now, classical time–frequency analysis techniques mainly include STFT, WVD, wavelet transform (WT), even wavelet packet transform (WPT), and HHT, etc. These techniques can reflect the change of a signal from both time domain and frequency one. Moreover, they can disclose intrinsic nature of a time-varying signal, for example, the wavelet transform. The principles of these classical techniques are introduced and interpreted in this section.

2.5.1 Short-Time Fourier Transform

For a given signal $x(t)$, the natural "stationary transform" is the well-known FT in Eq. (2.26), which is rewritten as

$$X(f)=\int_{-\infty}^{+\infty} x(t)\exp(-j2\pi ft)dt \tag{2.33}$$

where $X(f)$ defines the conception of global frequency f in a signal. This classical linear time–frequency representation, which is called the STFT, is first proposed by Gabor (Gabor, 1946; Rabiner and Schafer, 1978). This technique has been widely used in nonstationary signal analysis. Suppose that it is stationary when a signal $x(t)$ is seen through a window $h(t)$ with finite length and central location at time τ. The FT of the windowed signal $x(t)h^*(t-\tau)$ yields the STFT

$$STFT_x(\tau,f)=\int_{-\infty}^{+\infty} x(t)\,h^*(t-\tau)\exp(-j2\pi ft)dt \tag{2.34}$$

which maps the signal into a two-dimensional function in a time–frequency plane (τ, f). $h^*(t-\tau)$ denotes the complex conjugate of function $h(t-\tau)$ and the parameter f in Eq. (2.34) is equivalent to the Fourier frequency. Since the STFT is derived from traditional FT, so many properties of the FT are carried over to the STFT. However, the analysis performance of STFT is determined critically by selecting the window $h(t)$. Since the FT of a Gaussian function is also Gaussian, the "optimal" window for time-localization can be achieved by applying any Gaussian function

$$h_G(t)=\frac{1}{2\sqrt{\pi\alpha}}\exp\left(-\frac{t^2}{4\alpha}\right) \tag{2.35}$$

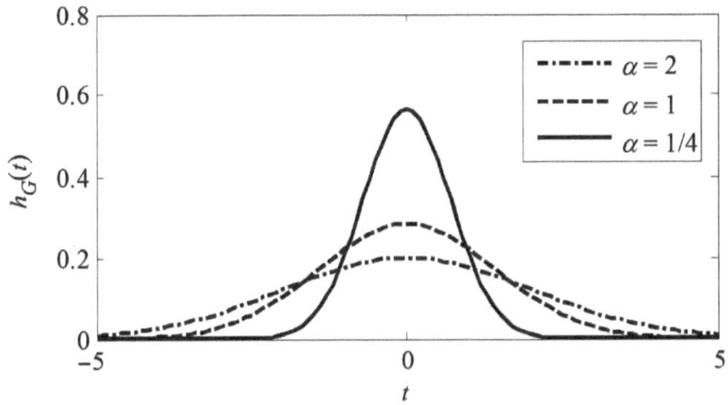

Figure 2.15. Gaussian function with different α.

where $\alpha > 0$ is fixed and the window function is shown in Fig. 2.15.

Considering the capability of the STFT to discriminate two pure sinusoids, a window function and its FT are written as $h(t)$ and $H(f)$, respectively. The corresponding bandwidth Δf is defined as

$$(\Delta f)^2 = \frac{\int f^2 |H(f)|^2 \, df}{\int |H(f)|^2 \, df} \tag{2.36}$$

where the denominator is the energy of $h(t)$. If they are only more than Δf apart, two sinusoids will be distinguished in the frequency domain. Therefore, the frequency resolution of the STFT analysis is defined by Δf. Similarly, the time resolution is defined by Δt as

$$(\Delta t)^2 = \frac{\int t^2 |h(t)|^2 \, dt}{\int |h(t)|^2 \, dt} \tag{2.37}$$

where the denominator is again the energy of $h(t)$. Two pulses in the time domain can be distinguished only if they are more than Δf apart.

However, the resolution in the time and frequency domains cannot be arbitrarily small simultaneously because their product has a lower boundary.

$$\text{Time-Bandwidth product} = \Delta t \Delta f \geq \frac{1}{4\pi} \tag{2.38}$$

The equality holds if and only if Gaussian windows are applied. The Eq. (2.38) is generally referred to as the uncertainty principle or Heisenberg inequality. It indicates that one can only make a tradeoff between time resolution and frequency resolution.

More importantly, once a window has been selected for the STFT, the time–frequency resolution determined by Eqs. (2.36) and (2.37) is fixed in the entire time–frequency plane since the same window is applied in all frequencies. If the signal consists of some small bursts associated with long quasi-stationary components, each type of components can be analyzed by using STFT with good time resolution or frequency one.

2.5.2 Wigner–Ville Distribution

WVD (Newland, 1994, Claasen, 1980) is a powerful time–frequency analysis tool, which focuses on the total energy distribution of a signal at a particular time and frequency. It was developed in quantum mechanics by Wigner (1932) and applied to harmonic analysis in signal processing by Ville (1948). Furthermore, the WVD has a high resolution of instantaneous energy density in both time and frequency domains. It is based on the following ideas. Assuming that an instantaneous correlation function of a time–domain signal $x(t)$ at time t is

$$R_x(t,\tau) = E\left[x\left(t+\frac{\tau}{2}\right)x^*\left(t-\frac{\tau}{2}\right)\right] \tag{2.39}$$

where τ denotes the time delay, the operator E represents the average of a statistical ensemble or collection of time–domain signal $x(t)$ and "*" denotes complex conjugate in Hilbert space. FT of this correlation function is then computed as

$$S_x(t,\omega) = \frac{1}{2\pi}\int_{-\infty}^{\infty} R_x(t,\tau)\exp(-j\omega\tau)d\tau \tag{2.40}$$

where the instantaneous spectral density $S_x(t, \omega)$ is a function associated with time t and frequency parameter ω ($\omega = 2\pi f$). It is impossible to calculate an ensemble average indeed. If that averaging operation is canceled and Eq. (2.39) is substituted into Eq. (2.40), we can obtain

$$WD_x(t,\omega) = \frac{1}{2\pi}\int_{-\infty}^{\infty} x\left(t+\frac{\tau}{2}\right)x^*\left(t-\frac{\tau}{2}\right)\exp(-j\omega\tau)d\tau \tag{2.41}$$

where $WD_x(t, \omega)$ is named as the WVD function associated with time t and angular frequency ω. The integral in Eq. (2.41) is centered at time t. It covers an infinite range of τ and is closely related with the characteristics of signal $x(t)$ far away from the local time t. In practice, we often apply a finite time window to solve infinite problem. Therefore, we obtain the pseudo Wigner–Ville distribution (PWD)

$$PWD_x(t,\omega)=\frac{1}{2\pi}\int_{-\infty}^{+\infty} WD_x(t,\Omega)W_h(0,\ \omega-\Omega)d\Omega \tag{2.42}$$

where W_h is the WVD with window $h(t)$. $PWD_x(t,\ \omega)$ is an approximation of $WD_x(t,\ \omega)$ as a result of $WD_x(t,\ \omega)$ windowed by window $h(t)$. It is an energy distribution of a signal on time–frequency plane.

For a discrete signal $x(n)$, Eq. (2.42) can be written as

$$PWD_x\left(n,\frac{k\pi}{N}\right)=2\sum_{m=M/2+1}^{M/2} x(n+m)x^*(n-m)|h(m)|^2\exp\left(\frac{-j2\pi mk}{N}\right) \tag{2.43}$$

where $k\pi/N$ denotes the angular frequency, N denotes the data length, and $h(m)$ denotes the time window with length m and its center located at n. The window $h(m)$ can be selected as those windows which are applied in FT. Eq. (2.43) can be computed by employing FFT algorithm.

WVD is a time–frequency representation, which overcomes the tradeoff problem between time and frequency resolution. Moreover, WVD has various distinct properties, such as preservation of time and frequency support, instantaneous frequency, group delay. These properties make the WVD a powerful tool for nonstationary signal analysis and processing.

2.5.3 Wavelet Analysis

Wavelet analysis provides multiresolution analysis and signal decomposition in independent frequency bands. In this kind of data decomposition, low-pass and band-pass filters are applied simultaneously at each operation. The signal energy in these frequency bands is conservative due to orthonormal decomposition without redundancy or omissions.

As we know, both Mallat's pyramid wavelet transform algorithm and Wickerhause wavelet packet algorithm make the previous data half at each decomposition operation, which suggests that after decompositions the number of data samples is manipulated shrinks by 2^j. Although the decomposed signals in certain frequency band are stationary, the FFT spectral analysis technique cannot be applied to process these signals since the data shrinkage causes that frequency resolution is too rough to analyze.

Generally, any signal can be decomposed as the sum of wavelet series with different wavelet coefficients. Therefore, we can construct orthogonal wavelets by adopting optimal method to optimize the wavelet coefficients based on convergence condition. This method is extensively used in the diagnosis of mechanical faults. It is significant to reprocess the signal decomposed by wavelet techniques. Therefore, we should organically combine wavelet techniques with some conventional methods to acquire much more information included in mechanical vibration signals.

2.5.3.1 Wavelet Transform

Let R represent real numbers, and $L^2(R)$ represent the vector space of measurable, square-integrable one-dimensional signal $x(t)$. For $x(t) \in L^2(R)$ and $g(t) \in L^2(R)$, the inner product between $x(t)$ and $g(t)$ is expressed as

$$\langle x(t), g(t) \rangle = \int_{-\infty}^{+\infty} x(t)\, g(t)\, dt \tag{2.44}$$

From Eq. (2.33), the FT is achieved by utilizing the basis function $g(t) = \exp(-j2\pi ft)$, which can be written as the inner product form

$$X(f) = \int_{-\infty}^{+\infty} x(t)\, \exp(-j2\pi ft) dt = \langle x(t), g(t) \rangle \tag{2.45}$$

Similarly, the STFT method is that basis function $g(t) = \exp(-j2\pi ft)$ in Eq. (2.45) is substituted by basis function $h^*(t-\tau)\exp(-j2\pi ft)$. So the Eq. (2.45) can be written as

$$\begin{aligned} STFT_x(\tau, f) &= \int_{-\infty}^{+\infty} x(t)\, h^*(t-\tau)\exp(-j2\pi ft) dt \\ &= \langle x(t), h^*(t-\tau)\exp(-j2\pi ft) \rangle \end{aligned} \tag{2.46}$$

Since the space $L^2(R)$ of measurable functions $x(t)$ is defined on the real line R, it satisfies

$$\int_{-\infty}^{+\infty} |x(t)|^2\, dt < \infty \tag{2.47}$$

Particularly, since the local average value of each function in space $L^2(R)$ must decay to zero at $\pm\infty$, the sinusoidal functions $g(t) = \exp(-j2\pi ft) = \cos 2\pi ft - \sin 2\pi ft$ cannot decay to zeros at $\pm\infty$ and do not belong to space $L^2(R)$. In fact, if we search for "waves" that produce $L^2(R)$, these waves should decay to zero at $\pm\infty$. For all practical purposes, the decay should be very fast. Thus, we search for small wave, or "wavelets," to generate space $L^2(R)$ (Chui, 2014). We define a single function as $\psi(t)$ to produce all of space $L^2(R)$. However, a new problem occurs that if the wavelet $\psi(t)$ decays very fast, it will not cover the whole real line. The feasible scheme is to shift $\psi(t)$ with b along real line R and further $\psi(t-b)$ is obtained.

Frequencies of these waves should be partitioned into consecutive frequency bands, thus they can be obtained by dilation, that is, dilation by a. Assuming that function $\psi(t)$ has unit length, that is, $\|\psi\|_2 = 1$ and the norm $\|\psi\|_2 = \langle \psi, \psi \rangle^{1/2}$, we can obtain

$$\psi_{a,b}(t) = |a|^{-1/2}\psi\left(\frac{t-b}{a}\right),\ a, b \in R \tag{2.48}$$

where $\psi_{a,b}(t)$ denotes a family function named wavelets. The $\psi(t)$ is named mother wavelet, and each $\psi_{a,b}(t)$ is a basic wavelet. If $\psi(t) \in L^2(R)$ satisfies the "admissibility" condition (Chui, 2014)

$$C_\psi = \int_{-\infty}^{+\infty} \frac{|\hat{\psi}(\omega)|^2}{|\omega|} d\omega < \infty \tag{2.49}$$

then $\psi(t)$ is a "basic wavelet." Relative to every basic wavelet $\psi_{a,b}(t)$, the integral wavelet transform on $L^2(R)$ is defined as

$$WT_x(b,a) = |a|^{-1/2} \int_{-\infty}^{+\infty} x(t)\psi^*\left(\frac{t-b}{a}\right)dt = \langle x, \psi_{a,b} \rangle,\ x(t) \in L^2(R) \tag{2.50}$$

where $a, b \in R$ and $a \neq 0$, and $\hat{\psi}(\omega)$ is FT of $\psi(t)$.

The basic wavelet $\psi(t)$ produces a time–frequency window with finite area $4\Delta_\psi \Delta_{\hat{\psi}}$. Additionally, $\hat{\psi}(\omega)$ is continuous under this additional assumption, so the finiteness of C_ψ in Eq. (2.49) suggests $\hat{\psi}(0) = 0$ or equivalently

$$\int_{-\infty}^{+\infty} \psi(t)dt = 0 \tag{2.51}$$

which means that the $\psi(t)$ must have some oscillation and it is the reason why $\psi(t)$ is named as "wavelet" (Chui, 2014).

When a function $x(t)$ is scaled: $x(t) \rightarrow x(at)$, where $a > 0$, it is contracted if $a > 1$ and expanded if $a < 1$. Now, the Eq. (2.50) is rewritten as

$$WT_x(b,a) = a^{-1/2} \int_{-\infty}^{+\infty} x(t)\psi^*\left(\frac{t-b}{a}\right)dt \tag{2.52}$$

or by a substitution of variable as

$$WT_x(b,a) = a^{1/2} \int_{-\infty}^{+\infty} x(at)\psi^*\left(t - \frac{b}{a}\right)dt \tag{2.53}$$

The explanation of Eq. (2.52) is that with the increase of the scale, the impulse response of the filter $\psi((t-b)/a)$ is spreading out in time when we consider only long-time behavior. Equivalently, Eq. (2.53) implies that with the increase of the scale, an increasingly contracted version of the signal is seen through a filter with constant length. That is the interpretation of the scale factor a in maps. Very large scales indicate global views, while small scales indicate detailed views.

Let Z represent the set of integers: $Z = \{\ldots, -2, -1, 0, 1, 2, \ldots\}$. To improve computational efficiency, integral powers of 2 is generally applied to frequency partitioning and time shift, which means that let $a = 2^j$, $b = 2^j k$, $j, k \in Z$. Through a binary dilation (i.e., dilation by 2^j) and a dyadic translation (of $k/2^j$), we can obtain the dyadic wavelet $\psi_{j,k}(t)$ as follow

$$\psi_{j,k}(t) = 2^{-j/2}\psi(2^{-j}t - k), \quad j, k \in Z \tag{2.54}$$

The Kronecker symbol is

$$\delta_{j,k} = \begin{cases} 1 & j = k \\ 0 & j \neq k \end{cases}, \quad j, k \in Z \tag{2.55}$$

A function $\psi(t) \in L^2(R)$ is named orthogonal wavelet, if the family $\{\psi_{j,k}(t)\}$ defined in Eq. (2.48) is an orthonormal basis of $L^2(R)$, it satisfies

$$\langle \psi_{j,k}, \psi_{l,m} \rangle = \delta_{j,l} \cdot \delta_{k,m}, \quad j, k, l, m \in Z \tag{2.56}$$

and each $x(t) \in L^2(R)$ is expressed as

$$x(t) = \sum_{j,k=-\infty}^{+\infty} c_{j,k}\psi_{j,k}(t) \tag{2.57}$$

where the convergence of the series in Eq. (2.57) satisfies

$$\lim_{M_1, N_1, M_2, N_2 \to \infty} \left\| x - \sum_{j=-M_2}^{N_2} \sum_{k=-M_1}^{N_1} c_{j,k}\psi_{j,k} \right\|_2 = 0 \tag{2.58}$$

The series expression of in Eq. (2.57) is named a wavelet series. The wavelet coefficients $c_{j,k}$ are calculated as

$$c_{j,k} = \langle x(t), \psi_{j,k}(t) \rangle \tag{2.59}$$

According to the definition of an integral wavelet transform $WT_x(b, a)$ on $L^2(R)$ in Eq. (2.52), the wavelet coefficients in Eqs. (2.57) and (2.59) can be written as

$$c_{j,k} = WT_x\left(\frac{k}{2^j}, \frac{1}{2^j}\right) \tag{2.60}$$

The linear transformation WT_x is named "integral wavelet transform" related to the "basic wavelet" $\psi_{a,b}(t)$. Therefore, the (j, k) th wavelet coefficient of $x(t)$ can be obtained by the integral wavelet transform of $x(t)$ estimated at the dyadic position $b = 2^j k$ with $a = 2^j$, $j, k \in Z$, where the same orthogonal wavelet $\psi_{a,b}(t)$ is applied to produce the wavelet series in Eq. (2.57) and define the integral wavelet transform in Eq. (2.50). It is noticed that the Fourier series and the FT are basically irrelevant. However, it is closely relevant between the corresponding wavelet series in Eq. (2.57) and the integral wavelet transform in Eq. (2.50), which is described in Eq. (2.60).

2.5.3.2 Wavelet Basis and Fast Pyramidal Algorithm

According to the scaling function $\varphi(t)$, wavelet bases in $L^2(R)$ can be further constructed. Assuming that $\varphi(t)$ is a scaling function of wavelet transform, which can produce multiresolution analysis $\{V_j\}_{j\in Z}$ and satisfy the following two conditions.

1. $\{\varphi_{0,n}\}_{n\in Z}$ is an orthonormal basis in the approximation subspace V_0.
2. Suppose that existing $\{h_n\} \in l^2$ can make $\varphi(t) = \sum_{n\in Z} h_n \varphi(2t-n)$, here $\{h_n\}_{n\in Z} \in l^2$. Let $g_n = (-1)^n h(1-n)$, $\psi(t) = \sum_{n\in Z} g_n \varphi(2t-n)$ and $W_j = \text{span}_{L^2(R)} \{\psi_{j,n}(t) = 2^{j/2}\psi(2^j t - n), n \in Z\}$ such that
   a. $W_j \perp V_j$, $W_j \oplus V_j = V_{j+1}$, and $L^2(R) = \oplus_{j\in Z} W_j$, $W_j \perp W_k$, $j \neq k$,
   b. $\{\psi_{j,n}\}_{j,n\in Z}$ is orthonormal basis in detail space W_j, so $\{\psi_{j,n}\}_{j,n\in Z}$ is orthonormal basis in $L^2(R)$.

According to the fact that $\varphi(t) \in V_0$, $\psi(t) \in W_0$ and $W_0 \subset V_1$, there exist two sequences $\{h_n\}_{n\in Z}$, $\{g_n\}_{n\in Z}$ and can satisfy the following two-scale equations

$$\begin{cases} \varphi(t) = \sum_n h_n \varphi(2t-n) \\ \psi(t) = \sum_n g_n \varphi(2t-n) \end{cases} \tag{2.61}$$

and decomposition equation

$$\varphi(t-n) = \sum_k \{h_{n-2k}\varphi(t-k) + g_{n-2k}\psi(t-k)\}, \quad \varphi(t) \in V_0, \psi(t) \in W_0 \tag{2.62}$$

The Fourier interpretation of Eq. (2.61) can be expressed as

$$\begin{cases} \hat{\varphi}(\omega) = H\left(\dfrac{\omega}{2}\right)\hat{\varphi}\left(\dfrac{\omega}{2}\right) \\ \hat{\psi}(\omega) = G\left(\dfrac{\omega}{2}\right)\hat{\varphi}\left(\dfrac{\omega}{2}\right) \end{cases} \tag{2.63}$$

where $H(\omega)=\sum_{n=-\infty}^{+\infty} h_n \exp(-jn\omega)$ and $G(\omega)=\sum_{n=-\infty}^{+\infty} g_n \exp(-jn\omega)$.

The two sequences $\{h_n\}_{n\in Z}$, $\{g_n\}_{n\in Z}$ in the time domain are called Quadrature Mirror Filters (QMF) in engineering and satisfy two equations $\sum h_{n-2j} h_{n-2k} = \delta_{j,k}$ and $\sum h_n = 1$. The corresponding $H(\omega)$ and $G(\omega)$ are the two expressions of QMF in the frequency domain.

The Eqs. (2.61) and (2.62) establish the relations among the scaling function, wavelet function, and QMF. It is helpful for us to construct the wavelet function. According to Eq. (2.61), the scaling function $\varphi(t) = \sin \pi t/(\pi t)$ can be constructed as corresponding wavelet function

$$\psi(t)=\frac{\sin\pi\left(t-\frac{1}{2}\right)-\sin 2\pi\left(t-\frac{1}{2}\right)}{\pi\left(t-\frac{1}{2}\right)} \tag{2.64}$$

According to the polynomial splines of order $2p+1$, we can build the following multiresolution approximations (Mallat, 1989a,b)

$$\begin{cases} \hat{\varphi}(\omega)=\dfrac{1}{\omega^n\sqrt{\sum_{2n}(\omega)}} \\ \hat{\psi}(\omega)=\dfrac{e^{-j(\omega/2)}\sqrt{\sum_{2n}\left(\frac{\omega}{2}+\pi\right)}}{\omega^n\sqrt{\sum_{2n}(\omega)\sum_{2n}\left(\frac{\omega}{2}\right)}} \end{cases} \tag{2.65}$$

where $n = 2p + 2$ and $\sum_n(\omega)=\sum_{k=-\infty}^{+\infty}\frac{1}{(\omega+2k)^n}$.

Based on Eq. (2.65), there exists $\hat{\varphi}(2\omega)=H(\omega)\hat{\varphi}(\omega)$. Then we can obtain

$$H(\omega)=\sqrt{\frac{\sum_{2n}(\omega)}{2^{2n}\sum_{2n}(2\omega)}} \tag{2.66}$$

where $H(\omega)$ is the transfer function of the quadrature mirror filter. Moreover, the filter is symmetrical. Therefore, the impulse response coefficients of the QMF, $\{g_n\}_{n\in Z}$, are able to be calculated by the equation $g_n = (-1)^{1-n} h_{1-n}$. According to multiresolution approximation and cubic splines, we can further obtain the functions $\hat{\varphi}(\omega)$ and $\hat{\psi}(\omega)$ with $n = 4$.

Suppose that A is a linear projection operator on multiresolution vector space $V_j \in L^2(R)$, which approximates a signal at the resolution 2^j and the original signal $x(t)$ is measurable with finite energy $x(t) \in L^2(R)$. Among all the approximated functions, the function $Ax^j(t)$ is the most similar to $x(t)$.

$$\forall y(t) \in V_j, \quad \|y(t) - x(t)\| \geq \|Ax^j(t) - x(t)\| \tag{2.67}$$

where $Ax^j(t)$ is named approximation signal. Since $V_j = V_{j-1} \oplus W_{j-1}$, the unique expression of $Ax^j(t)$ can be obtained as

$$Ax^j(t) = Ax^{j-1}(t) + Dx^{j-1}(t) \tag{2.68}$$

where $Dx^{j-1}(t)$ is named the detail signal at the resolution 2^j. It includes the difference of information between Ax^j and Ax^{j-1}. For convenience, let $Ax^i(t) = Ax^i$ and $Dx^i(t) = Dx^i$, then Eq. (2.68) can be further expressed as

$$\begin{aligned} Ax^j &= Ax^{j-1} + Dx^{j-1} \\ &= Ax^{j-2} + Dx^{j-2} + Dx^{j-1} = \cdots\cdots \\ &= Ax^{j-M} + Dx^{j-M} + Dx^{j-M+1} + \cdots + Dx^{j-1} \end{aligned} \tag{2.69}$$

where $Ax^i \in V_i$ and $Dx^i \in W_i$. $Ax^i(t)$ and $Dx^j(t)$ can be written as

$$\begin{cases} Ax^j(t) = \sum_k c_k^j \varphi(2^j t - k) \\ Dx^j(t) = \sum_k d_k^j \psi(2^j t - k) \end{cases} \tag{2.70}$$

According to the decomposition relation between $\varphi(t)$ and $\psi(t)$ in Eq. (2.61) and $Ax^j = Ax^{j-1} + Dx^{j-1}$, $\{c_k^{j-1}\}$ and $\{d_k^{j-1}\}$ can be calculated from $\{c_k^j\}$ as

$$\begin{cases} c_k^{j-1} = \sum_n h_{n-2k} c_k^j \\ d_k^{j-1} = \sum_n g_{n-2k} c_k^j \end{cases} \tag{2.71}$$

Similarly, based on the scaling relation in Eq. (2.61) and $Ax^j = Ax^{j-1} + Dx^{j-1}$, the reconstruction algorithm for $\{c_k^j\}$ from $\{c_k^{j-1}\}$ and $\{d_k^{j-1}\}$ can be expressed as

$$c_k^j = \sum_n \left\{ h_{k-2n} c_k^{j-1} + g_{k-2n} d_k^{j-1} \right\} \tag{2.72}$$

Suppose that $\{x_n^0\}$ is a sequence of discrete sampling data of $x(t)$ and $Ax^0 = \{x_n^0\}$ where $n \in Z$, according to Eq. (2.71) we can obtain

$$\begin{cases} Ax_k^{-1} = \sum_n h_{n-2k}\left(Ax_k^0\right) \\ Dx_k^{-1} = \sum_n g_{n-2k}\left(Ax_k^0\right) \end{cases} \tag{2.73}$$

For simplicity, discrete operators of wavelet decomposition can be expressed as two operators H and G, so Eq. (2.73) can be rewritten as $Ax^{j-1} = H(Ax^j)$ and $Dx^{j-1} = G(Ax^j)$, which is the famous Mallat's pyramid decomposition algorithm. The corresponding reconstruction algorithm is written as

$$Ax^0 = \sum_k \left\{h_{n-2k}\left(Ax^{-1}\right) + g_{n-2k}\left(Dx^{-1}\right)\right\} \tag{2.74}$$

Similarly, discrete operators of reconstruction are expressed as H^* and G^*, then Eq. (2.74) is rewritten as

$$Ax^0 = H^*\left(Ax^{-1}\right) + G^*\left(Dx^{-1}\right) \tag{2.75}$$

From Eqs. (2.73) and (2.74), it is noted that the Mallat's pyramid algorithm for signal decomposition and reconstruction is calculated by these coefficients of low-pass and band-pass filters $\{h_n\}_{n\in Z}$ and $\{g_n\}_{n\in Z}$ rather than scaling function $\varphi(t)$ and wavelet function $\psi(t)$. Moreover, the operation amount of the algorithm is proportional to $O(N\log_2 N)$ where N is the data length. In the process of signal decomposition, these filters are generally included in downsampling operations. Thus, steps of the algorithm are that the data is halved each time and every other sample is thrown away at each operation. This characteristic can effectively improve the algorithm's speed because the downsampling reduces the computation at each iteration geometrically, after J iterations, for example, the number of samples manipulated shrinks by 2^J. For signal reconstruction, the pyramid algorithm simply inverts the decomposition process. It combines two filters with upsampling operations. The upsampling operation in essence attempts to insert a zero between data values at every iteration, thereby doubling the data. This algorithm to reconstruct the original signal is also very fast. It also requires roughly the same number of computer's CPU cycles as the algorithm does for signal decomposition (Bruce et al., 1996). The status of the pyramidal algorithm in the wavelet transform is equivalent to that of FFT algorithm in the FT. This characteristic can make wavelet transform a very powerful tool in the practical engineering applications.

2.5.3.3 *Wavelet Packet Transform*

From Eq. (2.71) it is observed that high frequency sequences $\left\{d_k^j\right\}$ cannot be further decomposed by the Mallat's algorithm of wavelet decomposition. Here, the signal is only decomposed in subspaces V_j rather than in subspaces W_j. Therefore, it has ignored the high frequency resolution for signal processing. In order to solve this problem, later wavelet packet transform is developed by Coifman et al. (1994).

Assuming that $\{h_n\}_{n\in Z}$ is a conjugate quadrature mirror filter and satisfies

$$\sum_n h_{n-2k}\, h_{n-2l} = \delta_{k,l},\ \sum_n h_n = 2^{1/2} \tag{2.76}$$

where we can let

$$g_n = (-1)^k h_{1-k} \tag{2.77}$$

Now a family of wave function $w_n \in L^2(R)$, $n = 1, 2, \ldots$, is defined recursively from $\varphi(t)$ and $\psi(t)$ as follows (Wickerhauser, 1991, Qin and Yang, 1994):

$$w_0(t) = \varphi(t),\quad w_1 = \psi(t) \tag{2.78}$$

and

$$\begin{cases} w_{2n}(t) = 2^{1/2}\sum_k h_k w_n(2t-k) \\ w_{2n+1}(t) = 2^{1/2}\sum_k g_k w_n(2t-k) \end{cases} \tag{2.79}$$

where $\{w_n(t)\}_{n\in Z}$ is named wavelet packets based on scaling function $w_0(t) = \varphi(t)$ (Qin and Yang, 1994).

Similar to wavelet transform, we can define two operators H and G as

$$\begin{cases} H\{s_k\}_{(j)} = \sum_k s_k h_{k-2j} \\ G\{s_k\}_{(j)} = \sum_k s_k g_{k-2j} \end{cases} \tag{2.80}$$

and their adjoining operator H^* and G^* as

$$\begin{cases} H^*\{s_j\}_{(k)} = \sum_j s_j h_{k-2j} \\ G^*\{s_j\}_{(k)} = \sum_j s_j g_{k-2j} \end{cases} \tag{2.81}$$

where H, G, H^*, and G^* satisfy $H^*H + G^*G = I$, $HG^* = 0$, and $l^2 = H^*l^2 \oplus G^*l^2$. Therefore, Eq. (2.79) is rewritten as

$$\begin{cases} w_{2n}(t-l) = 2^{1/2}\sum_j h_{j-2l} w_n(2t-j) = H\{2^{1/2} w_n(2t-j)\}_{(l)} \\ w_{2n+1}(t-l) = 2^{1/2}\sum_j g_{j-2l} w_n(2t-j) = G\{2^{1/2} w_n(2t-j)\}_{(l)} \end{cases} \tag{2.82}$$

Then we can obtain

$$w_n(t-j) = 2^{-1/2}\sum_i h_{j-2i} w_{2n}\left(\frac{t}{2}-i\right) + 2^{-1/2}\sum_i g_{j-2i} w_{2n+1}\left(\frac{t}{2}-i\right) \tag{2.83}$$

when $n = 0$, it can be expressed as

$$w_0(t-k) = 2^{-1/2}\sum_i h_{k-2i} w_0\left(\frac{t}{2}-i\right) + 2^{-1/2}\sum_i g_{k-2i} w_1\left(\frac{t}{2}-i\right) \tag{2.84}$$

Assuming that there is a signal $x(t) \in U^0 \subset L^2(R)$, we can obtain

$$\begin{aligned} x(t) &= \sum_k s_k^0 w_0(t-k) \\ &= \sum_i H\{s_k^0\}_{(i)} 2^{-1/2} w_0\left(\frac{t}{2}-i\right) + \sum_i G\{s_k^0\}_{(i)} 2^{-1/2} w_1\left(\frac{t}{2}-i\right) \end{aligned} \tag{2.85}$$

Similarly, for $x(t) \in U^n \subset L^2(R)$, we can obtain

$$x(t) = \sum_i H\{s_k^n\}_{(i)} 2^{-1/2} w_{2n}\left(\frac{t}{2}-i\right) + \sum_i G\{s_k^n\}_{(i)} 2^{-1/2} w_{2n+1}\left(\frac{t}{2}-i\right) \tag{2.86}$$

where Eq. (2.86) indicates the decomposition of signal $x(t)$ is dependent on wavelet packets. For convenience, two symbols are introduced as follows.

$$\begin{aligned} U_j^0 &= V_j, \quad j \in Z \\ U_j^1 &= W_j, \quad j \in Z \end{aligned} \tag{2.87}$$

According to the relation of $V_{j+1} = V_j \oplus W_j$, we can obtain

$$U_{j+1}^0 = U_j^0 \oplus U_j^1, \quad j \in Z \tag{2.88}$$

For $0 < n \in Z_+$, the following relation can be obtained

$$U_{j+1}^n = U_j^{2n} \oplus U_j^{2n+1}, \quad j \in Z \tag{2.89}$$

Comparing Eq. (2.89) with Eq. (2.79), it is clear that U_j^{2n} corresponds to $w_{2n}(t)$ and U_j^{2n+1} to w_{2n+1}. Based on multiresolution analysis $L^2(R) = \oplus W_j$ where $j \in Z$, U_{j+1}^n is substituted by W_{j+1}^n, then we can obtain the relation of decomposition in wavelet subspaces W_{j+1}^n

$$W_{j+1}^{n} = U_{j+1}^{n} = U_{j}^{2n} \oplus U_{j}^{2n+1}, \quad j \in Z \tag{2.90}$$

It can be observed from Eq. (2.90) that wavelet subspace W_j is further decomposed step by step using wavelet packets. Let $n = 1, 2, \ldots; j = 1, 2, \ldots$, we can obtain the general representation of wavelet packet decomposition.

$$\begin{aligned} W_j &= W_j^1 = U_j^1 \\ &= U_{j-1}^2 \oplus U_{j-1}^3 \\ &= U_{j-2}^4 \oplus U_{j-2}^5 \oplus U_{j-2}^6 \oplus U_{j-2}^7 \\ &= \cdots \qquad \cdots \qquad \cdots \\ &= U_0^{2^j} \oplus U_0^{2^j+1} \oplus \cdots \oplus U_0^{2^{j+1}-1} \end{aligned} \tag{2.91}$$

This kind of decomposition representation of subspace W_j is rewritten as $U_{j-k}^{2^k+m}$ where $m = 0, 1, 2, \ldots, 2^k - 1$, $k = 1, 2, \ldots, j$ and $j = 1, 2, \ldots$. The corresponding orthonormal bases are

$$\left\{2^{(j-k)/2}\, w_{2^k+m}\left(2^{j-k}t - l\right), l \in Z\right\} \tag{2.92}$$

Obviously, when $k = 0$ and $m = 0$, the subspace $U_{j-k}^{2^k+m}$ becomes $U_j^1 = W_j$. The Eq. (2.92) is simplified as $\left\{2^{j/2} w_1\left(2^j t - l\right), l \in Z\right\}$, which is just wavelet basis $\psi_{j,l}(t)$. Signal decompositions of wavelet transform (the Mallat's algorithm) and wavelet packets are illustrated in Fig. 2.16A and B, respectively. In Fig. 2.16A, symbols A_j and D_j

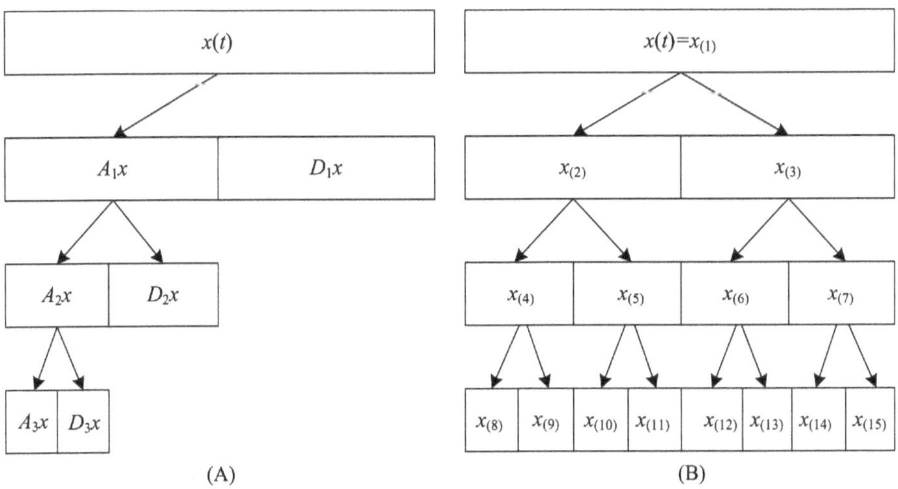

Figure 2.16. Signal decomposition comparison between wavelet and wavelet packet transforms. (A) Signal decomposition using wavelet transform and (B) signal decomposition using wavelet packet transform.

are applied to represent $Ax^j(t)$ and $Dx^j(t)$ in Eq. (2.70), respectively. Similarly, $U_{j-k}^{2^k+m}$ is represented as x_{2^k+m} in Fig. 2.16B for simplicity, which can be seen in Eq. (2.92).

2.5.4 Hilbert–Huang Transform

The combination of both the well-known Hilbert spectra analysis and the recently developed empirical mode decomposition (EMD) is designated as the HHT (Huang et al., 1998, 1999, Huang and Busalacchi, 2000). Indeed, it represents such a paradigm shift of data analysis methodology. The HHT is designed specifically for analyzing nonlinear and nonstationary signals. The key part of HHT is EMD which can decompose any complicated signals into finite and often small number of intrinsic mode functions (IMFs). The instantaneous frequency defined by the HT can reveal the physical meaning of local phase change better for IMFs than for any other non-IMF time series. Moreover, this decomposition method is adaptive and therefore highly efficient. As the decomposition is based on the local characteristic of the signal, it is very applicable to nonlinear and nonstationary signals.

The performance of HHT is mainly attributed to EMD technique. It provides a potentially viable method for nonlinear and nonstationary signal analysis, especially for time–frequency–energy representations. It has been tested and verified exhaustively, though only empirically. In almost all the cases studied, EMD can present the results much sharper than most of the traditional signal analysis methods. In addition, it has been proved that EMD has unprecedented prowess in revealing the hidden physical meanings in signals (Lei et al., 2015).

The HT can lead to an apparent time–frequency–energy representation of a signal. However, this representation may not be consistent with physically meaningful definitions of instantaneous frequencies and instantaneous amplitudes. The EMD can produce components of the signal whose HT can lead to physically meaningful definition of these two instantaneous quantities, and hence the combination of EMD and HT provides a more physically meaningful time–frequency–energy representation of a signal.

2.5.4.1 Empirical Mode Decomposition

Empirical mode decomposition (EMD) is one of the most popular time–frequency analysis techniques. Generally, a time–domain signal can be decomposed into a series of oscillatory functions in the time domain, which are called IMFs. The IMFs denote the internal oscillatory modes imbedded in the signal. Moreover, the number and selection of these IMFs depend on the signal itself rather than basis function. Therefore, the EMD method is a self-adaptive signal processing technique and is suitable for nonlinear and nonstationary statistic processes.

The EMD method was firstly proposed by Wu and Huang (Huang et al., 1998, Huang et al., 1999, Huang and Busalacchi, 2000). It is able to decompose a signal into some IMFs by the sifting process. Each IMF should satisfy the following two conditions: (a) the number of extremum and the number of zero-crossings must be equal or differ at most by one, and (b) at any point, the mean value of the upper envelope defined by the local maxima and the lower envelope defined by the local minima is zero. Though the EMD method is well known now, the theory is added here for completeness. The EMD process of a signal $x(t)$ is summarized in Table 2.3.

A collection of IMFs $c_i(i=1,2,\ldots,I)$ and a residue r_I are obtained at the end of the decomposition procedure of EMD. The linear sum of all IMFs and final residue satisfies this equality $x(t)=\sum_{i=1}^{I}c_i+r_I$. The residue reflects the trend of the signal $x(t)$ and is similar to the trend in the signal preprocessing. Therefore, it commonly does not carry any useful information. However, the IMFs, $c_1, c_2, \ldots, c_I$, represent different frequency bands ranging from high to low. It is different that the frequency components are included in each frequency band, which are determined by the signal itself $x(t)$. Although the EMD technique performs powerful performance

Table 2.3 The EMD Technique (Lei et al., 2013b,c)

(1) Initialize $r_0 = x(t)$ and $i = 1$;
(2) Detect the ith IMF c_i;
 (a) Initialize $h_{i(k-1)} = r_{i-1}$ and $k = 1$;
 (b) Detect the local maxima and minima of $h_{i(k-1)}$;
 (c) Obtain the upper and lower envelops of $h_{i(k-1)}$ by employing cubic spline interpolation scheme;
 (d) Compute the mean value $m_{i(k-1)}$ of the upper and lower envelops of $h_{i(k-1)}$;
 (e) Compute the mean value $m_{i(k-1)}$ of the upper and lower envelops of $h_{i(k-1)}$;
 (f) Let $h_{ik} = h_{i(k-1)} - m_{i(k-1)}$;
 (g) Compute the mean value $m_{i(k-1)}$ of the upper and lower envelops of $h_{i(k-1)}$;
(3) Define the remainder $r_{i+1} = r_i - c_i$;
(4) If r_{i+1} still has least 2 extrema then go to step (2) with $i = i + 1$ else the decomposition process is finished and r_{i+1} is the residue of the signal.

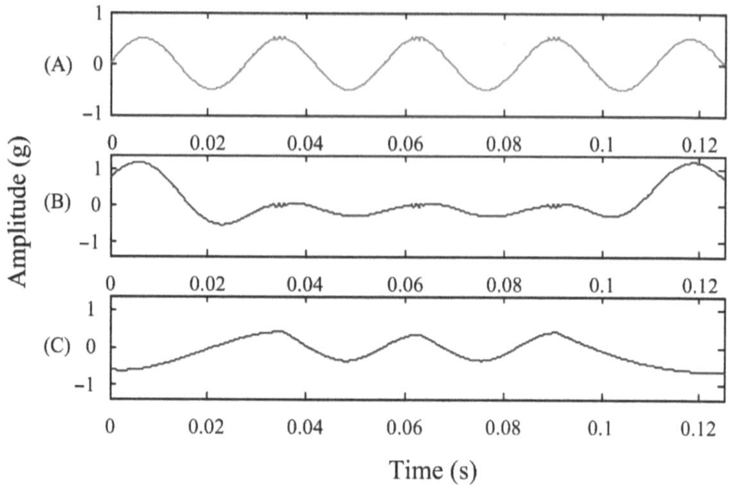

Figure 2.17. The decomposition results using EMD. (A) The simulated signal, (B) IMF c_1, and (C) IMF c_2.

in the process of processing nonlinear and nonstationary signals, the method itself still has some weakness mainly including mode mixing and end effects. Since there is no theoretical foundation for the EMD, it is very difficult to theoretically explain the mode mixing and end effects. For an intuitionistic explanation of these drawbacks of EMD, therefore, a simulated signal and its decomposition results using EMD are illustrated in Fig. 2.17. It is noticed that the simulated signal in Fig. 2.17A is decomposed into two IMFs c_1 and c_2, but the mode mixing occurs in c_1 which consists of an impulse component and a sinuous component.

2.5.4.2 Ensemble Empirical Mode Decomposition

To solve the problem of mode mixing in EMD, ensemble empirical mode decomposition (EEMD) was presented by Wu and Huang (2009), which defines the true IMF components as the mean of an ensemble of trials. Each trial consists of the decomposition results of the signal plus a white noise of finite amplitude. This new technique is based on the insight from recent researches on the statistical properties of white noise, which indicates that the EMD method is an effectively self-adaptive dyadic filter bank when it is applied to the white noise. In addition, the result studied by Flandrin et al. (2004) demonstrated that noise could help data analysis in the EMD method. All these investigations promote the advent of the EEMD technique.

Table 2.4 The EEMD Technique (Lei et al., 2009)

(1) Initialize the number of ensemble M, the amplitude of the added white noise, and $m = 1$;
  (a) Add a white noise series with the given amplitude to the investigated signal, that is, $x_m(t) = x(t) + n_m(t)$ where $n_m(t)$ indicates the mth added white noise series, and $x_m(t)$ denotes the noise-added signal of the mth trial;
  (b) Decompose the noise-added signal $x_m(t)$ into I IMFs $c_{i,m}$ ($i = 1, 2, \ldots, I$) utilizing the EMD method, where $c_{i,m}$ represnts the ith IMF of the mth trial, and I is the number of IMFs;
  (c) If $m < M$ then go to step (a) with $m = m + 1$. Repeat steps (a) and (b) again and again, but with different white noise series each time;

(2) Compute the ensemble mean a_i of the M trials for each IMF.

$$a_i = \frac{1}{M}\sum_{m=1}^{M} c_{i,m}, \quad i = 1, 2, \ldots, I, \; m = 1, 2, \ldots, M$$

(3) Report the mean a_i ($i = 1, 2, \ldots, I$) of each of the I IMFs as the final IMFs.

The principle of the EEMD technique is illustrated as follows (Lei et al., 2011, Lei and Zuo, 2009). The added white noise would populate the whole time–frequency space uniformly with the constituting components of different scales. When a signal is added to this uniformly distributed white noise background, the components in different scales of the signal are automatically projected onto proper scales of the reference built by the white noise in the background. As each of the noise-added decompositions consists of the signal and the added white noise, each individual trial may certainly produce very noisy results. But the noise in each trial is different in separate trials. Thus it can be decreased or even completely eliminated in the ensemble mean of enough trails. The ensemble mean is treated as the true answer because the only persistent part is the signal as more and more trials are added to the ensemble. The detail procedure of EEMD technique is shown in Table 2.4.

To verify the EEMD performance on overcoming the mode mixing problem, the simulated signal in Fig. 2.17A is decomposed again by using EEMD technique with the ensemble number 100 and the added noise amplitude 0.01. The decomposition results are plotted in Fig. 2.18B and C. It is observed that the two components included in the signal are divided into two different IMFs perfectly. IMF a_1 represents the impulse component and IMF a_2 represents the sinuous component. Therefore, the EEMD technique is able to solve the problem of mode mixing and achieve an improved decomposition with physical meaning.

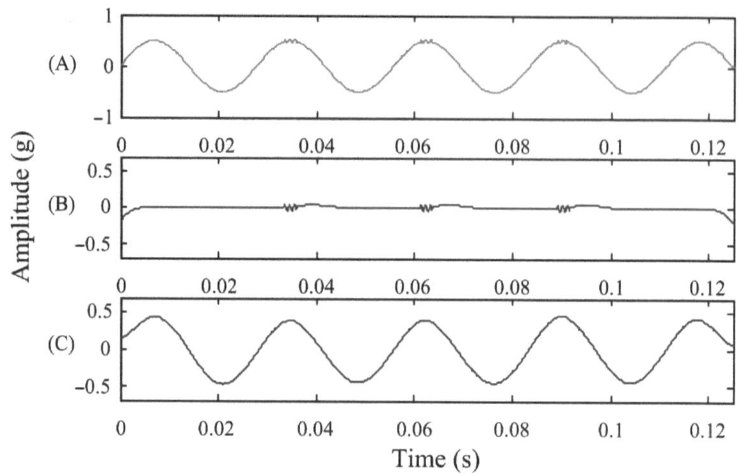

Figure 2.18. The decomposition results using EEMD. (A) The simulated signal, (B) IMF a_1 and (C) IMF a_2.

2.5.4.3 Hilbert Transform

Since HT is an indispensable part of Hilbert Huang transform, it is simply illustrated in this subsection. Generally, the HT for a given signal $x(t)$ is defined as (Quyen et al., 2001)

$$\hat{x}(t) = H[x(t)] = \frac{\int_{-\infty}^{\infty} \frac{x(t)}{t-\tau} d\tau}{\pi} = x(t) * \frac{1}{\pi t} \tag{2.93}$$

where the symbol "*" denotes convolution operation. According to the HT, the analytic signal can be defined as

$$z(t) = x(t) + \mathrm{j}\hat{x}(t) = a(t)\exp[\mathrm{j}\theta(t)] \tag{2.94}$$

where $\mathrm{j} = \sqrt{-1}$ and

$$a(t) = \sqrt{x^2 + \hat{x}^2} \quad \text{and} \quad \theta(t) = \arctan\left(\frac{\hat{x}}{x}\right) \tag{2.95}$$

Here, $a(t)$ is the instantaneous amplitude and θ is the instantaneous phase. The instantaneous frequency is simply defined as $\omega = \frac{d\theta}{dt}$. Then the Hilbert envelope spectrum can be given as follows:

$$h(f) = \int_{-\infty}^{\infty} \sqrt{x^2(t) + \hat{x}^2(t)} \exp(-\mathrm{j}2\pi f t) dt \tag{2.96}$$

HT is also widely used in the signal demodulation. When the mechanical faults occur, the collected vibration signals are generally modulated. Therefore, signal demodulation is able to separate the carrier component and modulation component. Moreover, the fault features are always hidden in the modulation component. It is necessary for us to implement the signal demodulation during signal preprocessing. To intuitively illustrate the signal demodulation using HT, the demodulation result for a given modulation signal is illustrated in Fig. 2.19, where a modulation signal is given in Fig. 2.19A (solid black curve) and the corresponding envelope obtained by HT is plotted in red dotted curve. The corresponding frequency spectra in Fig. 2.19B shows that modulation frequency can be released from the carrier frequency and its sidebands. It is also noticed that the obtained envelope contains a direct current component which should be

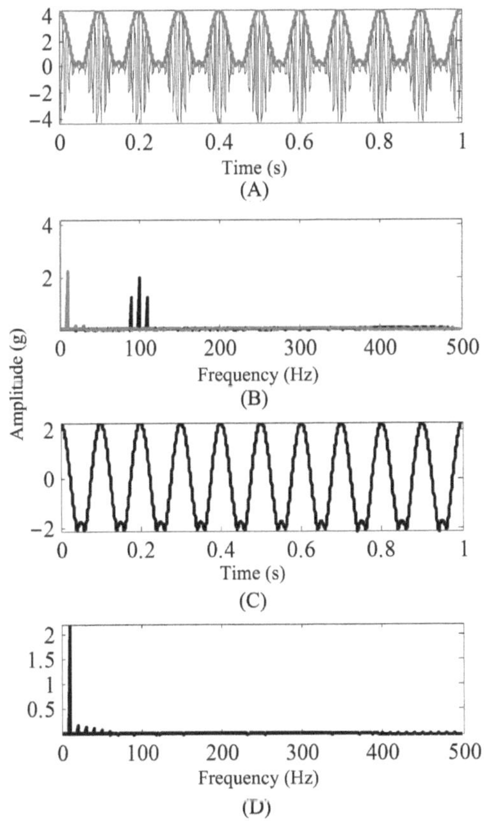

Figure 2.19. Envelope spectrum. (A) Modulated signal, (B) frequency spectrum of (A), (C) envelope, and (D) envelope spectrum preprocessed.

removed. The envelope is presented in Fig. 2.19C and the corresponding frequency spectrum is in Fig. 2.19D. Obviously, HT is an effective method for signal demodulation. In addition, the demodulated signal should be preprocessed through employing these methods in this chapter.

2.5.5 Common Feature Extraction in the Time-Frequency Domain

With the advent of time–frequency analysis techniques, such as typical wavelet transform and empirical mode decomposition, some effective features can be extracted from time–frequency distribution of a signal by employing different signal processing techniques. In this subsection, therefore, two common time–frequency features are introduced according to different time–frequency analysis techniques: one is frequency-band energy ratio of wavelet decomposition and even wavelet packet decomposition and the other is IMF energy ratio of EMD and even EEMD.

A signal can be decomposed into an "approximation" and a detail by employing wavelet or wavelet packet transform. The approximation reflects the low-frequency components and the detail represents the high-frequency components of the signal. The approximation itself is then decomposed into a second-level approximation and detail and the decomposition process can be repeated. It can be noticed from the structure of wavelet decomposition that with the increase of decomposition levels, the frequency–domain resolution of the frequency-band signals in high frequency domain decreases and corresponding time–domain resolution raises, while frequency-band signals in low frequency domain have high frequency–domain and low time–domain resolution powers. In wavelet packet analysis, however, the details as well as the approximations can be further decomposed simultaneously. The wavelet packet decomposition can adaptively select corresponding frequency band to match the spectrum of the signal on the basis of signal characteristics. In addition, it has high frequency resolution power in both high frequency and low frequency domains. This is a better method than wavelet decomposition because the wavelet packet decomposition is developed from rigorous mathematical theory and supported by numerical calculation verification.

Eq. (2.90) can be rewritten as follows

$$W_{j+1}^{n} = U_{j+1}^{n} = U_{j}^{2n} \oplus U_{j}^{2n+1}, \ j \in Z \tag{2.97}$$

and general representation of wavelet packet decomposition can be expressed as

$$W_j = U_{j-k}^{2^k} \oplus U_{j-k}^{2^k+1} \oplus \cdots \oplus U_{j-k}^{2^{k+1}-1}, \ j,k \in Z \tag{2.98}$$

The Eq. (2.98) can be further expressed as

$$W_j = \bigoplus_{m=0}^{2^k-1} U_{j-k}^{2^k+m}, \quad j,k,m \in Z \tag{2.99}$$

where $j = 1, 2, \ldots; k = 1, 2, \ldots j; m = 0, 1, 2, \ldots, 2^k - 1$.

Suppose that there is a given signal $x(t)$, and its wavelet packet decomposition is illustrated in Fig. 2.16B. Based on Eq. (2.99), the decomposed signal x_{2^k+m} belongs to subspace $U_{j-k}^{2^k+m}$. If $k = 0$ and $m = 0$, this indicates that the original signal $x(t)$ noted as x_1 is not decomposed. For example, after the first decomposition, that is, $k = 1$, $m = 0, 1$, two frequency-band signals x_2 and x_3 can be obtained from x_1. According to energy distribution, we can know that wavelet packet decomposition obeys the energy conservation principle (Yen and Lin, 2000, Peng and Chu, 2004), that is,

$$E_n(x(t)) = \sum_{m=0}^{2^k-1} E_n\left(U_{j-k}^{2^k+m}\right) = \sum_{m=0}^{2^k-1} E_n\left(x_{2^k+m}\right) \tag{2.100}$$

where $E_n(\circ)$ denotes the energy of the signal. Eq. (2.100) suggests that the original signal can be decomposed into orthogonal and independent frequency bands. Moreover, these decomposed signals' energy is conservative. The energy in each frequency band can reflect the operating conditions of machinery and further monitor it because dynamic information of various mechanical elements exists in certain frequency band. Generally, mechanical fault diagnosis can be monitored by selecting the amplitudes of specific frequencies in FFT spectrum. In other words, it takes advantage of the energy of sinusoids to monitor the mechanical faults. Many mechanical fault signals excited by looseness, friction, shock, and creep, which are nonstationary and nonlinear, are unsuitable to be decomposed as sinusoids by FFT technique. Aiming at the reasonableness of condition monitoring, we take the energy of the signal decomposed by wavelet packets as time–frequency feature to diagnose the mechanical faults, which includes not only the sinusoidal energy but also energy of any waveforms.

Assume that the data length of original signal $x(t)$ is N. The data length of decomposed signal x_{2^k+m} $\left(\text{i.e., } x_{2^k+m}(i)\right)$ is shortened as $N/2^k$, its energy can be expressed as

$$E_n\left(x_{2^k+m}\right) = \frac{1}{2^{-k}N-1} \sum_{i=1}^{2^{-k}N} \left\{x_{2^k+m}(i)\right\}^2 \tag{2.101}$$

where $m = 0, 1, 2, \ldots, 2^k - 1$, it indicates the position of frequency bands. In independent frequency band, the energy of the decomposed signal can be used to monitor the operating conditions of machinery. This method is more reasonable than only utilizing amplitudes of harmonics in FFT spectrum because the method takes the whole energy in every independent frequency band into account. In this way, the energy of vibration signals including nonlinear and nonstationary signals is able to be all monitored.

In view of common applications, the relative energy is adopted for condition monitoring. It is obvious that the relative energy in mth frequency band can be written as

$$E_n(m) = \frac{E_n\left(x_{2^k+m}\right)}{\sum_{m=0}^{2^k-1} E_n\left(x_{2^k+m}\right)} \tag{2.102}$$

The relative energy in mth frequency band is equal to $E_n(m)$, obviously it must satisfy this condition

$$\sum_{m=0}^{2^k-1} E_n(m) = 1 \tag{2.103}$$

The relative energy $E_n(m)$ obtained by utilizing WPT is called as frequency band energy of wavelet packets. It is also called time–frequency feature because it is extracted from time-frequency domain or by utilizing time–frequency analysis method.

In addition, it is well known that EMD can also adaptively decompose a signal into several IMFs. Therefore, the original signal $x(t)$ is equal to linear sum of all IMFs and final residue r_L

$$x(t) = \sum_{i=1}^{L} c_i + r_L \tag{2.104}$$

According to the EMD theory, we can know that IMFs, $c_1, c_2, \ldots, c_L$, contain different frequency bands ranging from high to low. The frequency components included in each IMF are different and they vary with the variation of the vibration signal $x(t)$, while r_L represents the central tendency of vibration signal $x(t)$. From the comparison between WPT and EMD, it is noted that WPT decomposes a signal into independent frequency bands with equal frequency widths and without overlap or gap such that they do not have redundant or omitted information. However, WPT has the shortcoming that the analysis performance is determined by the wavelet basis function. The number of frequency bands can be easily controlled in WPT through setting the decomposition level. On the other hand, EMD is based on the local characteristic imbedded in the signal and may decompose the signal into a set of complete and almost orthogonal IMFs. During the decomposition, there is no need to select a basis function. In addition, both the number of IMFs and the actual scaling are dependent on the signal itself instead of subjective selection. Thus, it is a self-adaptive signal processing method. However, the number of IMFs cannot be controlled beforehand because of the self-adaptive characteristic of EMD. Obviously, WPT and EMD process the signal from different views and have their own characteristics including advantages as well as shortcomings. Therefore, we can also take relative energy to monitor the health conditions of machinery by virtue of EMD. The energy of IMF c_i can be expressed as

$$E(c_i)=\frac{1}{N}\sum_{j=1}^{N}\{c_i(j)\}^2,\quad i=1,2,\cdots,L \tag{2.105}$$

where N is the data length and L is number of IMF. Then the relative energy in IMF c_i can be expressed as

$$E_m(c_i)=\frac{E(c_i)}{\sum_{i=1}^{L}E(c_i)},\quad m=1,2,\cdots,L \tag{2.106}$$

Similarly, we can also extract the energy features of wavelet transform and EEMD. In order to achieve higher diagnosis accuracy, we can often combine time-domain and frequency–domain features with time–frequency features to make an intelligent fault diagnosis. Therefore, we should extract proper features from time–, frequency–, or time-frequency domains by virtue of different signal processing techniques.

2.6 Conclusions

In order to achieve intelligent fault diagnosis and remaining useful life prediction, it is essential for us to preprocess the collected vibration signals by signal processing techniques and methods. This chapter introduces common signal processing methods from traditional time domain to frequency one, even to abundant time–frequency one. In addition, the common statistical features in every domain are listed in corresponding ends.

Autocorrelation and cross-correlation analyses are the representation of signal processing methods in the time domain. Autocorrelation can capture periodic components from a noisy signal and further achieve mechanical fault diagnosis because mechanical vibration signals are periodic when a fault occurs. However, this kind of technique is just applied to the vibration signals with high SNR. Cross-correlation can estimate the similarity between two different signals and therefore is widely used to speed measurement and information extraction. Finally, some common statistical features in the time domain are summarized and their physical natures are stated. These features can reflect the mechanical operating conditions from their waveforms of vibration signals, but they are not robust enough to completely reflect the mechanical health conditions.

FT reveals the nature of a signal from frequency spectrum and it is a powerful tool for stationary signal analysis. The collected vibration signals are always transient or nonstationary. Therefore, the FT is powerless in this case. Even so, for finite data length we think it is approximately stationary and therefore some common frequency–domain statistical features are extracted to reflect the mechanical operating conditions from the frequency spectrum.

To reflect the variation of a signal from both time and frequency domains, signal processing techniques in the time-frequency domain are developed rapidly, such as WVD and short time FT. When a window is chosen, the time–frequency resolution is then fixed over the entire time–frequency plane. If the signal is composed of small bursts associated with long quasi-stationary components, it can be analyzed with good time resolution or frequency resolution. Another important time–frequency representation is wavelet theory, which aims at extracting the localized contributions of signals. It can be interpreted as a decomposition of the original signal in an orthonormal wavelet basis or as a decomposition of the signal in a set of independent frequency channels. The independence is generated by the orthogonality of the wavelet functions. The decomposed signal in the independent frequency channel carries about vivid information of operating conditions and faults of specific mechanical elements. Based on the information we can monitor and diagnose different faults more conveniently and efficiently. The concept of multiresolution analysis in wavelet transform is very useful for signal processing. We can choose different resolutions to obtain information for condition monitoring and fault diagnosis. Wavelet transform performs best when a signal is composed of high frequency components of short duration plus low frequency components of long duration, which is often the case with signals encountered in practice. Since high frequency sequences cannot be decomposed by the Mallat's algorithm, wavelet packet decomposition is developed to provide better frequency resolution at high frequencies. Moreover, independent frequency bands link up mutually without redundant or omitted information. Wavelet transform, even wavelet packet transform, however, needs to select subjectively suitable basis functions, whereas HHT based on EMD and HT is able to decompose adaptively a signal into independent components. It overcomes the shortcoming of wavelet transform. For different signals, therefore, a suitable signal processing method should be selected to achieve the condition monitoring and fault diagnosis of rotating machinery.

References

Bruce, A., Donoho, D., Gao, H.-Y., 1996. Wavelet analysis [for signal processing]. IEEE Spectr. 33, 26–35.

Chui, C.K., 2014. An Introduction to Wavelets. Academic press, New York.

Claasen, T., 1980. The Wigner distribution-a tool for time–frequency signal analysis, part II: discrete-time signals. Philips J. Res. 35, 276–300.

Coifman, R.R., Meyer, Y., Quake, S., Wickerhauser, M.V., 1994. Signal processing and compression with wavelet packets. Wavelets and Their Applications. Springer, Netherlands.

Cooley, J.W., Tukey, J.W., 1965. An algorithm for the machine calculation of complex Fourier series. Math. Comp. 19, 297–301.

Flandrin, P., Rilling, G., Goncalves, P., 2004. Empirical mode decomposition as a filter bank. IEEE Signal Process. Lett. 11, 112–114.

Gabor, D., 1946. Theory of communication. part 1: the analysis of information. J. Inst. Electr. Eng. Part III 93, 429–441.

Huang, N., Busalacchi, A.J., 2000. A new view of earthquake ground motion data: the Hilbert spectral analysis. Proceedings of International Workshop on Annual Commemoration of Chi-Chi Earthquake, 2, 64–75.

Huang, N.E., Shen, Z., Long, S.R., 1999. A new view of nonlinear water waves: the Hilbert Spectrum 1. Ann. Rev. Fluid Mech. 31, 417–457.

Huang, N.E., Shen, Z., Long, S.R., Wu, M.C., Shih, H.H., Zheng, Q., Yen, N.-C., Tung, C.C., Liu, H.H., 1998. The empirical mode decomposition and the Hilbert spectrum for nonlinear and non-stationary time series analysis. Proc. Royal Soc. Lond. A, 903–995.

Jia, F., Lei, Y., Shan, H., Lin, J., 2015. Early fault diagnosis of bearings using an improved spectral kurtosis by maximum correlated kurtosis deconvolution. Sensors 15, 29363–29377.

Lei, Y., Zuo, M.J., 2009. Fault diagnosis of rotating machinery using an improved HHT based on EEMD and sensitive IMFs. Meas. Sci. Technol. 20, 125701.

Lei, Y., Han, D., Lin, J., He, Z., 2013b. Planetary gearbox fault diagnosis using an adaptive stochastic resonance method. Mech. Syst. Signal Process. 38, 113–124.

Lei, Y., He, Z., Zi, Y., 2009. Application of the EEMD method to rotor fault diagnosis of rotating machinery. Mech. Syst. Signal Process. 23, 1327–1338.

Lei, Y., He, Z., Zi, Y., 2011. EEMD method and WNN for fault diagnosis of locomotive roller bearings. Expert Syst. Appl. 38, 7334–7341.

Lei, Y., Li, N., Lin, J., Wang, S., 2013c. Fault diagnosis of rotating machinery based on an adaptive ensemble empirical mode decomposition. Sensors 13, 16950–16964.

Lei, Y., Lin, J., Han, D., He, Z., 2014. An enhanced stochastic resonance method for weak feature extraction from vibration signals in bearing fault detection. Proc. Inst. Mech. Eng. Part C 228, 815–827.

Lei, Y., Lin, J., He, Z., Zuo, M.J., 2013a. A review on empirical mode decomposition in fault diagnosis of rotating machinery. Mech. Syst. Signal Process. 35, 108–126.

Lei, Y., Liu, Z., Ouazri, J., Lin, J., 2015. A fault diagnosis method of rolling element bearings based on CEEMDAN. Proc. Inst. Mech. Eng. Part C.

Mallat, S.G., 1989a. Multiresolution approximations and wavelet orthonormal bases of $\mathbf{L}^2(\mathbf{R})$. Trans. Am. Math. Soc. 315, 69–87.

Mallat, S.G., 1989b. A theory for multiresolution signal decomposition: the wavelet representation. IEEE Trans. Pattern Anal. Mach. Intell. 11, 674–693.

Newland, D.E., 1994. Wavelet analysis of vibration: part 1—theory. J. Vib. Acoust. 116, 409–416.

Oppenheim, A.V., Schafer, R.W., Buck, J.R., 1989. Discrete-time signal processing, Prentice hall Englewood Cliffs, NJ.

Oxford, O.E., 2009. The Oxford English Dictionary. Oxford University Press, Oxford.

Peng, Z.K., Chu, F.L., 2004. Application of the wavelet transform in machine condition monitoring and fault diagnostics: a review with bibliography. Mech. Sys. Signal Process. 18, 199–221.

Qin, Q.-Q., Yang, Z.-K., 1994. Practical Wavelet Analysis. Publish House of Xian Electronic Science and Technology University, Xi'an.

Quyen, M.L.V., Foucher, J., Lachaux, J.-P., Rodriguez, E., Lutz, A., Martinerie, J., Varela, F.J., 2001. Comparison of Hilbert transform and wavelet methods for the analysis of neuronal synchrony. J. Neurosci. Methods 111, 83–98.

Rabiner, L.R., Schafer, R.W., 1978. Digital Processing of Speech Signals. Prentice Hall, Englewood Cliffs, NJ.

Randall, R.B., 1987. Frequency Analysis. Brül & Kjor, Copenhagen.

Randall, R.B., 2011. Vibration-Based Condition Monitoring: Industrial, Aerospace And Automotive Applications. John Wiley & Sons, Chichester, UK.

Ville, J.D., 1948. Théorie et applications de la notion de signal analytique. Cables et transmission 2, 61–74.

Wickerhauser, M.V., 1991. Lectures on wavelet packet algorithms. Lecture notes, INRIA, Citeseer.

Wigner, E., 1932. On the quantum correction for thermodynamic equilibrium. Phys. Rev. 40, 749.

Wu, Z., Huang, N.E., 2009. Ensemble empirical mode decomposition: a noise-assisted data analysis method. Adv. Adapt. Data Anal. 1, 1–41.

Yen, G.G., Lin, K.C., 2000. Wavelet packet feature extraction for vibration monitoring. IEEE Trans. Indu. Electron. 47, 650–667.

3

INDIVIDUAL INTELLIGENT METHOD-BASED FAULT DIAGNOSIS

3.1 Introduction to Intelligent Diagnosis Methods

To fully inspect health conditions of rotating machinery, condition monitoring systems are applied to collect real-time data from the machines, and the amount of data collected has grown exponentially by multiple sensors after the long-time operation (Jia et al., 2016). Such explosion of data makes mechanical fault diagnosis endorse the concept of big data. The power of mechanical big data not only promises to challenge the traditional methods in mechanical fault diagnosis, but also brings this field new perspectives for processing and discovering valuable information from such big data. Therefore, it becomes a new research subject to effectively extract characteristics from big data and accurately identify the machinery health conditions using advanced techniques (Lei et al., 2016).

The main purpose of fault diagnosis is to analyze the relevant external information to judge the conditions of the inaccessible internal components, by which one could decide if the machine needs to be dismantled or not. One of the tools for diagnosing faults of rotating machinery is vibration analysis (Villa et al., 2012). Through the vibration analysis, it is possible to obtain vital diagnosis information from the vibration signals. The results of traditional fault diagnosis methods are filtered signals and their spectra with signal processing techniques, which need a visual inspection of the diagnosis results to compare them with the healthy ones and find the fault characteristics, depending on the expertise of the diagnosticians and their ability to eyeball the results. So these diagnosis methods require a great deal of expertise to apply them successfully and diagnosticians to analyze the monitored signal one after another. Nowadays, as the data are

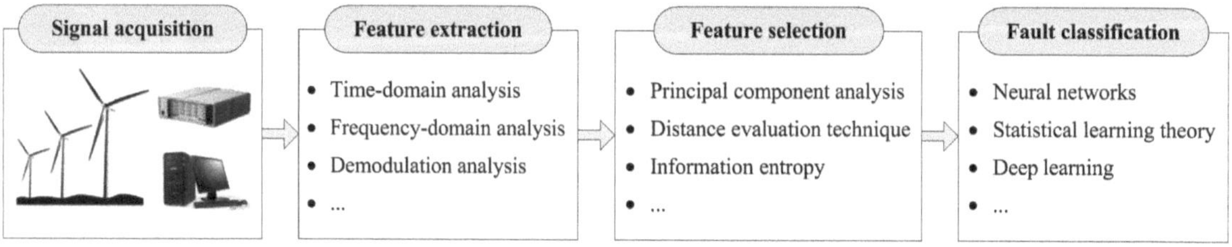

Figure 3.1. Flowchart of intelligent fault diagnosis.

generally collected faster than diagnosticians can analyze them, traditional diagnosis possibly is not suitable for dealing with mechanical big data. In contrast, intelligent fault diagnosis is able to replace diagnosis specialists with artificial intelligent techniques so as to rapidly and efficiently process collected signals and provide accurate diagnosis results (Lei et al., 2015). Without the need for a diagnosis specialist to examine data and diagnose problems, intelligent fault diagnosis (Lei et al., 2008) can be a potential tool to handle massive data of machinery fault diagnosis in Big Data Era.

Traditionally, the framework of intelligent fault diagnosis includes four main steps: signal acquisition, feature extraction, feature selection, and fault classification (Worden et al., 2011; Lei et al., 2007), as shown in Fig. 3.1. In the signal acquisition step, vibration signals have been extensively utilized since they provide the most intrinsic information about mechanical faults. The feature extraction step aims to extract representative features from the collected signals based on signal processing techniques, like time–domain statistical analysis, Fourier spectral analysis, and demodulation analysis. Although these features characterize the mechanical health conditions, they may contain useless or insensitive information and affect the diagnosis results, as well as computational efficiency. So in the third step, feature selection is used to select sensitive features through dimension reduction strategies, such as principal component analysis (PCA) and distance evaluation technique. In the fault classification step, the selected features are used to train artificial intelligence techniques (Michalski et al., 2013) like artificial neural networks (ANN) and support vector machine (SVM), and the mechanical health conditions are determined by these techniques.

In this chapter, we focus on various artificial intelligence techniques in intelligent fault diagnosis of rotating machinery, and introduce the intelligent diagnosis methods based on these techniques.

3.2 Artificial Neural Networks

3.2.1 Introduction to Artificial Neural Networks

An artificial neural network (ANN) is an information processing paradigm inspired by the way in which the human brain processes information (Hagan et al., 1996). Due to the ability of ANNs in learning and generalizing nonlinear relationships between input data and output data, they provide a flexible mechanism for learning and recognizing system faults. They have been established as a powerful intelligence technique in the fault diagnosis of rotating machinery. Fault diagnosis using ANN classifiers may largely increase the reliability of fault diagnosis methods. In this section, the basic concepts, notation, and basic learning algorithms are described for ANNs, which will be of great use for solving pattern recognition problems in the following sections of this book. We give a brief review of the basic concepts of neural networks and the backpropagation (BP) learning algorithm, which is commonly utilized to train an effective network. We consider this material necessary to understand the new networks for intelligent fault diagnosis.

3.2.1.1 Architecture of Neural Networks

The basic ANN is multilayer perceptron (MLP). It is a feedforward neural network consisting of the units (we also call them neurons) that are connected to each other using directed edges. Fig. 3.2 shows the architectural graph of a MLP with four layers. It can be seen that a unit in any layer of the network connects to all the units in the next layer. Data flow through the network in a forward direction, from left to right, and the last layer outputs the results of the networks (Haykin et al., 2009).

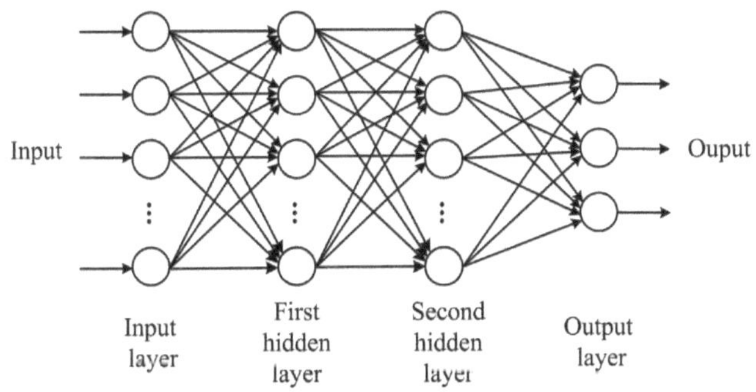

Figure 3.2. Architectural graph of multilayer perceptron (MLP) with two hidden layers.

In the applications of MLP for intelligent fault diagnosis of machinery, the inputs are the features extracted from mechanical vibration signals, where a feature concentrates the discriminative information for the signals of different health conditions and discards the irrelevant variations (Lei et al., 2013). The desired outputs are the labels that indicate the corresponding health conditions of the inputs. For each input, the aim of MLP is to minimize the discrepancy between the output results and the desired labels. To achieve this aim, we need to train MLP to adjust its parameters by algorithms.

3.2.1.2 Backpropagation Algorithm

The most popular algorithm for the training of MLP is backpropagation algorithm. The first description of the original backpropagation algorithm was contained in the thesis of Werbos (1974). But this thesis was not widely disseminated. Until 1980s, the backpropagation algorithm was rediscovered by David Rumelhart, Geoffery Hinton, and Ronald Williams (Rumelhart et al., 1986), Yann LeCun (LeCun et al., 1989), and David Parker (Parker, 1985) independently. The backpropagation algorithm was popularized after the book *Parallel Distributed Processing* by David Rumelhart and James McClelland was published (Rumelhart et al., 1987). This book spurred a great number of researches on neural networks. So the backpropagation algorithm became the most widely used algorithm for training the networks.

The main idea of the backpropagation algorithm is that the parameters, that is, weights and biases, of the MLP should be updated to minimize a prescribed error, which measures the discrepancy between the network outputs and the desired outputs. The training process has two phases, as shown in Fig. 3.3.

1. In the forward phase, the parameters of the MLP are fixed and the input data is propagated through the network, layer by layer, until it reaches the last layer. So in this phase, changes

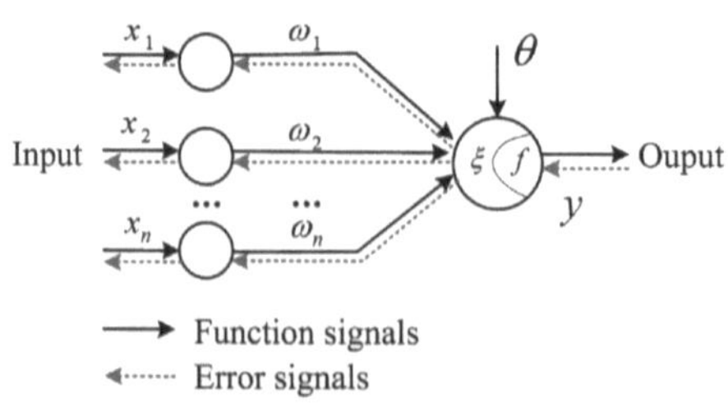

Figure 3.3. Illustration of the directions of two basic signal flows in a MLP: forward propagation for function signals and backpropagation of error signals.

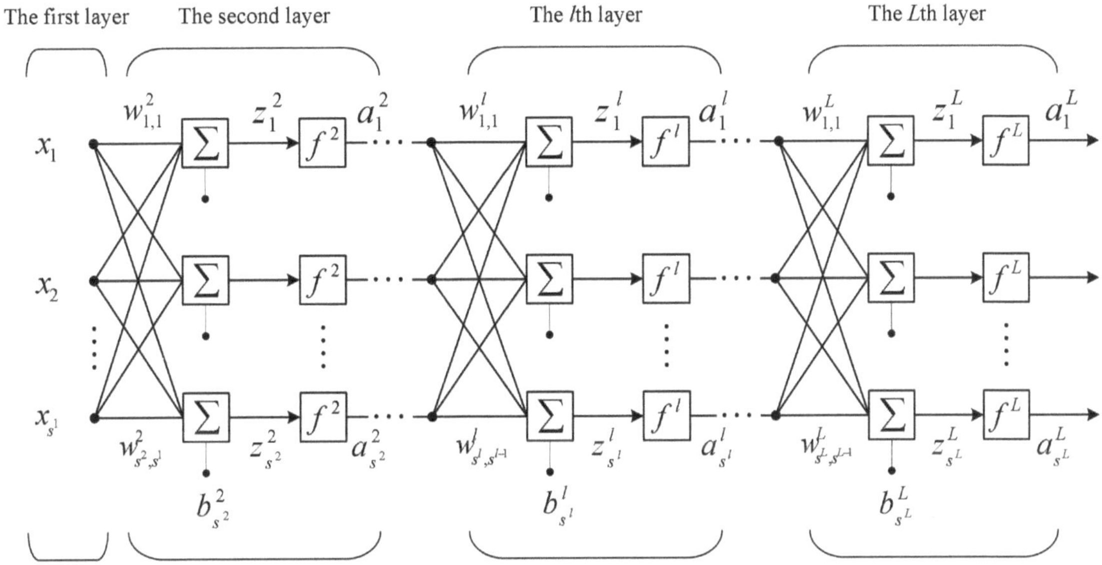

Figure 3.4. Detailed architectural graph of the MLP.

are confined to the activation functions and the outputs of the units in the MLP.

2. In the backward phase, an error signal is calculated by comparing the output result of the network with its desired label. This error signal is first produced in the output layer and then propagated through the network layer by layer again, but at this time the propagation is performed in the backward direction. In this phase, successive adjustments are performed to the weights and biases of the MLP. Calculation of these adjustments is straightforward in the output layer, but it is challenging for the middle layers.

To show the forward phase of the MLP clearly, we reproduce the architectural graph of the MLP and display it in Fig. 3.4. It can be seen that the MLP has L layers and the layer $l(l=1,\cdots,L)$ has s^l nodes, where we use superscripts to identify the layer number. The output of the lth layer is the input to the $l+1$th layer. It is noted that the input layer is considered as the first layer, which may be slightly different from other references. Thus, the weight matrix starts from the $\boldsymbol{W}^2$ instead of $\boldsymbol{W}^1$, and the bias matrix $\boldsymbol{b}^2$ is the same. The elements of the weight matrix $\boldsymbol{W}^l$ in the lth layer are represented by $w^l_{i,j}$ $(i=1,\cdots,s^l;j=1,\cdots,s^{l-1})$ and elements of the bias matrix $\boldsymbol{b}^l$ are represented by b^l_i. The nonlinear function of unit i in the lth layer can be represented by f^l.

Suppose that we have a training dataset $\left\{\boldsymbol{x}_q,\boldsymbol{y}_q\right\}_{q=1}^{Q}$ where $\boldsymbol{x}_q \in \Re^{s^1\times 1}$ is the qth feature vector (we also call it a training sample) as the input data and $y_q \in \Re^{s^L \times 1}$ is the corresponding label as the output data. For x_q, the output data of MLP

$$z_i^l = \sum_{j=1}^{s^{l-1}} w_{i,j}^l \cdot a_j^{l-1} + b_i^l \tag{3.1}$$

$$a_i^l = f^l\left(z_i^l\right) \tag{3.2}$$

where $2 \le l \le L$ and $a_j^1 = x_j$.

For the qth input vector x_q, we can define the sum of the squared errors to measure the discrepancy between the output vector a_q^L and the desired labels y_q.

$$E_q = \frac{1}{2}\left(y_q - a_q^L\right)^2 \tag{3.3}$$

where $a_q^L = \left[a_1^L, \cdots a_{s^L}^L\right]$ for x_q. Obviously, when E_q is equal to zero, the network is able to reproduce exactly the desired output vector for x_q. The backpropagation algorithm aims to find the set of parameters that minimizes the error over all training samples.

To solve this optimization problem, gradient descent is used in the backpropagation algorithm. One iteration of gradient descent updates the parameters $w_{i,j}^l$ and b_i^l as follows:

$$w_{i,j}^l = w_{i,j}^l - \alpha \frac{\partial E(w,b)}{\partial w_{i,j}^l} \tag{3.4}$$

$$b_i^l = b_i^l - \alpha \frac{\partial E(w,b)}{\partial b_i^l} \tag{3.5}$$

where α is the learning rate. The key step in Eqs. (3.4) and (3.5) is computing their partial derivatives. The backpropagation algorithm gives an efficient way to compute the partial derivatives above. We write these partial derivatives as follows:

$$\frac{\partial E(w,b)}{\partial w_{i,j}^l} = \frac{\partial E(w,b)}{\partial z_i^l} \cdot \frac{\partial z_i^l}{\partial w_{i,j}^l} \tag{3.6}$$

$$\frac{\partial E(w,b)}{\partial b_i^l} = \frac{\partial E(w,b)}{\partial z_i^l} \cdot \frac{\partial z_i^l}{\partial b_i^l} \tag{3.7}$$

The second term in each of the equations can be easily computed, since the input of the lth layer is an explicit function of the weights and bias in that layer, as described in Eq. (3.1).

Therefore, the results of the second term are:

$$\frac{\partial z_i^l}{\partial w_{i,j}^l} = a_j^{l-1}, \ \frac{\partial z_i^l}{\partial b_i^l} = 1. \tag{3.8}$$

But the first terms in Eqs. (3.6) and (3.7) also need to be calculated. So we define

$$s_i^l = \frac{\partial E(w,b)}{\partial z_i^l} \tag{3.9}$$

where s_i^l is the sensitivity of $E(w, b)$ to the changes in the ith units in the lth layer. Eqs. (3.6) and (3.7) can be simplified to

$$\frac{\partial E(w,b)}{\partial w_{i,j}^l} = s_i^l \cdot a_j^{l-1} \tag{3.10}$$

$$\frac{\partial E(w,b)}{\partial b_i^l} = s_i^l. \tag{3.11}$$

Now, it remains for us to calculate the sensitivities s_i^l. This calculation will give us the term backpropagation, because it describes a recurrence relationship where the sensitivity at the l–1th layer could be computed from the sensitivity at the lth layer.

To deduce the recurrence relationship for these sensitivities, we will use the following Jacobian matrix:

$$\frac{\partial \mathbf{z}^{l+1}}{\partial \mathbf{z}^l} = \begin{bmatrix} \partial z_1^{l+1}/\partial z_1^l & \partial z_1^{l+1}/\partial z_2^l & \cdots & \partial z_1^{l+1}/\partial z_{s^l}^l \\ \partial z_2^{l+1}/\partial z_1^l & \partial z_2^{l+1}/\partial z_2^l & \cdots & \partial z_2^{l+1}/\partial z_{s^l}^l \\ \vdots & \vdots & & \vdots \\ \partial z_{s^{l+1}}^{l+1}/\partial z_1^l & \partial z_{s^{l+1}}^{l+1}/\partial z_2^l & \cdots & \partial z_{s^{l+1}}^{l+1}/\partial z_{s^l}^l \end{bmatrix} \tag{3.12}$$

For each element in this Jacobian matrix:

$$\begin{aligned} \frac{\partial z_i^{l+1}}{\partial z_i^l} &= \frac{\partial\left(\sum_{j=1}^{s^l} w_{i,j}^{l+1} \cdot a_j^l + b_i^{l+1}\right)}{\partial z_j^l} = w_{i,j}^{l+1} \cdot \frac{\partial a_j^l}{\partial z_j^l} \\ &= w_{i,j}^{l+1} \frac{\partial f^l\left(z_j^l\right)}{\partial z_j^l} = w_{i,j}^{l+1} \dot{f}^l\left(z_j^l\right) \end{aligned} \tag{3.13}$$

where $\dot{f}^l\left(z_j^l\right) = \frac{\partial f^l\left(z_j^l\right)}{\partial z_j^l}$.

Based on Eq. (3.13), the Jacobian matrix can be rewritten as

$$\frac{\partial z^{l+1}}{\partial z^l} W^{l+1}\dot{F}^l(z^l) \tag{3.14}$$

where $\dot{F}^l(z^l)=\begin{bmatrix} \dot{f}^l(z_1^l) & 0 & \cdots & 0 \\ 0 & \dot{f}^l(z_2^l) & \cdots & 0 \\ \vdots & \vdots & & \vdots \\ 0 & 0 & \cdots & \dot{f}^l(z_{s^l}^l) \end{bmatrix}$.

We can represent the recurrence relation for the sensitivities in matrix form:

$$s^l = \frac{\partial E(w,b)}{\partial z^l} = \left(\frac{\partial z^{l+1}}{\partial z^l}\right)^{\mathrm{T}} \frac{\partial E(w,b)}{\partial z^{l+1}} = \dot{F}^m(z^l)(W^{l+1})^{\mathrm{T}} \frac{\partial E(w,b)}{\partial z^{l+1}} \tag{3.15}$$

Because $s^{l+1} = \frac{\partial E(w,b)}{\partial z^{l+1}}$, Eq. (3.15) is rewritten as

$$s^l = \dot{F}^l(z^l)(W^{l+1})^{\mathrm{T}} s^{l+1}\ s^l = \dot{F}^l(z^l)(W^{l+1})^{\mathrm{T}} s^{l+1} \tag{3.16}$$

Now, it can be seen that the sensitivities computed in the backpropagation algorithm derived from its name. These sensitivities are propagated backward from the last layer to the first layer through the network:

$$s^L \to s^{L-1} \to \cdots \to s^l \to \cdots \to s^2. \tag{3.17}$$

If we can compute the starting sensitivity s^L, all of the terms in Eqs. (3.6) and (3.7) can be calculated. So we still have one more step to complete the backpropagation algorithm. We need the starting sensitivity s^L and it is obtained at the output layer based on Eq. (3.3).

$$s_i^L = \frac{\partial E(w,b)}{\partial z^L} = -\left(y_i - a_i^L\right)\frac{\partial a_i^L}{\partial z_i^L} = -\sum_{q=1}^{Q}\left(y_i - a_i^L\right)\dot{f}^L\left(z_{s^L}^L\right) \tag{3.18}$$

Here, we summarize the backpropagation algorithm in the matrix form. The first step is to propagate the input forward through the network:

$$a^1 = x \tag{3.19}$$

$$a^l = f^l(W^l a^{l-1} + b^l) \quad \text{for} \quad l = 2,\cdots,L \tag{3.20}$$

The next step is to propagate the sensitivities backward through the network:

$$s^L = -(y - a^L)\dot{F}^L(z^L) \quad (3.21)$$

$$s^l = \dot{F}^l(z^l)(W^{l+1})^T s^{l+1} \quad (3.22)$$

Finally, the weights and biases are updated by the following equations:

$$W^l = W^l - \alpha s^l (a^{l-1})^T \quad (3.23)$$

$$b^l = b^l - \alpha s^l. \quad (3.24)$$

3.2.1.3 Speeding up the Backpropagation

The backpropagation algorithm gives approximations to the trajectories in the weight and bias space, which are computed by the method of gradient descent. The smaller the learning rate in Eqs. (3.4) and (3.5) we used, the smaller the changes to the weights and biases of the network will be in one iteration, as well as the smoother the trajectories in the weight and bias space will be. However, the update process will be achieved slowly using a slower learning rate. If we use a too large learning rate in order to speed up the rate of learning, the resulting changes in the weights and biases of the network may become unstable. Therefore, a simple method of speeding up the training process and simultaneously avoiding the problem of instability is developed to modify the backpropagation algorithm. The main change is that a momentum term is introduced to the update rules in Eqs. (3.4) and (3.5). The modified rules of the updating the weights are expressed as:

$$\Delta w_{i,j}(k) = \eta \Delta w_{i,j}(k-1) + \alpha \frac{\partial E(w,b)}{\partial b_i^l} \quad (3.25)$$

$$w_{i,j}^l(k) = w_{i,j}^l(k-1) - \Delta w_{i,j}(k) \quad (3.26)$$

where η is called the momentum constant and is usually set as a value between 0 and 1. The modified rules of updating the biases are as same as those of updating the weights. The addition of the momentum term not only smoothes the weight and bias updating but also tends to resist erratic weight changes because of the gradient noise or high spatial frequencies in the weight and bias space. This modified backpropagation algorithm is the mostly used algorithm for training MLP in intelligent fault diagnosis. There are also some modified strategies but they are not commonly used, so we have not included them in this book.

One reason of the popularity of the MLP lies on its nonlinear ability, which actually originates from the function f^l shown in Fig. 3.4. This function is usually a differentiable nonlinear activation function. There are two types of the most widely used activation functions in the MLP, that is, sigmoid function and tanh function.

3.2.1.4 Epilog

MLP, which is the most popular and successful neural network architecture, is suitable for a wide range of applications, such as identification and classification. It comprises a number of identical units organized in layers, with which on one layer connected to those on the next layer, the output of one layer is considered as the input to the next layer. The backpropagation algorithm for the design of a MLP with multilayers is commonly used and may be viewed as the application of an optimization method known in statistics as stochastic approximation. However, MLP suffers from a series of problem in practice.

1. MLP doesn't guarantee the global minimum of the error function, but only a local minimum. This issue is a major drawback caused by the nonconvexity of error functions in neural networks.
2. The vanishing gradient problem is a difficulty found in training a MLP with multiple layers, where the propagated errors shrink exponentially with the increasing number of layers as the errors propagate from layer to layer.
3. MLP needs plenty of training samples, as well as a long training time to train an effective network for practical problems.

Therefore, a new neural network needs to be used in intelligent fault diagnosis to improve the diagnosis accuracy and reliability. Many neural networks are developed to deal with the drawbacks of MLP, such as radial basis function (RBF) network, wavelet neural network (WNN) and adaptive neuro-fuzzy inference system (ANFIS). Each of these developed networks has its advantages in intelligent fault diagnosis of rotating machinery. We will introduce these networks respectively in the following sections.

3.2.2 Radial Basis Function Network-Based Fault Diagnosis

3.2.2.1 Introduction

The RBF method was employed by Micchelli and Powell to study interpolation problems of a high-dimension data space individually (Micchelli, 1984; Powell, 1987). Inspired by the work,

Broomhead applied the RBF method to improve the performance of ANNs and the new network was called as RBF network. The RBF network is a single hidden layer feedforward network and the hidden layer is called radial basis layer and its activation function is defined as a Gaussian RBF. It provides good approximation for nonlinear inputs, and also can automatically determine the number of neurons of the radial basis layer in the training process. The RBF network is alternative to other multilayer networks by regarding the design of a neural network as a classification problem of a high-dimensional space, with faster convergence, higher reliability, and smaller extrapolation errors. Due to these advantages, the RBF network has attracted interests increasingly in intelligent fault diagnosis of rotating machinery, as well as other engineering applications.

The fault diagnosis method based on the RBF network consists of the steps, that is, feature extraction and selection, and fault recognition, which are described in Section 3.1. Traditionally, to acquire fault features from vibration signals in which the fault characteristics are immersed in heavy noise, some advanced signal processing techniques are needed to preprocess the vibration signals before feature extraction, especially for the signals of faults at the early stage. Among the signal processing techniques, wavelet packet transform (WPT) and empirical mode decomposition (EMD) are proved to be highly powerful for mechanical fault diagnosis in many applications. Thus, both WPT and EMD are employed to preprocess the vibration signals to highlight the fault characteristics. In order to obtain the salient feature set, time–domain features are first extracted from not only the vibration signals but also the preprocessed signals by WPT and EMD. Then a few features that obviously characterize the mechanical health conditions are selected from all the features. There are some feature selection methods, such as principal components (PCs) analysis, conditional entropy, distance evaluation technique, and so on. Due to the simpleness and reliability of distance evaluation technique, it is generally applied in intelligent fault diagnosis.

In view of the analysis aforementioned, an intelligent diagnosis method of rotating machinery is presented based on the RBF network. In the method, WPT and EMD are first used to preprocess the original vibration signals collected from rotating machinery to obtain more faulty characteristics. Then, the time–domain features are extracted from both the original and preprocessed signals to generate a feature set with rich fault information. Next, the distance evaluation technique is performed on all features in the feature set to evaluate the features, and sensitive features are selected for diagnosis. Finally, the RBF network is employed to recognize the health conditions of the rotating machinery by inputting the sensitive features.

3.2.2.2 Radial Basis Function Network

Cover's theorem (Cover, 1965) theorem proves that if a set of training samples cannot be separated linearly, one can project the training set into a higher-dimensional space using some nonlinear transformation so as to separate it linearly with high probability. Cover's theorem aims at the separability of patterns, namely a nonlinear pattern classification problem in a high-dimensional space is more likely to be linearly separable than in a low-dimensional space. Based on Cover's theorem, the RBF network is able to perform complex classification tasks by transforming the classification tasks into a high-dimensional space in a nonlinear manner to make them linear. It also indicates that the output layer of the RBF network could be used as a linear layer.

Given a training set $\{x_i, y_i\}_{i=1}^{M}$, the goal of the RBF network is to find a smooth function that fits every label exactly. A general architecture of the RBF network is displayed in Fig. 3.5. It can be seen that RBF network contains three layers as follows:

1. *Input layer*, which consists of N source nodes, where N is the dimension of the input vector $\boldsymbol{x}_i$.
2. *Radial basis layer* (also called hidden layer), which consists of the neurons using RBFs as the activation function, and the number of the neurons is as same as the number of training samples M. Typically, Gaussian function is used as the RBFs. The Gaussian function has the property that each basis

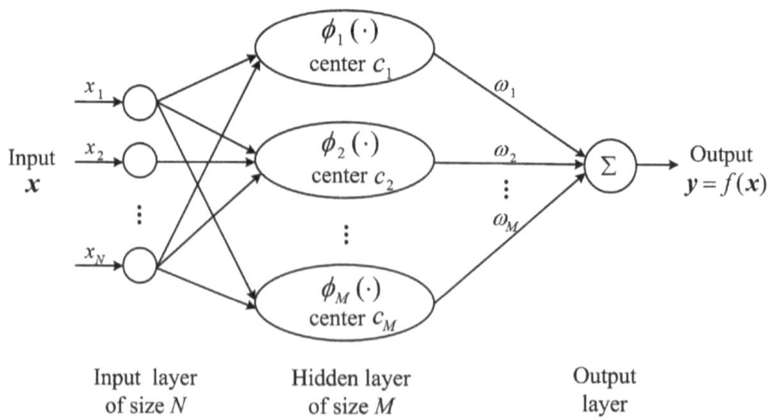

Figure 3.5. A general architecture of the radial basis function (RBF) network.

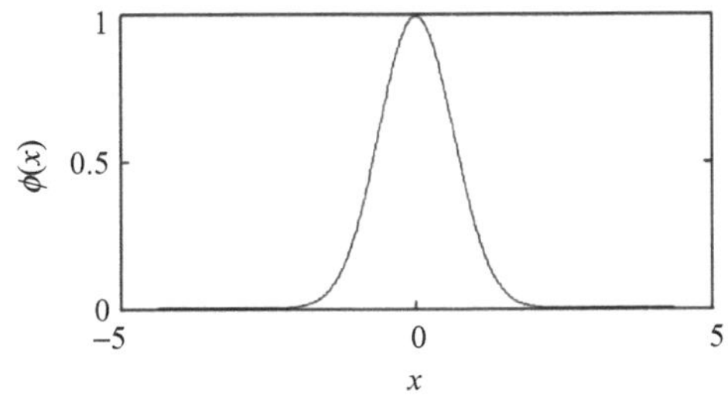

Figure 3.6. Gaussian function.

function relies on the Euclidean distance to a center $\boldsymbol{c}_j$, and it is expressed as

$$\phi_j(\boldsymbol{x}_i) = \exp\left(-\frac{(\boldsymbol{x}_i - \boldsymbol{c}_j)^2}{2b_j^2}\right) \tag{3.27}$$

where b_j is the width of the jth RBF with center c_j and $j = 1,2,\cdots,M$. The Gaussian function is shown in Fig. 3.6. It is seen that the function only responds to a small region of the input space since the Gaussian is centered. The value of the Gaussian function is peak at the zero distance and its values become smaller when the distances to the center increase. Therefore, the neuron in the radial basis layer produces an output of one when the input is a center, and its output value decreases along with the increase of distance between the input and the center.

3. *Output layer*, which is connected by the radial basis layer using weights. Typically, the size of the output layer is much smaller than that of the hidden layer. It is determined by the number of health conditions in the diagnosis tasks, namely, the dimension of $\boldsymbol{y}_i$. The value of the output layer is a linear combination of RBFs.

$$f(\boldsymbol{x}_i) = \sum_{j=1}^{M} w_j \phi_j(\boldsymbol{x}_i) \tag{3.28}$$

In practice, however, there is an issue that needs attention. When dealing with large training samples, it is wasteful for computational resources that the RBF network has a hidden layer of the same size as the number of the training samples. As the radial basis layer is described in Eq. (3.27), it is found that any correlation existing between adjacent data of the training sample is correspondingly transplanted into adjacent neurons. In other words, it may be redundancy of neurons in the radial basis layer. Additionally, the training set $\{\boldsymbol{x}_i, \boldsymbol{y}_i\}_{i=1}^{M}$ is typically noisy, especially in

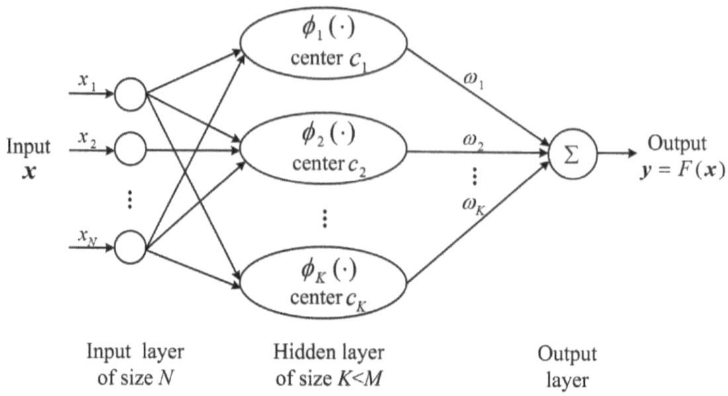

Figure 3.7. Architecture of the RBF network.

mechanical intelligent fault diagnosis. Selecting noisy samples as the centers could lead to misleading results. So it is necessary to design an RBF network in a different approach. Based on these reasons, it can reduce the size of the radial basis layer to a fraction of the number of the training samples, as shown in Fig. 3.7.

Although the RBF networks in Figs. 3.5 and 3.7 are different, they share a common feature: the training of an RBF network does not involve the BP of error signals, which is unlike the case for multilayer perceptron. Moreover, the value of the output layer becomes

$$F(\boldsymbol{x}_i) = \sum_{j=1}^{K} w_j \phi(\boldsymbol{x}_i, \boldsymbol{c}_j) \tag{3.29}$$

where each neuron of the radial basis layer is characterized by the function $\phi(\boldsymbol{x}_i, \boldsymbol{c}_j)$ and $j=1,2,\cdots,K$ with K being smaller than M. So the difference between the structure of Fig. 3.5 and that of Fig. 3.7 is that all the training samples are used as the centers in Fig. 3.5 whereas parts of training samples are used as the centers in Fig. 3.7.

So the key of a successful implementation of the RBF network is to find the suitable centers for the Gaussian functions. A simple way of choosing the centers is to randomly choose a subset of the training samples. A more systematic approach is named as orthogonal least squares (Chen et al., 1991). This is a sequential selection process in which, at each step the next data point chosen as a basis function center corresponds to the one that gives the greatest reduction in the sum-of-squares error. Values for the expansion coefficients are determined as part of the algorithm. Clustering algorithms can also be used, and they may give a set of centers that no longer coincide with the training samples.

3.2.2.3 Fault Diagnosis Method Based on RBF Network

The time–domain features are not only irrespective of loads and speeds of rotating machinery but also able to indicate early faults occurring in the rotating machinery effectively. Here, six of them are used: skewness, kurtosis, crest indicator, clearance indicator, shape indicator, and impulse indicator. First, the six features are extracted from the original vibration signals to characterize the general change of the mechanical health conditions. Then, WPT and EMD are used to preprocess each of the original vibration signals to obtain additional fault characteristic information, and the six time–domain features are extracted from the frequency-band signals of WPT and the intrinsic mode functions (IMFs) of EMD. Suppose that the decomposition level of WPT is I and the IMF number of EMD is L. We can get a combined feature set with $6 + 6 \times 2^I + 6L$ features for each original vibration signal.

Although the features of the combined feature set may reflect the mechanical faults from different aspects, they show different sensitivity to these faults. The sensitive features mean that samples from the same class have similar feature values and those from different classes have significantly different values. So feature selection is important before the feature set is fed into a classifier. By eliminating insensitive features, the feature selection is able to reduce the computational burden and improve the generalization capability of the trained classifier. Here, a feature evaluation technique is applied to select the sensitive features from the combined feature set.

Suppose that the combined feature set of C health conditions is

$$\left\{q_{m,c,j}, m=1,2,\cdots,M_c; c=1,2,\cdots,C; j=1,2,\cdots,J\right\}, \tag{3.30}$$

where, $q_{m,c,j}$ is the jth feature extracted from the mth signal under the cth health condition, M_c is the signal number of the cth health condition, and J is the number of the features of each condition. The distance evaluation technique can be described as follows.

1. Get the average distance of the signals with the same health condition

$$d_{c,j} = \frac{1}{M_c \times (M_c - 1)} \sum_{l,m=1}^{M_c} |q_{m,c,j} - q_{l,c,j}|, \quad l,m=1,2,\cdots,M_c, l \neq m. \tag{3.31}$$

2. Calculate the average distance of C health conditions

$$d_j^{(w)} = \frac{1}{C}\sum_{c=1}^{C} d_{c,j}. \tag{3.32}$$

3. Obtain the average value of the features for all signals with the same health condition

$$u_{c,j} = \frac{1}{M_c}\sum_{m=1}^{M_c} q_{m,c,j}. \tag{3.33}$$

4. Calculate the average distance between different health condition signals

$$d_j^{(b)} = \frac{1}{C\times(C-1)}\sum_{c,e=1}^{C}\left|u_{e,j} - u_{c,j}\right|, \quad c,e=1,2,\cdots,C, c\neq e. \tag{3.34}$$

5. Calculate the ratio $d_j^{(b)}$ and $d_j^{(w)}$ so as to get the distance evaluation factor

$$\alpha_j = \frac{d_j^{(b)}}{d_j^{(w)}}. \tag{3.35}$$

We use the distance evaluation factor to indicate the sensitivity of the features of various mechanical health conditions. After the distance evaluation factors of all the features are calculated, these features are ranked from large to small according to their evaluation factors. Therefore, we can obtain a ranked feature set, from which the first D features are selected and applied to diagnose the faults of rotating machinery.

To make sense of the above feature evaluation technique, Fig. 3.8 shows a simple classification example. This is a three-class classification problem and each of the classes has 10 signals characterized by two features (f1 and f2), respectively. The sensitivities of f1 and f2 are evaluated by the distance evaluation technique. From Fig. 3.8, it can be obviously seen that 10 signals of each class

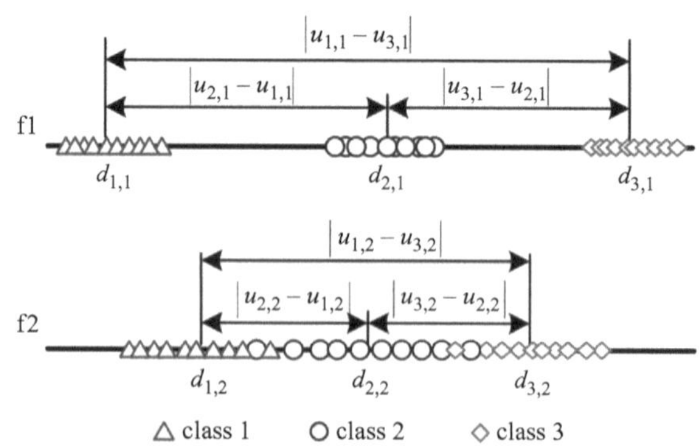

Figure 3.8. Sensitivity comparison between two features in a three-class problem.

characterized by f1 are clustered closer with each other than f2. So f1 is able to separate these three classes easier than f2 and the classification accuracy with f1 will be higher than that with f2. Therefore, the f1 is better or more sensitive than f2 in this classification problem. Based on the distance evaluation technique, the value of the average distance within classes of f1 $\left(d_1^{(w)} = \frac{1}{3}\sum_{c=1}^{3} d_{c,1}\right)$ is smaller than that of f2 $\left(d_2^{(w)} = \frac{1}{3}\sum_{c=1}^{3} d_{c,2}\right)$. But the value of the average distance between classes of f1 $\left(d_1^{(b)} = \frac{1}{6}\sum_{c,e=1}^{3} |u_{e,1} - u_{c,1}|\right)$ is bigger than that of f2 $\left(d_2^{(b)} = \frac{1}{6}\sum_{c,e=1}^{3} |u_{e,2} - u_{c,2}|\right)$. Therefore, the evaluation factor of f1 $\left(\alpha_1 = d_1^{(b)} / d_1^{(w)}\right)$ is larger than that of f2 $\left(\alpha_2 = d_2^{(b)} / d_2^{(w)}\right)$. This analysis indicates the reason that the distance evaluation technique can select sensitive features for the classification problem of rotating machinery and improve the diagnosis accuracy.

In summary, the intelligent diagnosis method using the RBF network is described in Fig. 3.9. It involves four procedures as follows:

1. Acquire vibration signals from rotating machinery, and preprocess them using WPT with db10 and EMD to obtain the frequency-band signals and IMFs, respectively.
2. Extract the six time–domain features from both the original and preprocessed signals.
3. Combine all the features into a feature set and evaluate these features by the distance evaluation technique. According to the evaluation factors, the features are rearranged to form the ranked feature set.
4. Increase the number of the selected sensitive features from one to the number of features, and feed the selected features into the RBF network for training. If one of the following two criteria is satisfied, the training process will stop. One is the training accuracy increasing to 100%, and the other is that the training accuracy hardly increases as the number of sensitive features increases to F (in this section $F = 5$).

3.2.2.4 Intelligent Diagnosis of Bearing Faults: An Experimental Case Study

Rolling element bearings play an important role in rotating machinery. Bearing faults may lead to fatal damages of the machinery. Therefore, it is absolutely essential to be able to detect and diagnose the faults of the bearings accurately and

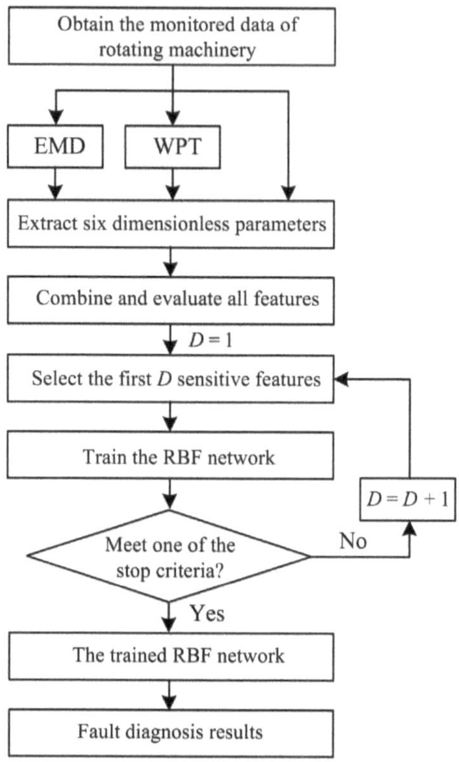

Figure 3.9. Flowchart of the intelligent diagnosis method based on RBF network.

automatically. In this section, a rolling element bearing experiment is performed to verify the effectiveness of the proposed intelligent diagnosis method.

The experimental system illustrated in Fig. 3.10A and B displays the structure sketch of the test rig. The tested bearings are fixed on a shaft driven by the motor. The motor drives the shaft by a transmission. An accelerometer is installed on the shaft box to collect the vibration signals from the bearings. The Advanced Data Acquisition and Analysis System by Sony EX is applied to collect the data. Some parameters in the experiment are listed in Table 3.1.

The faults are introduced into the outer race, the inner race, and the roller of the tested bearings using the electrical discharge machining (EDM) technique, respectively. The defect depth is 0.05–0.15 mm and the width is 0.3–0.7 mm. These faulty bearings can be seen in Fig. 3.11. The bearing data set is acquired from the experiment system under four different operating conditions: (1) normal condition; (2) with outer race fault (OF); (3) with inner race fault (IF); (4) with roller fault. For each of the four operating conditions,

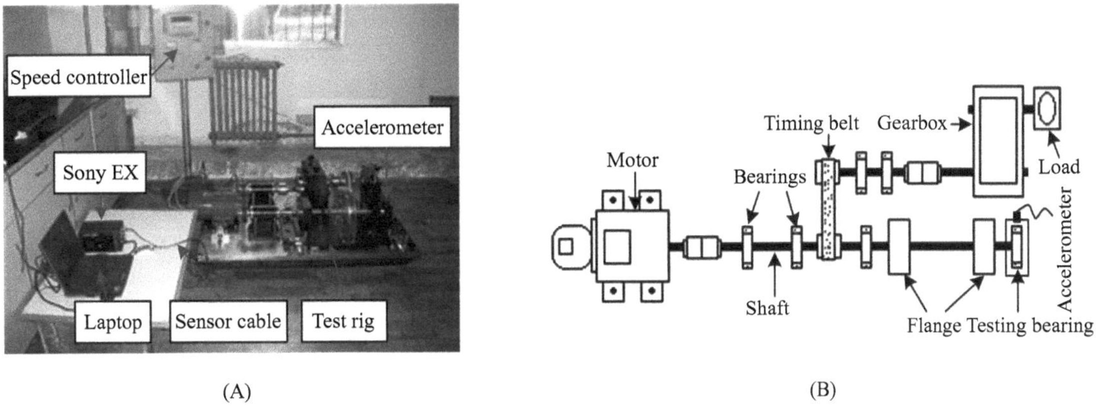

Figure 3.10. (A) Experimental system and (B) structure sketch.

95 data samples are collected. Therefore, it is a four-class classification problem that aims to classify different fault categories.

WPT is performed on each of the original vibration signals with level 3 and therefore eight frequency-band signals are obtained. In addition, each original signal is preprocessed by EMD. The first eight IMFs are selected and used to extract features for covering most of the information. The six dimensionless parameters are extracted from the original signal, each of the eight frequency-band signals derived by WPT and each of the eight IMFs achieved by EMD. Therefore, a data sample can be represented by 102 features.

Table 3.1 Parameters in the Experiment

Parameter	Value
Bearing specs	NU312 (NSK)
Outer race diameter	27 mm
Inner race diameter	14 mm
Roller diameter	6.35 mm
Roller number	10
Contact angle	0°
Rotating speed	1200 rpm
Sampling frequency	25.6 kHz
Sampling points	4096

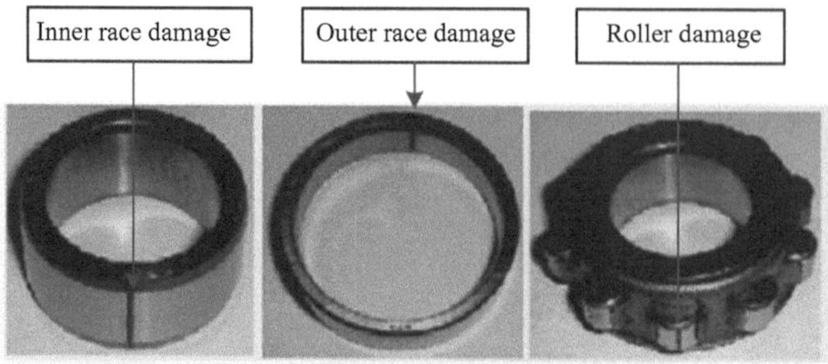

Figure 3.11. Faults in the rolling element bearings.

The 95 samples of each condition are divided into two sets: 80 samples used to train and 15 samples used to test. With the proposed method applied to diagnose faults of the bearings, the combined feature set with 102 feature values is extracted by the distance evaluation technique. Fig. 3.12 shows the evaluation factors of all the features. We add the number of the sensitive features one by one and use them to train the RBF network. The output layer of the RBF network contains four neurons corresponding to the four operating conditions of the bearings. If the second stop criterion in the proposed method is satisfied first (the training accuracy is not improved if the number of sensitive features increases to five), the training process of the RBF network is terminated. 20 sensitive features are selected, whose evaluation factors are over the horizontal line in Fig. 3.12. The horizontal line indicates the value of the extracted factor of the 20th selected sensitive feature. Table 3.2 clearly displays these 20 features.

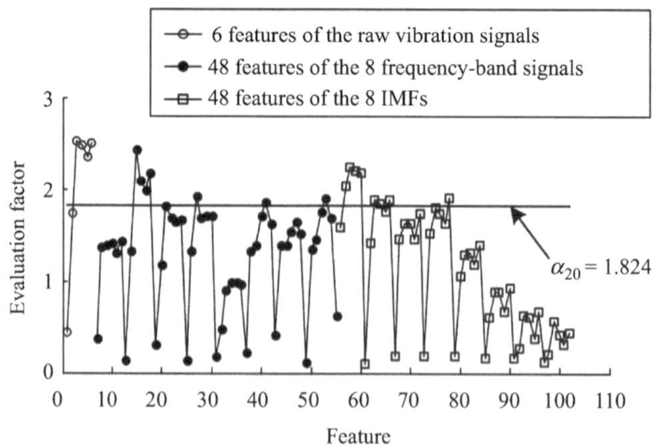

Figure 3.12. Distance evaluation factors of all the features.

Table 3.2 Selected 20 Sensitive Features in the Fault Diagnosis of the Bearings

	Original	Frequency-Band Signals								IMFs							
Parameter	**Signal**	**1**	**2**	**3**	**4**	**5**	**6**	**7**	**8**	**1**	**2**	**3**	**4**	**5**	**6**	**7**	**8**
SK	−	−	−	−	−	−	−	−	−	−	−	−	−	−	−	−	−
KU	−	−	−	−	−	−	−	−	−	−	−	−	−	−	−	−	−
CI	+	−	+	+	+	−	−	−	−	−	+	+	−	−	−	−	−
CLI	+	−	+	−	−	−	−	−	−	−	+	+	−	−	−	−	−
SI	+	−	+	−	−	−	+	−	+	−	+	−	−	−	−	−	−
IMI	+	−	+	−	−	−	−	−	−	−	+	+	−	+	−	−	−

+, Features selected; −, features not selected.

After the training process, the test samples are utilized to evaluate the generalization performance of the proposed method. The testing result is obtained with the 20 selected sensitive features extracted from each test sample and fed into the RBF network. Fig. 3.13 shows the testing results.

Two additional experiments are conducted to evaluate the performance of the proposed method. (1) The six parameter values extracted from the original vibration signals and the whole combined feature set (102 features) are input to the RBF network to

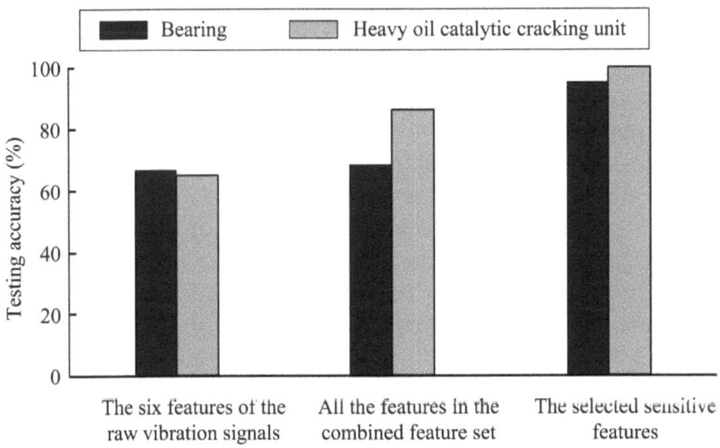

Figure 3.13. Testing accuracies of the experiment and application.

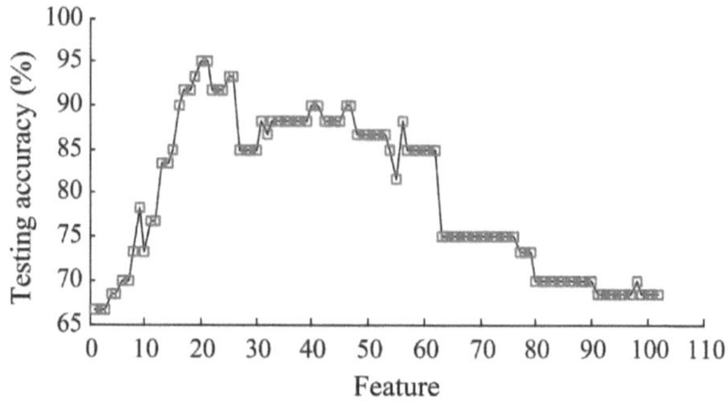

Figure 3.14. Testing accuracy varying with input features.

train and test, respectively. The testing results are displayed in Fig. 3.13. (2) According to the evaluation factors of the features from large to small, with the number of the sensitive features added one by one until all the features are input to the RBF network, the testing accuracy varies with the number of the input features as shown in Fig. 3.14.

From Fig. 3.13, it is seen that for the four health condition identification of the rolling element bearings, the lower classification accuracy equaling 66.67% is obtained with the six features extracted from the original vibration signals. With WPT and EMD used to decompose the vibration signals and all the features including the original signals, frequency-band signals, and IMFs utilized to identify conditions, the accuracy rate increases by a small improvement of 1.66%. It indicates that additional characteristic information is indeed mined using WPT and EMD, but some information unrelated to faults is also obtained. In other words, not all features extracted from the frequency-band signals and IMFs can make contributions to identifying the conditions of the bearings. Some of them are not sensitive to the four bearing conditions. These features would confuse the classifier based on the RBF network and therefore do not lead to an obvious increase in the classification accuracy. However, if the sensitivities of the features are evaluated by the distance evaluation technique and 20 sensitive features are chosen as the input of the RBF network, the testing accuracy increases greatly to 95%.

Observing that the testing accuracy varies with the feature number in Fig. 3.14, we may find that the features have the higher sensitivity at the beginning and the testing accuracy increases with the number of features fed into the RBF network increased. However, the accuracy begins to decrease with the number of the selected features added up to 20, because the features added subsequently are not sensitive to the four conditions of the rolling element bearings. This might indicate that with the increase of the feature's number, the selected features contain too much fault-unrelated information and a high degree of overlap exists in the values of these features. Although the number of input features increases, the accuracies present the opposite trend.

Clearly, the following conclusions may be drawn from the analysis results earlier. It is able to obtain more fault information by using WPT and EMD to decompose the original vibration signals. However, some features extracted from the signals decomposed by WPT and EMD also include much useless information. Only if sensitive features are selected and unrelated, even harmful features are discarded by the distance evaluation technique, can different operational conditions of the bearings be accurately identified by the RBF network with a smaller configuration. Therefore, the experiment results of fault diagnosis of the rolling element bearings verify the effectiveness of the proposed intelligent diagnosis method.

3.2.2.5 Intelligent Diagnosis of Rub Faults: A Heavy Oil Catalytic Cracking Unit Case Study

The heavy oil catalytic cracking unit is one of the key machine sets and the typical rotating machinery used in oil refineries. The breakdown of the heavy oil catalytic cracking unit because of machine faults will cause a great loss of production. Therefore, it is important to diagnose the faults occurring in the heavy oil catalytic cracking unit.

The proposed method is applied to slight rub fault diagnosis of the heavy oil catalytic cracking unit in an oil refinery. This machine set is composed of a gas turbo, a compressor, a gearbox, and a driving motor. Fig. 3.15 shows the structure sketch of the heavy oil catalytic cracking unit. The gas turbo transforms the heat energy

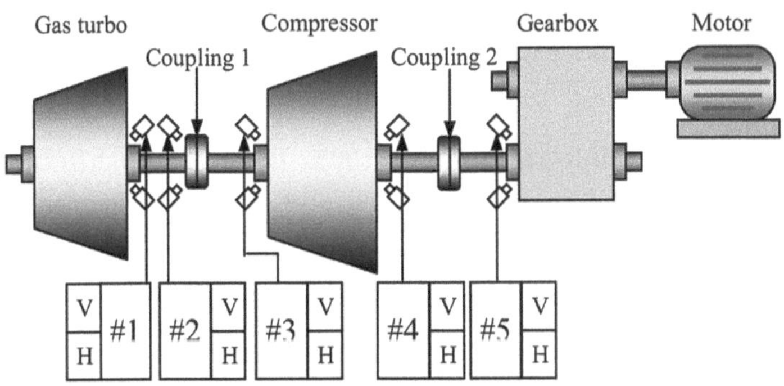

Figure 3.15. Structure sketch of the heavy oil catalytic cracking unit.

into the mechanics energy. Two bearings (#1 and #2) are used to support the gas turbo shaft, (#3 and #4) are used to support the compressor shaft, and #5 is to support the gearbox shaft. The hub and the laminas (left components of the gas turbo) on the shaft are a cantilever structure. The rotating speed of the machine set equals 5745 rpm (95.75 Hz). Bently 3300 system is used to monitor its operating condition. To collect vibration signals, eddy current transducers are fixed on each bearing case in vertical and horizontal directions respectively. A computer on-line monitoring system is applied for data acquisition and record.

After an overhaul, this machine set runs again. But the vibration of the machine set is beyond alarm limit and the machine set has to be stopped. When it is disassembled, it is found that there is a large area of rub between the hub (rotating with its shaft) and the gas seal (stationary). After the rub area is burnished, the rub becomes slight and the vibration weakens. However, the rub fault cannot be removed thoroughly. It is found that the vibration monitored in the vertical direction of bearing case #1 is larger and more complex than other monitoring points when the machine set is running. Therefore, we analyze it to recognize the different conditions of the machine set. Under each of the three operation conditions: before the overhaul (normal condition), after the overhaul (a large area of rub fault), and after burnishing (slight rub fault), 82 data samples are acquired with a sampling frequency equaling 2000 Hz and a data sample has 512 sampling points. Therefore, there are totally 246 samples collected from the machine set.

The proposed intelligent diagnosis method is applied to diagnose faults of the machine set. In order to identify the severe grades of faults, we solve the three-class classification problem. The 246 data samples are divided into 123 training samples and 123 testing samples. Each of the original vibration signals is decomposed by db10 WPT at level 3 and EMD. The first six IMFs which carry the major information are chosen. The six dimensionless parameters are obtained from each original signal, frequency-band signal, and IMF. Thus, 90 feature values are obtained to represent a data sample.

When the training accuracy is 100%, the first stop criterion of the proposed method is satisfied and the training process terminates. The three features selected in rub fault diagnosis are the skewness of the signal, the impulse indicator of the first frequency-band signal of WPT and the shape indicator of the first IMF of EMD. Fig. 3.13 shows the testing results of the proposed method, and it also displays the comparison results selecting different input features.

The results in Fig. 3.13 show that more fault characteristics are acquired because of using WPT and EMD, and thus the diagnosis accuracy rises from 65.04% to 86.18%. After evaluating the combined feature set extracted from both the original and the preprocessed signals and selecting the sensitive features, the computation burden of the RBF network is able to be reduced and the diagnosis accuracy of the RBF network is improved to 100%. Thus, the proposed intelligent diagnosis method is not only able to distinguish the normal condition from the large area rub fault, but also to recognize the slight rub fault from the normal condition in the machine set. It means that the method proposed is capable of diagnosing slight faults. These verify the effectiveness of the proposed fault diagnosis method which integrates WPT, EMD, the distance evaluation technique, and the RBF network.

3.2.2.6 Epilog

In order to improve the fault diagnosis accuracy of rotating machinery, WPT, EMD, the distance evaluation technique, and the RBF network are integrated into an intelligent diagnosis method. The method takes full advantage of each part. WPT and EMD are first employed to obtain rich fault information from the original vibration signals. Then, the six time–domain features are extracted from the original signals, the frequency-band signals of WPT and the IMFs of EMD. Third, all the features are used to generate a combined feature set, in which the sensitive features are selected by the distance evaluation technique. Finally, these sensitive features are fed into the RBF network to improve the fault diagnosis accuracy. The diagnosis results of the rolling element bearings and a heavy oil catalytic cracking unit demonstrate that the presented method is an effective method in fault diagnosis of rotating machinery.

3.2.3 Wavelet Neural Network-Based Fault Diagnosis

3.2.3.1 Introduction

In the previous section, EMD is used to decompose a complicated signal of rotating machinery into IMFs based on the local characteristic time scales of the signal (Huang et al., 1998). These IMFs work as the basis functions determined by the signal itself, and characterize the natural oscillatory modes embedded in the signal. Therefore, EMD is a self-adaptive signal processing technique and has been widely applied in fault diagnosis of rotating machinery. However, one drawback of EMD is known as the mode mixing problem. It is defined as either a single IMF consisting of components of widely disparate scales, or a component of a similar scale residing in different IMFs. To alleviate the mode mixing problem in EMD, a noise-assisted data analysis method called ensemble empirical mode decomposition (EEMD) was developed (Wu and Huang, 2009). By sifting an ensemble of white noise-added signal, the EEMD method can automatically eliminate the problem of mode mixing. Thus, the EEMD is a promising tool to replace EMD in intelligent fault diagnosis of rotating machinery.

In fault diagnosis of rotating machinery, it is a fact that some IMFs generated by EEMD are sensitive and related to the faults while others are not. So it encourages us to select the sensitive IMFs from all the IMFs, which may provide fault-related information, and discard irrelevant IMFs. Loutridis considered that the IMF sensitivity to mechanical faults could look through the energy content in each IMF of EMD. So he proposed an empirical law using the energy content to diagnose gear damages (Loutridis et al., 2005). Different from Loutridis's way, a new approach is proposed to select the sensitive IMFs from the IMFs decomposed by EEMD. The main idea behind this approach is to measure the sensitivity of IMFs through their kurtosis. As we know, the vibration signals of rotating machinery show a series of impulses when local defects occur, and Kurtosis is the right index for detecting impulse components caused by the local defects.

After obtaining the features from the sensitive IMFs, one needs to use an intelligent technique to make a decision on the health conditions of the rotating machinery automatically. In the previous section, RBF network is used to overcome the MLP's drawback of limiting to characterize local features. So the RBF network is able to be well suitable for mechanical fault classification. However, the activation function of the RBF network is Gaussian function, which is nonadaptive. Contrarily, wavelet bases have the adaptive spatial and frequency localization properties, so wavelet transform is effective in representing and detecting local characteristics of rotating machinery. By combining wavelet transform and MLP, Zhang and Benveniste (1992) proposed a new neural network named wavelet neural network (WNN). Since combining the characteristics of time–frequency location of the wavelet transform and self-learning capability of ANNs, WNN simultaneously harnesses the advantages of wavelet transform and ANNs.

Therefore, based on EEMD and WNN, an intelligent fault diagnosis method is presented in this section. In the method, the collected vibration signals are first decomposed by the EEMD method. Then, the most sensitive IMF is selected based on kurtosis index. Third, time– and frequency–domain features are extracted from the most sensitive IMF, its frequency spectrum and envelope spectrum. Finally, these extracted features are fed into WNN to recognize the mechanical health conditions.

3.2.3.2 *Wavelet Neural Network*

WNN has been successfully used in different applications. Zhang and Benveniste (1992) used WNN for approximating arbitrary nonlinear functions, and proved that WNN can provide good performance in nonlinear function learning. We will focus on its applications in intelligent fault diagnosis of rotating machinery after introducing the theories of the WNN.

The architecture of WNN is similar to that of MLP, consisting of input layer, one hidden layer, and output layer. The hidden layer composes of neurons, whose activation functions replace the ordinary basis function networks with a wavelet basis $\psi_{a,b}(\cdot)$. The model of a wavelet neuron is displayed in Fig. 3.16. It is seen that the inputs of the wavelet neuron include the sum of weighted inputs and parameters of the wavelet function. During the training process, both weights and wavelet parameters will be optimized using the BP algorithm.

The output of the wavelet neuron in Fig. 3.16 is expressed as,

$$c = \psi\left(\frac{u_k - b}{a}\right) \tag{3.36}$$

where a and b are the dilation and translation parameters, and the sum of weighted inputs u_k

$$u_k = \sum_{i=1}^{p} w_{ki} x_i \tag{3.37}$$

The schematic diagram of the three-layer WNN is shown in Fig. 3.17. It is seen that the output layer consists of a linear weighted combination of the wavelet activation functions.

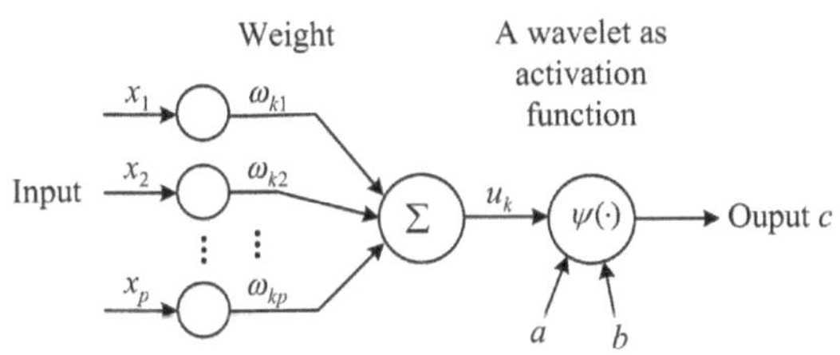

Figure 3.16. A wavelet neuron model.

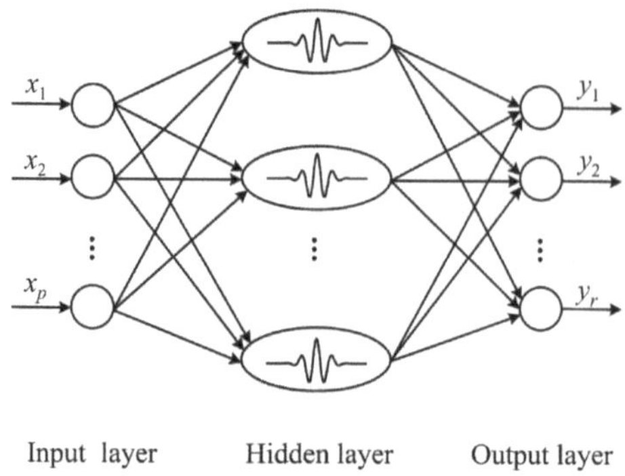

Figure 3.17. Structure of wavelet neural network.

The commonly used wavelet functions include Morlet, Haar, Mexican Hat, Shannon, and Meyer. Here, the Morlet wavelet is used as the activation function of the neurons in the hidden layer, because its explicit expression is simple and the partial derivatives of the corresponding error function can be easily calculated. This property is necessary for parameter updating in the training process.

The Morlet wavelet is expressed as follows

$$\psi(t)=\cos(5t)\exp\left(-\frac{t^2}{2}\right) \tag{3.38}$$

Its derivative is given by,

$$\dot{\psi}(t)=-5\sin\ (5t)\exp\left(-\frac{t^2}{2}\right)-t\cos(5t)\exp\left(-\frac{t^2}{2}\right) \tag{3.39}$$

The training of WNN is done by the backpropagation algorithm, minimizing an error function between the actual network outputs and the target outputs. The training continues until a specified value of the error function is met or the maximum number of iterations is reached.

3.2.3.3 Sensitive IMF Selection and Feature Extraction

EEMD not only inherits the merits of EMD in decomposing nonstationary and nonlinear signals, but also alleviates the mode mixing problem of EMD. We use EEMD to decompose the vibration signals of rotating machinery and obtain the IMFs. In these IMFs, some are sensitive and related to the faults while others are not. To avoid interference of fault-unrelated components, we can select the sensitive IMF to further process.

When local defects occur in rotating machinery, such as rub and flaking in outer race, inner race, and roller of bearings, the vibration signals show a series of impulses. Since kurtosis is able to detect impulse components caused by local defects, a new approach based on kurtosis is developed to select the sensitive IMF from the IMFs decomposed by EEMD.

1. For each health condition of the dataset, the kurtosis values of I IMFs of each data sample are computed. Let $k_{i,s}$ denote the kurtosis of the ith IMF of the sth data sample, and it can be expressed as:

$$k_{i,s} = \frac{N\sum_{n=1}^{N}(a_{n,i}-\bar{a}_i)^4}{\left(\sum_{n=1}^{N}(a_{n,i}-\bar{a}_i)^2\right)^2}, \quad n=1,2,\cdots,N, \quad i=1,2,\cdots,I, \quad s=1,2,\cdots,S \tag{3.40}$$

where $a_{n,i}$ is the nth data point in the ith IMF a_i, $\bar{a}_i$ is the average of a_i, N is the number of data points of a_i, I is the number of IMFs, and S is the number of the data samples.

2. Compute the mean and standard deviation of kurtosis values of each IMF

$$m_i = \frac{1}{S}\sum_{s=1}^{S}k_{i,s} \quad i=1,2,\cdots,I \tag{3.41}$$

where m_i is the mean of the kurtosis values.

$$std_i = \sqrt{\frac{1}{(S-1)}\sum_{s=1}^{S}(k_{i,s}-m_i)^2} \quad i=1,2,\cdots,I \tag{3.42}$$

where std_i is the standard deviation of the kurtosis values.

3. Construct the criterion of the sensitive IMF selection. For the signals of the normal health condition, the kurtosis values of IMFs are small. However, when the machines have local defects, its kurtosis values of IMFs will increase. Therefore, normal signals have smaller mean m_i than faulty signals. Furthermore, smaller standard deviation std_i is easier to distinguish different health conditions whether the health condition is normal or faulty. So we construct the selection criterion of the sensitive IMF as follows.

$$f_i = \begin{cases} m_i \cdot std_i, & \text{for normal bearings} \\ \dfrac{std_i}{m_i} & \text{for faulty bearings} \end{cases} \quad i=1,2,\cdots,I \tag{3.43}$$

4. Select the sensitive IMF of the data sample under each health condition. Based on the definition of the selection criterion, the smallest f_i in Eq. (3.43) indicates the sensitive IMF. Thus, the selected IMF will be further used to extract features.

3.2.3.4 WNN-Based Fault Diagnosis Method

For diagnosis purpose, time- and frequency–domain features are extracted from the sensitive IMF to identify various mechanical health conditions. They are standard deviation (x_{std}), kurtosis (x_{kur}), shape factor (*SF*), and impulse factor (*IF*) in the time domain, and mean frequency (x_{mf}), root mean square frequency (x_{rmsf}), standard deviation frequency (x_{stdf}) and spectrum peak ratio of bearing outer race (*SPRO*), inner race (*SPRI*), and roller (*SPRR*) in the frequency domain.

In the diagnosis process, four time–domain features are extracted from the sensitive IMF of each data sample, and three frequency–domain features x_{mf}, x_{rmsf}, and x_{stdf} are extracted from its frequency spectrum. Moreover, demodulation technique makes the diagnosis process independent of a particular machine since it focuses on the low-amplitude high-frequency broadband signals. So we also use *SPRO, SPRI,* and *SPRR* to extract features from the envelope spectrum of the sensitive IMF. Therefore, 10 features are produced from the sensitive IMF of each data sample, as displayed in Table 3.3. They will be fed into the WNN to automatically recognize the mechanical health conditions.

Fig. 3.18 shows the intelligent diagnosis method based on EEMD and WNN. It contains the following four procedures. First, each of the vibration signals acquired from rotating machinery is decomposed into a series of IMFs using EEMD. Second, the sensitive IMF is selected from these IMFs via the kurtosis-based approach described in Section 3.2.3.3. Third, 10 features are extracted from the most sensitive IMF, its frequency spectrum and its envelope spectrum. Finally, WNN is used to identify the health condition of rotating machinery and obtain the diagnosis results.

3.2.3.5 Intelligent Diagnosis of the Compound Faults: A Bearing Case Study

1. Experimental setup and data acquisition—Fig. 3.19 displays the test bench of a locomotive roller bearing. The test bench is composed of a hydraulic motor, two supporting pillow blocks, a tested bearing (52732QT) which is loaded on the outer race by a hydraulic cylinder, a hydraulic radial load application system, and a tachometer for shaft speed measurement. The bearing is fixed in a hydraulic motor driven mechanical system.

Table 3.3 The Extracted Features

1	Standard deviation	X_{std}	6	Root mean square frequency	X_{rmsf}
2	Kurtosis	X_{kur}	7	Standard deviation frequency	X_{stdf}
3	Shape factor	SF	8	Spectrum peak ratio of bearing outer race	$SPRO$
4	Impulse factor	IF	9	Spectrum peak ratio of bearing inner race	$SPRI$
5	Mean frequency	X_{mf}	10	Spectrum peak ratio of bearing roller	$SPRR$

$$SPRO=\frac{K\sum_{i=1}^{H}p_0(h)}{\sum_{k=1}^{K}s(k)},\quad SPRI=\frac{K\sum_{i=1}^{H}p_I(h)}{\sum_{k=1}^{K}s(k)},\quad SPRR=\frac{K\sum_{i=1}^{H}p_R(h)}{\sum_{k=1}^{K}s(k)}$$

where $s(k)$ is a spectrum for $k=1,2,\ldots,K$, K is the number of spectrum lines; f_k is the frequency value of the kth spectrum line; $p_0(h)$, $p_I(h)$ and $p_R(h)$ are respectively the peak values of the hth ($h=1,2,\ldots,H$, and H is the number of harmonics.) harmonics of the characteristic frequencies for bearing outer race (f_0), inner race (f_I) and roller (f_R), which can be calculated according to the following equations: $f_0=\frac{f_r}{2}N_R\left(1-\frac{B}{C}\cos\alpha\right)$, $f_I=\frac{f_r}{2}N_R\left(1+\frac{B}{C}\cos\alpha\right)$ and $f_R=\frac{f_rC}{2B}\left[1-\left(\frac{B}{C}\cos\alpha\right)^2\right]$. f_r is the shaft rotational frequency; N_R is the roller number; α is the contact angle; B and C are the roller and pitch diameters, respectively.

608A11-type ICP accelerometers are installed on the load module beside the outer race of the tested bearing to measure its vibrations. The Advanced Data Acquisition and Analysis System by Sony EX is applied to collect the data. Some parameters in the experiment are listed in Table 3.4.

A bearing data set containing nine subsets is acquired from the experimental system under nine different health conditions. The nine conditions are displayed in Table 3.5. Among the nine bearing conditions, the serious fault in the outer race (condition 2) is a serious flaking fault and other faults all belong to slight rub. Fig. 3.20 shows the faulty bearings. Each data subset consisting of 50 samples corresponds to one of the nine conditions. Each sample is a vibration signal and contains 8192 sampling points. Fig. 3.21A–I show the data samples for the nine bearing health conditions: (A) normal condition, (B) slight rub fault in the outer race, (C) serious flaking fault in the outer

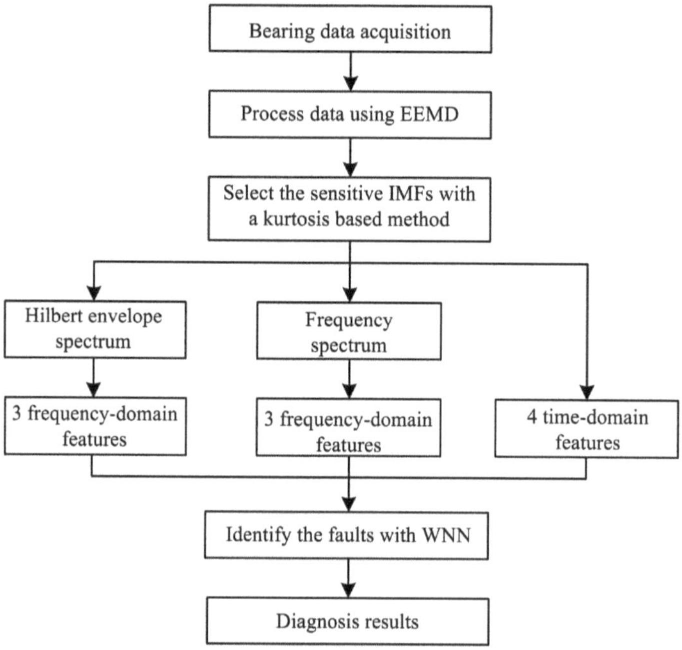

Figure 3.18. Flowchart of the WNN-based diagnosis method.

race, (D) slight rub fault in the inner race, (E) roller rub fault, (F) compound faults in the outer and inner races, (G) compound faults in the outer race and roller, (H) compound faults in the inner race and roller, and (I) compound faults in the outer and inner races and roller.

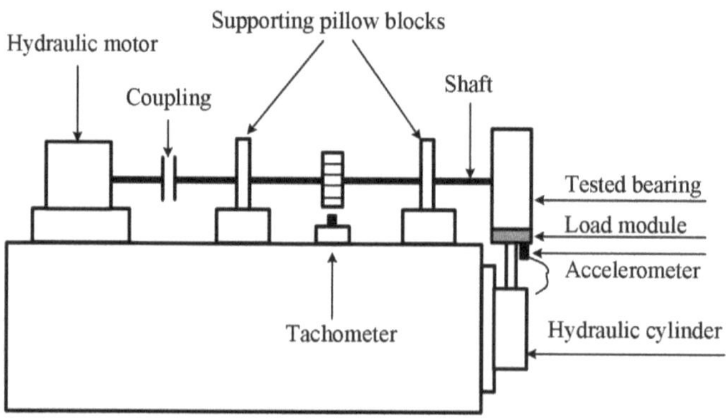

Figure 3.19. Test bench of the locomotive roller bearing.

Table 3.4 Parameters in the Experiment

Parameter	Value
Bearing specs	552732QT
Load	9800 N
Inner race diameter	160 mm
Outer race diameter	290 mm
Roller diameter	34 mm
Roller number	17
Contact angle	0°
Sampling frequency	12.8 kHz

To verify the effectiveness of the method proposed, three experiments are conducted over three different data sets chosen from the whole data set of the locomotive roller bearings. For comparison, MLP neural network with three layers and RBF neural network are also used to analyze the same data sets, respectively. MLP neural network, RBF network, and WNN have the same architecture. They have three layers, that is, input layer, hidden layer, and output layer, and the nodes in the layers are 10, 12, and 1, respectively.

Table 3.5 Description of the Faulty Bearings

Condition	Rotating Speed (rpm)	Label
Normal condition	490	1
Slight rub fault in the outer race	490	2
Serious flaking fault in the outer race	480	3
Slight rub fault in the inner race	500	4
Roller rub fault	530	5
Compound faults in the outer and inner races	520	6
Compound faults in the outer race and rollers	520	7
Compound faults in the inner race and rollers	640	8
Compound faults in the outer and inner races and rollers	550	9

Figure 3.20. Faults in the locomotive roller bearings.

2. Experiment 1: Diagnosis performance comparison of different fault categories—In this experiment, a bearing data set composed of the four data subsets acquired respectively under normal condition, serious fault in the outer race, inner race fault, and roller fault, is utilized to evaluate how the proposed method performs in recognizing different fault categories of the locomotive roller bearings. Each of the four data subsets contains 50 samples, and therefore the whole data set which corresponds to the four bearing health conditions totally includes 200 samples. The whole data set is separated into two sets: 100 samples for training and the remaining 100 for testing.

   Applying the 10 features extracted in Table 3.3 as the input features, the three classifiers based on MLP neural network, RBF neural network, and WNN are trained respectively and tested in recognizing the fault categories of the locomotive roller bearings. The testing results are displayed in Table 3.6 and Fig. 3.22, respectively. In this experiment, MLP neural network, RBF neural network, and WNN achieve the classification

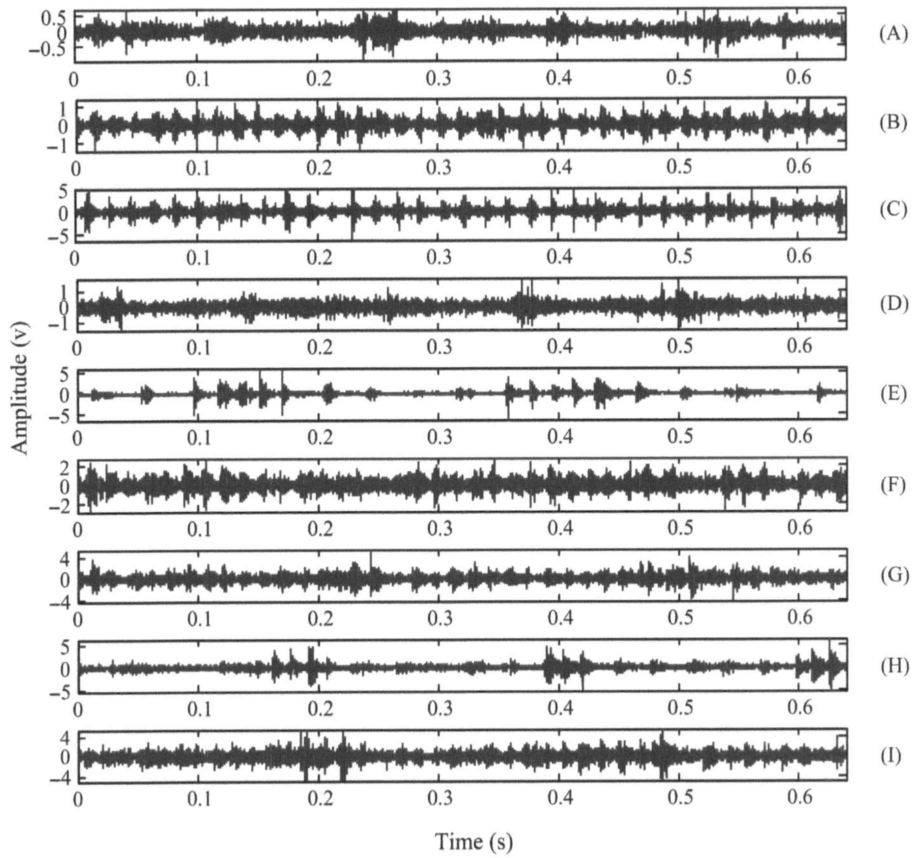

Figure 3.21. Data samples of the nine bearing conditions.

Table 3.6 Testing Accuracies (%) of MLP Neural Network, RBF Network and WNN

Methods	Experiment 1	Experiment 2	Experiment 3
MLP	89	100	67.56
RBF	90	100	78.67
WNN	100	100	91.56

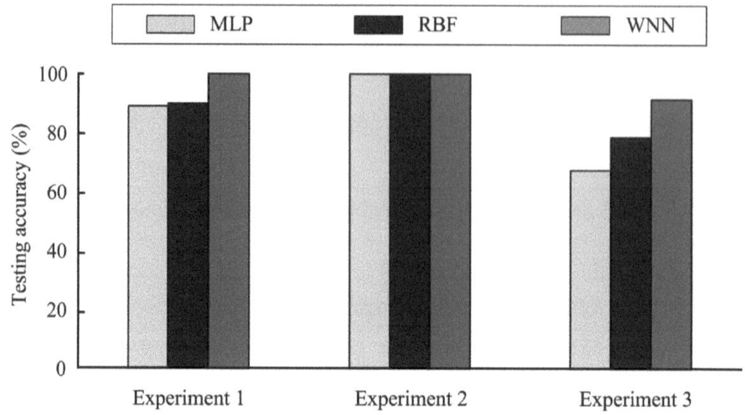

Figure 3.22. Testing accuracies of the three neural networks in the three experiments.

accuracies of 89, 90, and 100%, respectively. The proposed method based on WNN acquires the best identification result, which enhances the diagnosis accuracy by 11% compared with MLP and 10% compared with RBF. This comparison result indicates that the proposed method is obviously superior to the other two methods in recognizing different categories of bearing faults.

3. Experiment 2: Diagnosis performance comparison of different fault severities—To examine the effectiveness of the proposed method in recognizing different fault severities, the proposed method is applied on a data set containing different fault severities. The data set includes 150 samples where 50 samples are of the normal condition, 50 samples are with slight fault in the outer race, and the remaining 50 samples belong to serious fault in the outer race. 75 samples of the 150 samples act as the training set and the remaining 75 samples perform as the testing set.

   The testing results of the three neural networks on identifying different fault severities are listed in Table 3.6 and shown in Fig. 3.22, respectively. Since this experiment is a three-class classification problem and relatively simple, all the neural networks achieve a high testing accuracy equalling 100%.

4. Experiment 3: Diagnosis performance comparison of compound faults—Generally, in addition to the different categories of a single fault and different fault severities, compound faults also occur in rolling element bearings. If they cannot be detected as early as possible, a more serious consequence will usually

occur. Therefore, it is of great practical value to diagnose the compound faults in bearings correctly and timely.

In this experiment, we analyze a data set containing not only different fault categories and severities but also compound faults. The data set represents the nine conditions (described in Table 3.5) of the locomotive roller bearings and is composed of 450 samples. Half of the 450 data samples are used for training and the remaining 225 samples are used for testing.

The testing results of MLP neural network, RBF neural network, and WNN in this experiment are shown in Table 3.6 and Fig. 3.22. It can be seen from the table and the figure that testing accuracies of the three methods are 67.56, 78.67, and 91.56%, respectively. The diagnosis accuracies decrease because this is a ten-class classification problem and thus relatively difficult. However, the proposed method still achieves the highest testing accuracy. The aforementioned results indicate that the proposed method is more capable of recognizing not only different fault categories and severities but also compound faults.

5. Discussions. First, the diagnosis results of the above experiments verify that the proposed WNN-based method obtains high diagnosis accuracies and has a good diagnosis capability. It can recognize not only the different fault categories and severities but also the compound faults. The reasons of the good performance of the WNN-based method are as follows: (1) EEMD overcomes the mode mixing problem of EMD and increases the decomposition precision of vibration signals. (2) The sensitive IMF is selected to reflect the bearing fault characteristics, avoiding the interference of other fault-unrelated components. (3) The features from the time and frequency domain are able to represent the bearing health conditions well. (4) WNN has good learning ability so as to improve the diagnosis accuracies.

   Second, the EEMD method achieves better decomposition results than the EMD method because of solving the problem of mode mixing of EMD. Thus, EEMD is a powerful tool for mechanical signal analysis. However, the EEMD method needs to select the parameters about the magnitude of noise and number of ensembles. So different noise magnitudes and ensemble numbers should be tried and selected in the diagnosis method. The authors would like to investigate the automatic selection of these parameters in future.

   Finally, although WNN achieves the satisfactory results in diagnosing the health conditions of the locomotive roller bearings, the network initialization and the optimum structure selection affect the performance of WNN. These issues will be investigated in future work.

3.2.3.6 Epilog

Based on EEMD and WNN, an effective and automated diagnosis method of rotating machinery is presented. In the method, the collected vibration signals are first decomposed by EEMD. Then, the most sensitive IMF is selected using the kurtosis-based approach. Thirdly, time- and frequency–domain features are extracted from the most sensitive IMF, its frequency spectrum and Hilbert envelope spectrum. Finally, these features are input into WNN to diagnose the health conditions of the rotating machinery. The results indicate that the presented method can reliably distinguish not only the single faults and its different fault severities, but also the compound faults.

Different from the method in Section 3.2.2, feature selection is not utilized in this method. Since the features are extracted from the sensitive IMF, which has highlighted the potential fault characteristics, the feature selection is not necessary in the method. The selection of sensitive IMF meets the main purpose of the feature selection in obtaining salient features which obviously characterize the machine conditions from the whole feature set. So it is viewed as a different way to obtain salient features in intelligent fault diagnosis of rotating machinery.

3.2.4 Adaptive Neuro-Fuzzy Inference System-Based Fault Diagnosis

3.2.4.1 Introduction

The previously presented methods use the vibration information collected from individual sensors and obtain good performance. However, it is a fact that individual sensor can provide limited characteristic information for intelligent fault diagnosis of rotating machinery in a complex system, such as planetary gearboxes. So we would like to investigate the intelligent fault diagnosis of various gears in planetary gearboxes using the signals collected by multiple sensors.

The term of planetary gearbox refers to the gearboxes that include the transmission structure of a planetary gear set (Lei et al., 2014). This set is a compound gear system using several planet gears between a sun gear and an outer ring gear. The sun gear rotates around its own center, and the outer ring gear is either stationary or rotating. So the planet gears not only rotate around their own centers but also revolve around the sun gear. Due to high transmission efficiency and strong load-bearing capacity of planetary gearboxes, they are universally used in wind turbines and construction machinery. Generally, planetary gearboxes work under tough operating condition so that they always suffer different damage modes like crack, pitting, and so on. However, their complex gear transmission structures weaken the fault characteristics of the individual signals, which makes intelligent diagnosis of their gears difficult. Multiple sensors are able to provide complementary characteristic information for fault diagnosis, which may improve fault diagnosis accuracy of planetary gearboxes. Thus, how to fuse fault characteristics from multiple sensors becomes the vital issue.

We use ANFIS to deal with the problem above. ANFIS was developed by Jang (1993), aiming to solve real-world decision-making, modeling, and control problems. It is derived from the combination between fuzzy logic and neural networks. ANFIS not only transforms human knowledge or experience into the rule base and database of a fuzzy inference system, but also tunes the membership functions to minimize the output error measure or maximize performance index. Thus, it combines the advantages of the adaptive capability of neural networks and the qualitative approach of fuzzy logic. With these merits, ANFIS promises to play an important role in the design of intelligent diagnosis systems.

Therefore, an ANFIS based method is presented for intelligent diagnosis of planetary gearboxes based on the idea of multisensor data fusion. In this method, two features specially designed for planetary gearboxes are used to characterize the health conditions, and ANFIS is employed to fuse all of features extracted from different sensors. We will first introduce the two particular features for the planetary gearboxes and the algorithm of ANFIS, and then detail the method based on multisensor data fusion.

3.2.4.2 Adaptive Neuro-Fuzzy Inference System

ANFIS is an integration system in which neural networks are applied to optimize the fuzzy inference system. ANFIS constructs a series of fuzzy if–then rules with appropriate membership functions to produce the stipulated input–output pairs. The initial fuzzy rules and membership functions are first set by using human expertise about the outputs to be modeled. Then, ANFIS can modify these fuzzy if–then rules and membership functions to minimize the output error measure or explain the input–output relationship of a complex system. To describe the architecture of ANFIS, two fuzzy if–then rules are considered:

Rule 1: If (x is A_1) and (y is B_1) then ($z_1 = p_1x + q_1y+r_1$),
Rule 2: If (x is A_2) and (y is B_2) then ($z_2 = p_2x + q_2y+r_2$),

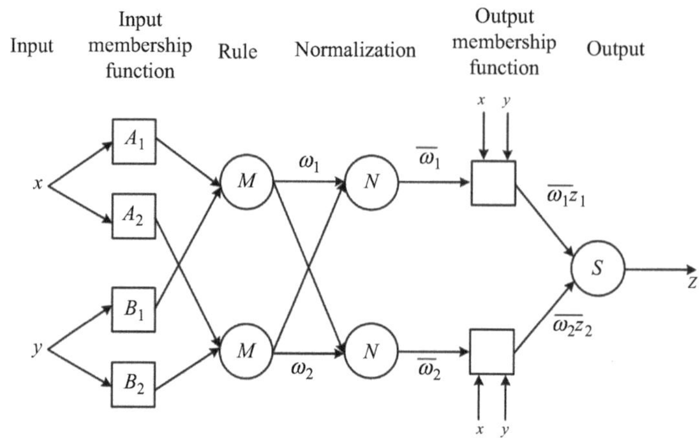

Figure 3.23. Architecture of the ANFIS.

where x and y are the inputs, A_i and B_i are the fuzzy sets, z_i ($i = 1,2$) are the outputs within the fuzzy region specified by the fuzzy rules, and p_i, q_i, and r_i are the parameters determined during the training process. To implement these two rules, the architecture of ANFIS has five layers, as shown in Fig. 3.23. It can be seen that a square stands for an adaptive node, and a circle indicates a fixed node.

Layer 1: Input membership function—The first layer is used to fuzzificate the inputs, and all the nodes of this layer are adaptive. Its outputs are the membership grade of the inputs.

$$o_i^1 = u_{A_i}(x), \quad i = 1,2, \tag{3.44}$$

$$o_i^1 = u_{B_{i-2}}(y), \quad i = 3,4, \tag{3.45}$$

where $u_{A_i}(x)$ and $u_{B_{i-2}}(y)$ are the fuzzy membership functions. Traditionally, the bell shaped membership function is used, and it is expressed by

$$u_{A_i}(x) = \frac{1}{1 + \left[\left(\frac{x - c_i}{a_i}\right)^2\right]^{b_i}}, \quad i = 1,2, \tag{3.46}$$

where a_i, b_i, and c_i are the parameters of the membership functions.

Layer 2: Rule—The nodes of this layer are fixed nodes. They are labeled with M, which indicates that they perform as multipliers. The outputs of this layer represent the fuzzy strengths ω_i of each rule and can be expressed as

$$o_i^2 = \omega_i = u_{A_i}(x)u_{B_i}(y), \quad i = 1,2. \tag{3.47}$$

Layer 3: Normalization—In this layer, the nodes are also fixed. These nodes are labeled with N, which means that they play a normalization role to the fuzzy strengths from the previous layer. The normalization factor is computed by the sum of the weight functions. The outputs of this layer are called normalized fuzzy strengths and are expressed as

$$o_i^3 = \bar{\omega}_i = \frac{\omega_i}{\sum_{i=1}^{2} \omega_i}, \quad i = 1,2. \tag{3.48}$$

Layer 4: Output membership function—The nodes of this layer are adaptive ones. Its outputs are represented by

$$o_i^4 = \bar{\omega}_i z_i = \bar{\omega}_i (p_i x + q_i y + r_i), \quad i = 1,2, \tag{3.49}$$

where p_i, q_i, and r_i are the parameters of the membership functions, respectively.

Layer 5: Output—Only one single fixed node, labeled with S, is in this layer. This node performs the sum of the incoming signals. Thus, the overall output is expressed as follows

$$o_i^5 = z = \sum_{i=1}^{2} \bar{\omega}_i z_i = \frac{\sum_{i=1}^{2} \omega_i z_i}{\sum_{i=1}^{2} \omega_i}. \tag{3.50}$$

In the training process of ANFIS, a hybrid learning algorithm of the gradient descent approach and the least-squares estimate is utilized to tune the parameters of the membership functions. In the forward pass, the node outputs advance until the output membership function layer and the consequent parameters are identified by the least-squares estimate. With the error signals propagated backward, the backward pass uses the BP gradient descent method for updating the premise parameters.

3.2.4.3 Diagnosis Method With Multisensor Data Fusion

Fig. 3.24 shows the diagram of the method using multisensor data fusion. It can be seen from the figure that the vibration signals are first acquired using multiple sensors, which are mounted on different locations of the housing of a planetary gearbox. Then the features, that is, the root mean square of the filtered signal (FRMS) and the normalized summation of positive amplitudes of the difference spectrum between the unknown signal and the healthy signal (NSDS), are extracted from these vibration signals of different sensors. Finally, all extracted features are fused using ANFIS so as to distinguish the health conditions of the planetary gearbox.

Here, we detail two diagnostic features specifically designed for fault diagnosis of planetary gearboxes.

1. FRMS: FRMS is developed by calculating the root mean square of the filtered signal instead of the original signal, and it is expressed as

$$\mathrm{FRMS} = \sqrt{\frac{1}{T}\sum_{t=1}^{T}(s(t))^2} \tag{3.51}$$

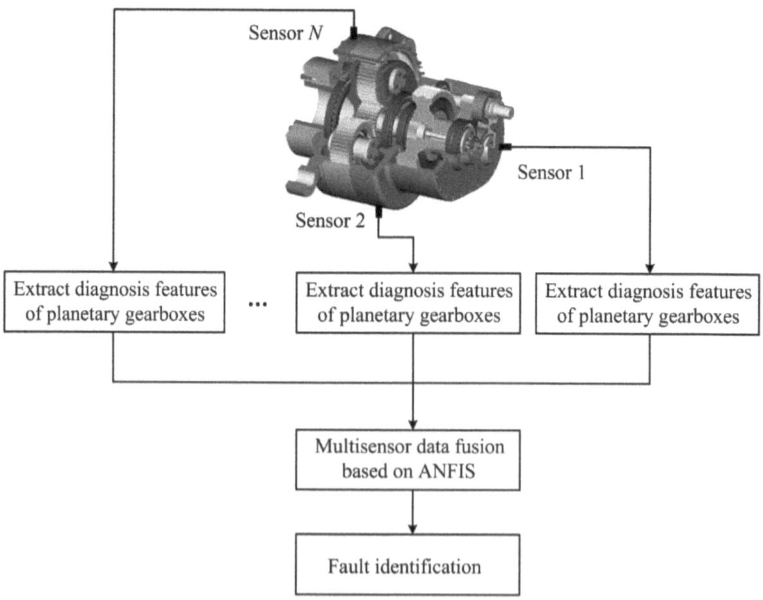

Figure 3.24. Diagram of the proposed method.

where $s(t)$ $(t=1,\cdots,T)$ is the tth data point of the filtered signal S, and T is the dimension of S.

The filtered signal $\boldsymbol{S}$ can be obtained from the original signal $\boldsymbol{X}$ using the following filtering process. (1) Get the frequency spectrum of $\boldsymbol{X}$. (2) In this spectrum, filter out the shaft frequency and its five order harmonics, the gear meshing frequency and its three order harmonics, and the modulation sidebands and its harmonics in the frequency spectrum. (3) Perform inverse Fourier transform on the filtered spectrum so as to obtain the filtered signal $\boldsymbol{S}$. The idea of this filtering process comes from the observation that the shaft frequency and its harmonics, the meshing frequency and its harmonics, and rich modulation sidebands are dominant components in both vibration spectra of a healthy and faulty planetary gearbox (McFadden, 1991). So these components cannot provide the information about the health condition of the planetary gearbox. This observation is fundamentally different from a healthy fixed-axis gearbox. Traditionally, the spectrum of the healthy fixed-axis gearbox only has the shaft frequency and its harmonics, and the meshing frequency and its harmonics. But rich modulation sidebands also dominate the vibration spectrum in the vibration spectrum of the healthy planetary gearbox. So the rich modulation sidebands are no longer the indication of faults in the case of planetary gearboxes.

2. NSDS: It is generated by normalizing the sum of the positive amplitudes of the difference spectrum between the measured signal and the healthy one. It is defined as

$$\text{NSDS} = \frac{\sum_{i=1}^{I} y_d(i)}{\sum_{i=1}^{I} y_m(i)} \tag{3.52}$$

$$\begin{cases} y_d(i) = y_m(i) - y_h(i) & \text{if } y_m(i) > y_n(i) \\ y_d(i) = 0 & \text{if } y_m(i) \le y_n(i) \end{cases} \tag{3.53}$$

where $i=1,\ldots,I$, y_m, and y_h represent the frequency spectra of the measured signal and the healthy one, y_d is the difference spectrum, and I is the total number of spectrum lines contained in y_d, y_m, and y_h. The motivation of using normalization, namely dividing the sum of the positive amplitudes in the difference spectra by the sum of the measured signal spectra, is to alleviate or even eliminate the effect of the various operation conditions like different rotating speeds and loads.

3.2.4.4 Intelligent Diagnosis of Gear Faults: A Planetary Gearbox Case Study

1. Experimental system for planetary gearboxes. In order to test the effectiveness of the proposed method, we established an experimental system of a planetary gearbox test rig and performed experiments on the test rig. Fig. 3.25 shows the schematic model of the test rig. The test rig consists of two gearboxes, a 3-hp motor to drive the gearboxes, and a magnetic brake used to load. The motor's rotating speed is controlled by a speed controller allowing the tested gear to operate under various speeds. The load is provided by the magnetic brake connecting to the output shaft and the torque can be adjusted using a brake controller.

   There are two gearboxes in the test rig: a two-stage planetary gearbox and a two-stage fixed-axis gearbox, as shown in Fig. 3.25, and the two-stage planetary gearbox is what we concern in our present study. An inner sun gear is surrounded by three or four rotating planet gears and a fixed outer ring gear in each stage of the planetary gearbox. The torque is transmitted through the sun gear to the planets mounted on a planetary carrier, and then transmitted to the output shaft through the carrier.

   In a planetary gearbox, sun gear teeth suffer faults more easily because their multiple meshes with the planet gears increase the potential of being damaged. Therefore, a cracked tooth and a pitted tooth are created on the sun gear in the first stage and a chipped tooth and a missing tooth are set in

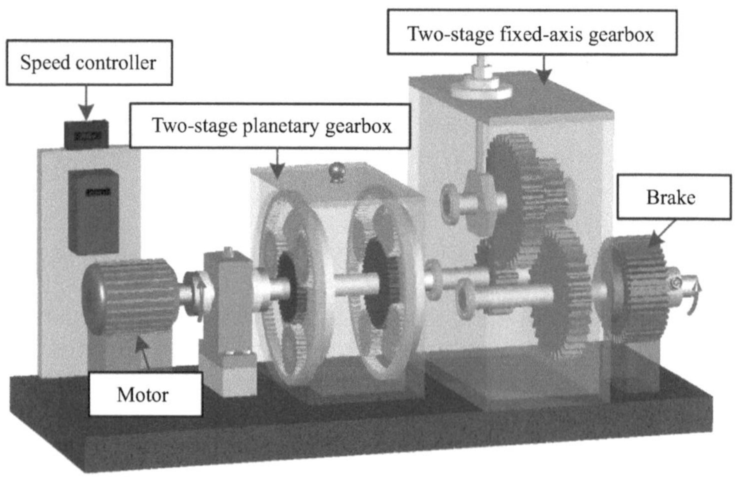

Figure 3.25. Schematic model of a two-stage planetary gearbox test rig.

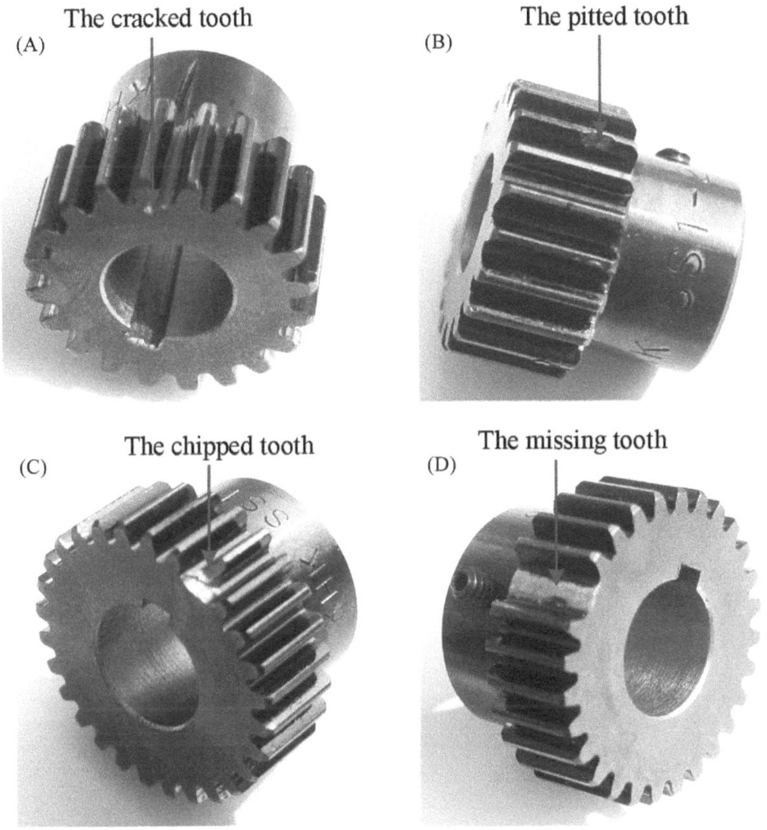

Figure 3.26. Damaged sun gears. (A) Having a cracked tooth, (B) having a pitted tooth, (C) having a chipped tooth, and (D) having a missing tooth.

the second stage, respectively. Fig. 3.26 displays the photos of the faulty sun gears, and the gear parameters of the two-stage planetary gearbox are listed in Table 3.7.

The experimental system consists of the test rig, the accelerometers, an NI data acquisition system and a laptop with the data acquisition software, as presented in Fig. 3.27A. Two accelerometers are used to collect the vibration signals, as shown in Fig. 3.27B. One is a triaxial accelerometer with a low sensitivity equaling 100 mv/g which is mounted at the location of the first stage bearing end cover. The other accelerometer is a unidirectional accelerometer with a high sensitivity of 500 mv/g which is mounted at the location of the second stage bearing end cover. All the vibration signals are collected with a sampling frequency equaling 5120 Hz at four different drive motor speed levels (2100, 2400, 2700, and 3000 rpm) and under two loading

Table 3.7 Gear Parameters in the Two-Stage Planetary Gearbox

	The First Stage				The Second Stage			
Gear	Sun	Planet	Ring	Number of Planets	Sun	Planet	Ring	Number of Planets
Number of teeth	20	40	100	3	28	36	100	4

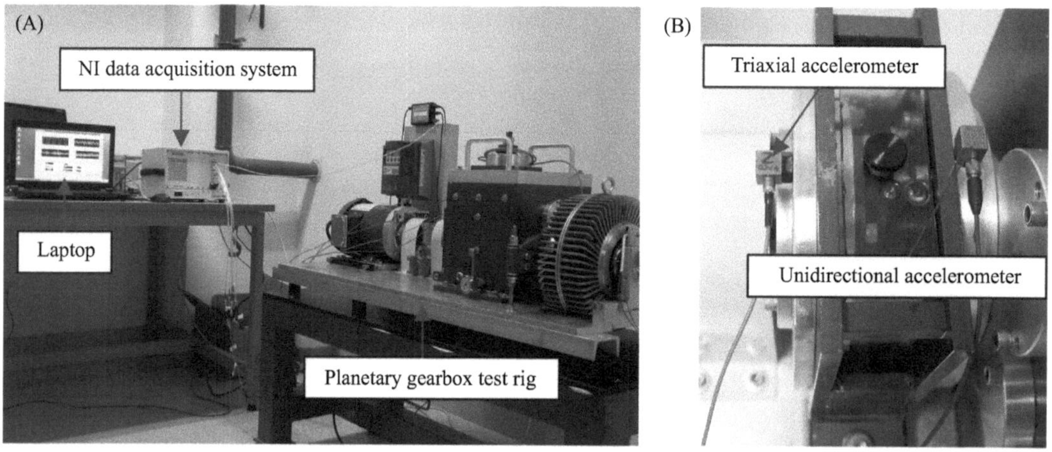

Figure 3.27. (A) Experimental system and (B) locations of the two accelerometers.

conditions (no load and the maximum load). The maximum load equals 13.5 Nm in our experiments which is calculated by using the gear material properties, the transmission structures, and so on. Collecting the data at varying motor speeds and loads level aims to simulate the general planetary gearbox operating conditions.

We conduct five experiments corresponding to the five gear health conditions which include normal, the cracked tooth on the first stage sun gear, the pitted tooth on the first stage sun gear, the chipped tooth on the second stage sun gear, and the missing tooth on the second stage sun gear. In each experiment, except the sun gear of different health conditions, all the other gears installed in the test rig are normal. The axial and radial vibrations at the first stage bearing end cover are measured by the triaxial accelerometer, and the radial vibrations at the second stage are acquired by the unidirectional accelerometer. We consider and use the vibration signals of the three channels including the axial and the vertical directions of the triaxial accelerometer and the vertical direction of the unidirectional accelerometer for multisensor data fusion.

The vibration signals collected from the planetary gearbox test rig are divided into several data samples and each of them is actually a data series which contains 20,224 data points. We obtain 30 data samples for each identical operating condition (identical motor speed, load, and gear health condition). Therefore, there are totally 240 data samples under two loading conditions and at 4 motor speed levels for each of the 5 gear health conditions.

2. Case 1: Fault identification of the sun gear in the first stage. In this case, we apply the proposed method to identify the health conditions of the sun gear in the first stage of the planetary gearbox. The sun gear of the first stage involves three health conditions: normal, a cracked tooth, and a pitted tooth. For each of the health conditions, 240 data samples are acquired, and thus the whole data set corresponding to the 3 conditions contains totally 720 samples. 360 samples are selected for training and the remaining 360 samples used to test. This is a three-class classification issue corresponding to the three gear conditions.

First, two diagnosis features, FRMS and NSDS, are extracted from each data sample. Consequently, six features are extracted from the data samples of the three channels, that is the vertical direction of the unidirectional accelerometer and the axial and the vertical directions of the triaxial accelerometer. Then, the six features are used as inputs of ANFIS by the data fusion strategy. Finally, the health conditions of the sun gear are recognized. The classification accuracy is the percentage that the number of the data samples identified correctly divides by the whole number of the data samples. The classification accuracies of the first case study are shown in Fig. 3.28.

To validate the improved performance of the proposed method using the multisensor data fusion, the methods with each individual channel or sensor are also used to process the same data. We name these methods for the three individual channels as Method 1, Method 2, and Method 3. The three methods only use diagnosis features extracted from individual channels. Thus, two features are extracted for each data sample in the three methods and also utilized as inputs of ANFIS. Their classification results are also displayed in Fig. 3.28.

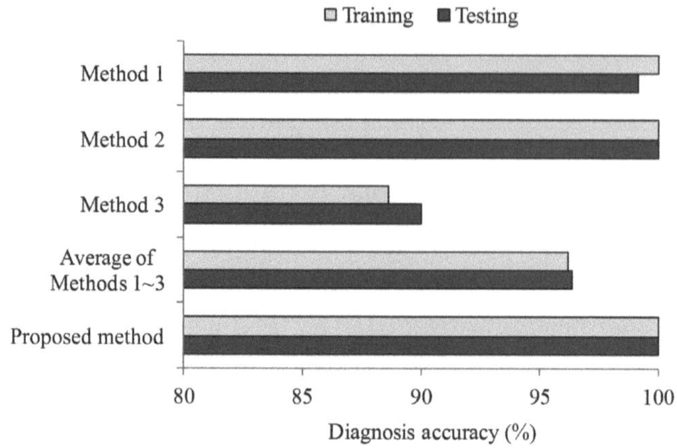

Figure 3.28. Accuracy comparison between Methods 1~3 and the proposed method for Case 1.

Since the first case is relatively simple, Methods 1~3 achieves acceptable accuracies in the training and testing processes. They range from 88.61% to 100% for training and from 90% to 100% for testing. As we can see, both the proposed method and Method 2 obtain the highest accuracy equaling 100%. However, as the classification errors of the two methods shown in Fig. 3.29, the classification error of the proposed method is much smaller than that of Method 2.

3. Case 2: Fault identification of the sun gear in the second stage. Different from Case 1, damage is created on the sun gear of the second stage in Case 2. Similarly, half of 720 data samples are used for training and the others are used for testing. They are used to test the classification accuracies of the proposed method as well as Methods 1~3. In this case, the faults to be identified are on the sun gear of the second stage in the planetary gearbox test rig, where the sun gear of the second stage operates under lower rotating speed than the sun gear of the first stage. As a consequence, the fault characteristics of the damaged sun

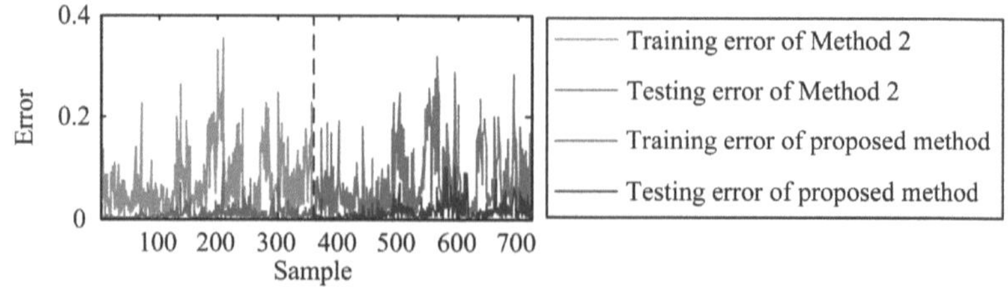

Figure 3.29. Classification errors of Method 2 and the proposed method.

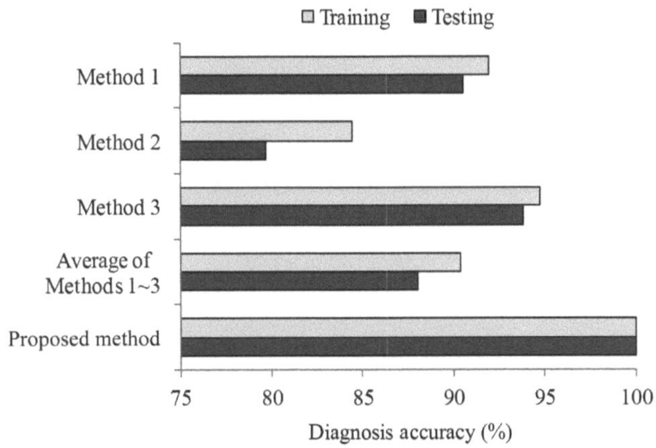

Figure 3.30. Accuracy comparison between Methods 1~3 and the proposed method for Case 2.

gear in the second stage are not excited evidently. Therefore, it is more difficult to recognize the faults in the second stage than those in the first stage. Fig. 3.30 gives the training and testing accuracies of the methods, respectively.

In Fig. 3.30, it is seen that the accuracies of Methods 1~3 for the second stage sun gear are obviously lower than those for the first stage sun gear in Fig. 3.28. The training accuracies of this case range from 84.44% to 94.72%, and the testing accuracies range from 79.72% to 93.88%. However, the proposed method still obtains the accuracy of 100% for the training and the testing. This result indicates that the proposed method performs robustly compared with Methods 1~3.

4. Case 3: Fault identification of the sun gears inside of all stages. The purpose of this case is to verify the performance of the proposed method in recognizing the health conditions on both stages of the planetary gearbox. Thus, 1200 data samples involving 5 gear health conditions, that is, the normal condition, the 2 damage modes in the first stage sun gear and the 2 damage modes in the second stage sun gear, are employed to test the proposed method. The 1200 data samples are divided into 2 sets: 600 samples for the training and 600 for the testing. Fig. 3.31 shows the final accuracies of the proposed method as well as Methods 1~3.

   It is seen from Fig. 3.31 that the training accuracies of all the methods decrease, ranging from 74% to 99.33%, because it is a five-class classification problem and more difficult than Case 1 and Case 2. However, among these methods, the proposed method still presents the highest training accuracy (99.33%).

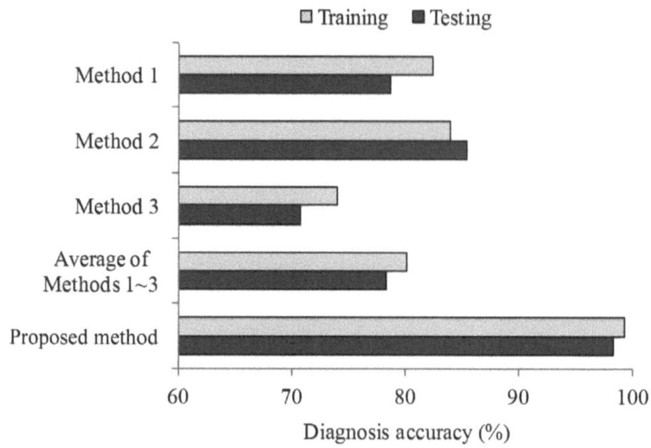

Figure 3.31. Accuracy comparison between Methods 1~3 and the proposed method for Case 3.

And the testing accuracy of the proposed method is much higher (98.33%) than those of Methods 1~3 (average 78.33%). So the results indicate that the method using multisensor data fusion can effectively recognize various fault modes and different fault locations.

All the results in the three cases demonstrate that fusing the data from different sensors is better than using the data from individual ones. The multisensor fusion based method is able to obtain obvious improvements in classification accuracies and performs robustly in intelligent diagnosis of planetary gearboxes.

3.2.4.5 Epilog

By mounting multiple sensors on different locations of a planetary gearbox, we can obtain complementary or different sensitivity information for intelligent fault diagnosis. Based on this fact, a method is presented using multisensor data fusion. In this method, ANFIS is employed as the fusion technique and two features designed for planetary gearboxes are used as the inputs of ANFIS. By comparing the diagnosis results of three cases, it is verified that the proposed method is superior to the method using individual sensor information.

Although the presented method is especially developed for fault diagnosis of gears in planetary gearboxes, the idea of using multisensor data fusion could be applied for other rotating machinery. If diagnosis accuracy with individual sensor information is insufficient, we can try the multisensor data fusion and improve the diagnosis accuracy.

3.3 Statistical Learning Theory

Artificial intelligence technologies aim at mining potential disciplines from the training samples or observable data, and applying the mined disciplines to predict the testing samples or future data. Most of the current artificial intelligence techniques, including ANNs, are to find out the disciplines that the samples are subject to when the number of samples is close to be infinite. This is named as the empirical risk minimization (ERM), which aims to minimize the error on the training samples (Zhang, 2010).

The neural network based on ERM is commonly used in plenty of engineering applications. In the training process, it is hoped to build up a neural network model of the system observed, namely to deduce responses of the system using the observed data. So the quantity and quality of the observed data may dominate the performance of this neural network model. Since the data obtained are finite and traditionally sampled nonuniformly, the neural network approaches suffer the problem of generalization, leading to the models that may overfit the data. This is a reason why the optimization algorithms are used for parameter selection so as to obtain the best model. Moreover, the limited number of data samples generally results in the insufficient performance in real applications. Therefore, support vector machine (SVM) based on the structural risk minimization (SRM) is used for dealing with the problems of the neural networks trained using ERM. Different from ERM, SRM is achieved by minimizing the upper bound of the expected risk and displays better generalization abilities, especially for limited sample cases. SRM is rooted in statistical learning theory (SLT), which has attracted more and more attention since 1990s. SLT is based on solid foundations, provides a uniform skeleton for limited sample learning tasks, and integrates some existing methods. Thus, it is beneficial for mining the unsolved challenges.

In Section 3.3, we first briefly introduce the SLT, and then focus on the application of SVM in intelligent fault diagnosis. Finally, we present the diagnosis method using relevance vector machine (RVM).

3.3.1 Introduction to Statistical Learning Theory

SLT is a machine learning framework that draws from the fields of statistics and functional analysis. It deals with the problem of finding a function describing the pattern of the data. The basic concepts of the SLT include the VC (Vapnik–Chervonenkis) dimension and the SRM (Burges, 1998).

1. VC Dimension: The VC dimension is a scalar value that measures the capacity of a set of functions. The VC dimension of a set of functions is p if and only if there are a set of points $\{x^i\}_{i=1}^{p}$ which can be separated into all 2^p possible configurations, and no set $\{x^i\}_{i=1}^{q}$ satisfies this property where $q > p$.

   Fig. 3.32 illustrates why three points in the plane can be shattered by a set of linear functions but four points cannot (Gunn, 1998). The VC dimension equals to the number of the free parameters for this case, but that is not always right. For example, the function $A\sin(bx)$ has an infinite VC dimension. The set of linear functions in n-dimension space will have a VC dimension equaling $n + 1$.
2. Structural risk minimization: Create a structure where S_h has VC dimension equaling h and

$$S_1 \subset S_2 \subset \cdots \subset S_\infty \tag{3.54}$$

SRM aims to solve the following problem

$$\min_{S_h} R_{emp}[f] + \sqrt{\frac{h\ln\left(\frac{2l}{h}+1\right) - \ln\left(\frac{\delta}{4}\right)}{l}} \tag{3.55}$$

For deterministic processes, a multiple output problem is usually regarded as a set of single output problems, which can be considered independently. Therefore, it is better to consider the processes with multiple inputs as a single output prediction.

The principle of SRM is summarized here. The VC dimension depends on the chosen class of functions, whereas the

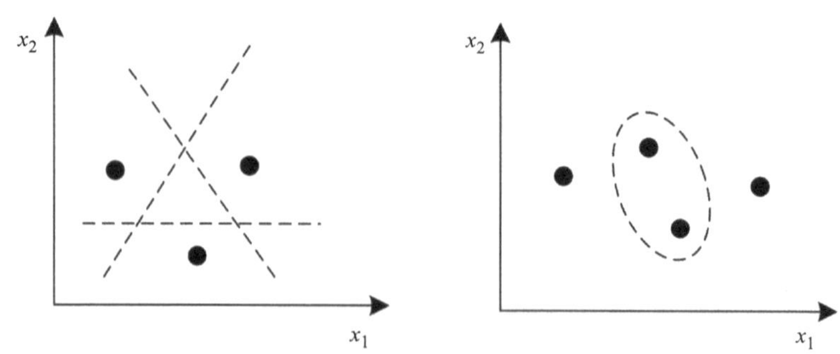

Figure 3.32. VC dimension illustration.

empirical and actual risk depend on the function chosen by the training processes. The subset of the chosen class of functions would be found so that the risk bound of the subset is minimized. Obviously, we cannot make the VC dimension h smoothly because the dimension is an integer. So we introduce a "structure" that divides the class of functions into the nested subsets. For each subset, we can calculate the dimension h or get its bound. Consequently, SRM are used to find the subset of functions by minimizing the bound of the actual risk. This can be done by training a series of machines. The trained machines for each subset aim to minimize the empirical risk for a given subset, and then we minimize the sum of the empirical risk and the VC confidence of the trained machines in the series (Burges, 1998). For a much deeper understanding of SLT, please refer to the Vapnik (1999) paper.

3.3.2 Support Vector Machine-Based Fault Diagnosis Method

3.3.2.1 Introduction

SVM is a supervised learning technique based on SLT and is widely used for classification problems in various fields (Cortes and Vapnik, 1995). It is well-known that SVM has the merits of simple structure, the global optimal, excellent generalization performance, as well as fast learning speed. SVM maps original sample space to high-dimensional feature space using the nonlinear transformation and finds the optimal classification plane to classify samples in the high-dimensional feature space. Even in the case of small sample sizes, SVM can also achieve high precision. Since the samples of the classification problem are often limited in intelligent fault diagnosis of rotating machinery, it is a good choice to use SVM as the fault recognition technique.

Since the vibration signals of rotating machinery display nonlinear and nonstationary properties, the features fed into SVM are expected to characterize these properties. Compared with the features extracted by the pervious diagnosis methods, the techniques for nonlinear dynamic parameter estimation may offer a potential alternative to extract fault-related features hidden in the complex and nonlinear fault signals. Nowadays, a number of nonlinear parameter extraction techniques have been investigated and introduced to fault diagnosis, such as fractal dimension, Lempel–Ziv complexity, and approximate entropy.

Among these feature extraction techniques, permutation entropy (PE) is a powerful tool for the analysis of mechanical signals, as it estimates the complexity of mechanical signals and allows assessing the temporal state of rotating machinery. PE is proposed by Bandt and Pompe (2002) and has the peculiarities of robustness, simplicity, and low computational cost. Because of these merits, it has been widely applied to analyze the complexity and dynamics of time series (Zanin et al., 2012), such as cardiac heart interbeat signal analysis, temperature fluctuation analysis of a basic electron heat transport experiment, and the electroencephalograph signal analysis. Although PE has performed well in these applications, it has limited ability in measuring the complexity of time series derived from physical and biological systems including mechanical systems. These series show structures on multiple spatiotemporal scales, so multiscale permutation entropy (MPE) is introduced to quantify the complexity of the mechanical signals over multiple scales (Aziz and Arif, 2005). MPE is expected to effectively measure dynamic characteristics of the mechanical signals with different scale factors and reveal the dynamic difference among the mechanical signals of different faults.

Therefore, a new intelligent diagnosis method of rotating machinery is developed based on MPE and SVM. In the method, MPE is used to extract features that measure the complexity of mechanical vibration signals, and SVM is used as a fault classifier to determine the health conditions of rotating machinery.

3.3.2.2 Support Vector Machine

Basically, given a set of training samples marked as two categories, a binary SVM is trained to classify new samples into one category or the other. For a training set in which a sample is viewed as a p-dimensional vector, we want to find a $(p-1)$-dimensional hyperplane that separates such a set into two categories. As shown in Fig. 3.33, there are many hyperplanes that might separate the

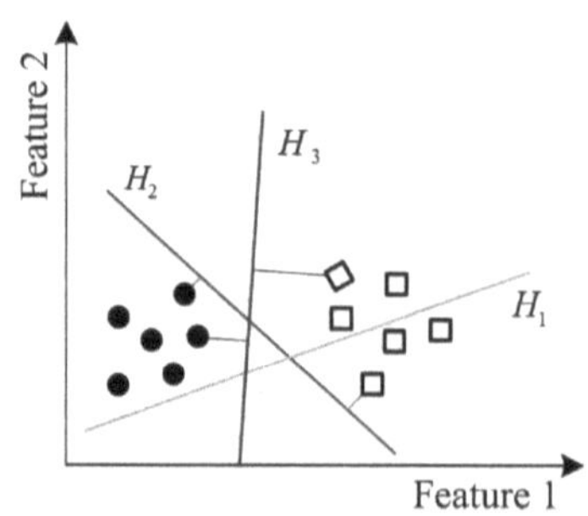

Figure 3.33. The hyperplanes: H_1 does not separate the classes. H_2 does, but only with a small margin. H_3 separates them with the maximum margin.

dataset. One reasonable hyperplane represents the largest separation or margin between the two categories. In other words, the distance from the chosen hyperplane to the nearest data point of each category is maximized. This hyperplane is known as the maximum-margin hyperplane and SVM aims to find this hyperplane to classify the dataset (Burges, 1998).

Traditionally, the classification of a dataset is a nonlinear problem. So the linear SVM can be extended to a nonlinear version, which efficiently implements the nonlinear classification using the kernel trick. The original samples of the dataset are mapped into a high-dimensional space with the kernel function, and SVM finds the optimal hyperplane to classify samples in the high-dimensional space. Here, we first describe linear and nonlinear SVM, and then introduce the SVM for multiclass classification.

1. Linear SVM: Given a set of samples $\boldsymbol{x}_i\,(i=1,2,\cdots,M)$, where M is the number of samples. The set has two classes, that is positive class and negative class. We denote $y_i = 1$ for the positive class and $y_i = -1$ for the negative class, respectively. It is possible to find a hyperplane $f(\boldsymbol{x}) = 0$ that classifies the given dataset

$$f(\boldsymbol{x})=\boldsymbol{w}^{\mathrm{T}}\boldsymbol{x}+b=\sum_{j=1}^{M}w_jx_j+b=0 \tag{3.56}$$

where $\boldsymbol{w}$ is a M-dimensional vector and b is a scalar, and they are used to define the hyperplane.

If the samples associated with labels can be separated linearly, we can find two hyperplanes that separate the samples and there are no points between them. These hyperplanes are called separating hyperplanes, and the region bounded by them is called "margin." SVM aims to maximize this margin so as to improve its generalization ability. The two separating hyperplanes can be described by the equations

$$f(\boldsymbol{x}_i)=1 \text{ if } y_i=1, \tag{3.57}$$

$$f(\boldsymbol{x}_i)=-1 \text{ if } y_i=-1, \tag{3.58}$$

or the equation in the following form

$$y_i f(\boldsymbol{x}_i)=y_i(\boldsymbol{w}^{\mathrm{T}}\boldsymbol{x}_i+b)\geq 1 \quad \text{for} \quad i=1,2,\cdots M. \tag{3.59}$$

The separating hyperplane that creates the maximum margin is called the optimal hyperplane. An example is presented in Fig. 3.34. In this figure, a series of samples with two different labels are shown, that is squares for the negative class and

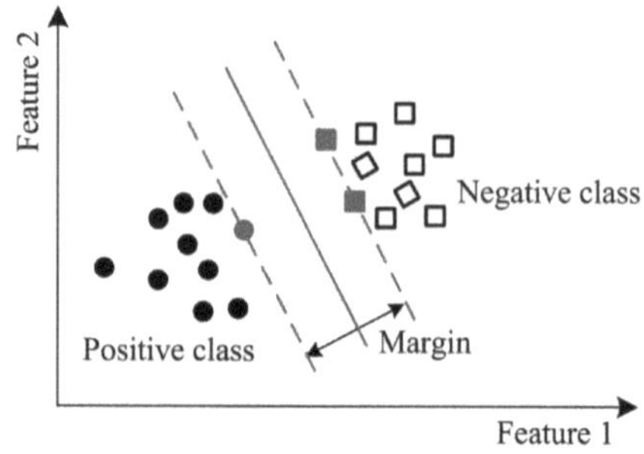

Figure 3.34. Classification of two classes using SVM.

circles for the positive class. SVM tries to find a hyperplane between the two different classes, and orientate it by maximizing the margin between the two hyperplanes (marked as dotted line). The samples used to define the two hyperplanes are called support vectors, marked as the grey squares and circles. Since the support vectors contain the information needed to define the classifier, the remaining samples are not required after the support vectors are selected.

In the geometry, the margin is $||\boldsymbol{w}||^{-2}$ according to Eq. (3.59). So the optimal separating hyperplane can be obtained by solving the following optimization problem:

$$\text{minimize } \frac{1}{2}||\boldsymbol{w}||^2 \tag{3.60}$$

$$\text{subject to } y_i(\boldsymbol{w}^{\mathrm{T}}\boldsymbol{x}_i + b) \geq 1, \quad i = 1, \cdots, M \tag{3.61}$$

This problem can be transformed into solving the equivalent Lagrangian dual problem with the help of the Karush–Kuhn–Tucker (KKT) condition. Its dual quadratic optimization problem is expressed as

$$\text{maximize } L(\boldsymbol{a}) = \sum_{i=1}^{M} a_i - \frac{1}{2}\sum_{i,j=0}^{M} a_i a_j y_i y_j \boldsymbol{x}_i^{\mathrm{T}} \cdot \boldsymbol{x}_j \tag{3.62}$$

$$\text{subject to } a_i \geq 0, \quad i = 1, \cdots, M, \quad \sum_{i=1}^{M} a_i y_i = 0. \tag{3.63}$$

By solving the above optimization problem, the coefficients a_i are obtained, leading to the decision function.

$$f(\boldsymbol{x}) = \operatorname{sign}\left(\sum_{i,j=1}^{\mathrm{M}} a_i y_i \boldsymbol{x}_i^T \cdot \boldsymbol{x}_j + b\right). \tag{3.64}$$

2. Soft margin: If there is no hyperplane that can separate the samples of two classes because of the noisy samples, the linear SVM will fail to find a feasible solution. So the soft margin method is proposed to handle this problem. A further optimization problem is introduced in this method to relax the constraints in Eqs. (3.57) and (3.58). By introducing positive slack variables ξ_i, the new optimization problem can be expressed as follows:

$$\text{minimize } \frac{1}{2}\|\boldsymbol{w}\|^2 + C\sum_{i=1}^{M}\xi_i \tag{3.65}$$

$$\text{subject to} \begin{cases} y_i(\boldsymbol{w}^{\mathrm{T}}\boldsymbol{x}_i + b) \geq 1 - \xi_i, & i = 1,\cdots,M \\ \xi_i \geq 0, & i = 1,\cdots,M \end{cases} \tag{3.66}$$

In the equations, ξ_i measures the distance between the margin and the examples $\boldsymbol{x}_i$ lying on the wrong side of the margin. Since the soft margin method is able to weaken the effect of outliers on SVM, it is widely used in classification problems.

3. Nonlinear classification: SVM can be extended to solve nonlinear classification tasks when the set of samples cannot be separated linearly. By applying kernel functions, the samples are mapped onto a high-dimensional feature space, in which the linear classification is possible. A kernel function $k\,(\boldsymbol{x}, \boldsymbol{y})$ is an inner product between the samples where $k(\boldsymbol{x},\boldsymbol{y}) = \langle\phi(\boldsymbol{x}),\phi(\boldsymbol{y})\rangle$. Valid kernel functions must satisfy the Mercer's conditions, namely $k\,(\boldsymbol{x}, \boldsymbol{y})$ must equal $k\,(\boldsymbol{y}, \boldsymbol{x})$.

   By using the kernel function, a nonlinear version of SVM can be developed and the expression of the optimization problem for such SVM can be written (in the dual form) as:

$$\text{maximize } L(\boldsymbol{a}) = \sum_{i=1}^{M} a_i - \frac{1}{2}\sum_{i,\,j=0}^{M} a_i a_j y_i y_j k(\boldsymbol{x}_i, \boldsymbol{x}_j) \tag{3.67}$$

$$\text{subject to} \begin{cases} 0 \leq a_i \leq C, \quad i = 1,\cdots,M \\ \sum_{i=1}^{M} a_i y_i = 0, \end{cases} \tag{3.68}$$

The decision function of the nonlinear SVM can be described by:

$$f(\boldsymbol{x}) = \text{sign}\left(\sum_{i,j=1}^{\mathrm{M}} a_i y_i k(\boldsymbol{x}_i, \boldsymbol{x}_j) + b\right) \tag{3.69}$$

Table 3.8 Formulation of Kernel Function

Kernel	$k(x, x_j)$
Linear	$\boldsymbol{x}^T\boldsymbol{x}_j$
Polynomial	$(\gamma \boldsymbol{x}^T\boldsymbol{x}_j+r)^d,\quad \gamma>0$
Gaussian RBF	$\exp(-\Vert\boldsymbol{x}-\boldsymbol{x}_j\Vert^2/2\gamma^2)$
Hyperbolic tangent	$\tanh(\kappa\boldsymbol{x}^T\boldsymbol{x}_j+c)\ \kappa>0$ and $c<0$

Note that the samples closest to the separating hyperplanes are those whose coefficients a_i are nonzero. These samples are called support vectors. The support vectors include the necessary information to construct the optimal hyperplane while other samples lay no effects on it. This is the reason why SVM could be used for the classification tasks whose number of samples is limited.

There are different kernel functions used in SVM, like linear, polynomial and Gaussian RBF, as shown in Table 3.8. Selecting the appropriate kernel function is crucial for the performance of SVM, since the kernel defines the high-dimensional space where the samples will be classified. Among the kernel functions, Gaussian RBF is the most commonly used in intelligent fault diagnosis.

3.3.2.3 Multiclass SVM

The SVM discussed earlier deals with binary classification and its labels take only two values, that is 1 and -1. In fault diagnosis of rotating machinery, however, the classification problem traditionally has more than two classes. For example, there are several fault classes for gears like pitting, crack, missing tooth, and so on. Therefore, the multiclass classification strategy of SVM should be investigated (Widodo and Yang, 2007).

1. One-Against-All (OAA): The commonly used method for multiclass classification of SVM is OAA method. It constructs k SVM models and k is the number of classes. The ith SVM ($i=1,\cdots k$) is trained with all samples from the ith class with positive labels, and other samples with negative labels. Thus, given M training samples $(\boldsymbol{x}_1,y_1),\cdots,(\boldsymbol{x}_M,y_M)$ where $\boldsymbol{x}_j\in\Re^n$, $j=1,\cdots M$ and $y_j\in\{1,\cdots,k\}$ is the label of $\boldsymbol{x}_j$, the ith SVM solves the problem as follows:

$$\begin{aligned}
&\text{minimize } \frac{1}{2}\|w^i\|^2 + C\sum_{i=1}^{M}\xi_j^i (w^i)^{\mathrm{T}} \\
&\text{subject to } (w^i)^{\mathrm{T}}\phi(x_j) + b^i \geq 1 - \xi_j^i \quad \text{if } y = i, \\
&\qquad (w^i)^{\mathrm{T}}\phi(x_j) + b^i \leq 1 - \xi_j^i \quad \text{if } y \neq i, \\
&\qquad \xi_j^i \geq 0, \qquad j = 1, \cdots, M,
\end{aligned} \tag{3.70}$$

where C is the penalty parameter.

2. One-Against-One (OAO): Another common method is called OAO method. This method constructs $k(k-1)/2$ SVMs, each of which is trained by samples from two classes. For training samples of the ith and the jth classes, we solve the following classification problem:

$$\begin{aligned}
&\text{minimize } \frac{1}{2}\|w^{ij}\|^2 + C\sum_{t}\xi_t^{ij} (w^{ij})^{\mathrm{T}} \\
&\text{subject to } (w^{ij})^{\mathrm{T}}\phi(x_t) + b^{ij} \geq 1 - \xi_t^{ij} \quad \text{if } y_t = i, \\
&\qquad (w^{ij})^{\mathrm{T}}\phi(x_t) + b^{ij} \leq 1 - \xi_t^{ij} \quad \text{if } y_t \neq i, \\
&\qquad \xi_t^{ij} \geq 0, \qquad t = 1, \cdots, M.
\end{aligned} \tag{3.71}$$

The decision of the training samples using OAO method is made by the following strategy. If sign$[(w^{ij})^{\mathrm{T}}\phi(x) + b^{ij}]$ indicates a sample x in the ith class, then the votes of x for the ith class increase by one. Otherwise, the ith is added by one. Then the label of x is predicted using the largest vote in the class.

3.3.2.4 Fault Diagnosis Method Based on SVM

Here, we first detail the technique of MPE and then describe the intelligent diagnosis method based on MPE and SVM.

1. Multiscale permutation entropy: As a kind of signal complexity metric, MPE first uses a coarse-grained procedure to acquire multiple scale time series from the original signals, and then measures the complexity of these time series. The details of MPE are described as follows.

   For a vibration signal $\{x_i, i = 1, 2, \cdots, n\}$, multiple coarse-grained time series are created by averaging the data within nonoverlapping window of length s. The coarse-grained process is described by the following equation:

$$y_j^s = \frac{1}{s}\sum_{i=(j-1)s+1}^{js} x_i, \quad 1 \leq j \leq n/s \tag{3.72}$$

where s represents the scale factor and y_j^s is the coarse-grained time series with multiple scales. It is worth pointing out that when the scale is 1, namely $s = 1$, the coarse-grained time series are the original time series.

After the coarse-grained process, PE of the coarse-grained time series y_j^s with different scale factors is calculated. This procedure is called MPE.

First, y_j^s is embedded to phase space

$$Y_t^s = \left[y_t^s, y_{t+\tau}^s, \cdots, y_{t+(m-1)\tau}^s \right], \quad t \in [1, n/s - m + 1] \tag{3.73}$$

where m is the embedding dimension and τ is the delay time. For each t, the m components of the vector Y_t^s can be arranged in an increasing order.

$$\left[y_{t+(k_1-1)\tau}^s \le y_{t+(k_2-1)\tau}^s \le \cdots \le y_{t+(k_m-1)\tau}^s \right] \tag{3.74}$$

Then we can find a vector $\boldsymbol{\pi}_t$ to identify the permutation pattern of Y_t^s

$$\pi_t = \left[k_1, k_2, \cdots, k_m \right] \tag{3.75}$$

In the phase space of embedding dimension m, the total number of the permutations of Y_t^s is $m!$. We use N_l to notate the occurrence times of the lth permutation for each t, where $1 \le l \le m!$. The occurrence probability of the lth permutation can be described as follows

$$p_l^s = \frac{N_l}{n/s - m + 1} \tag{3.76}$$

where n/s is the length of coarse-grained time series with scale factor s. Thereby the PE of coarse-grained time series with scale s is defined as follows

$$H_p^s = -\sum_{l=1}^{m!} p_l^s \ln p_l^s . \tag{3.77}$$

When all the permutations of time series have the same probability, the PE of the time series obtains the maximum value and their characteristics are similar to the noise characteristics. Contrarily, the more regular the time series are, the lower the PE value is. So when time series are collected from the linear predictive system, PE of the time series will reach the minimum value. In addition, since MPE is a statistical method, the length of the calculated time series should be long enough to ensure the precision of the calculation results.

2. Phase space reconstruction: In MPE method, mechanical vibration signals are embedded to phase space using the embedding dimension m and the delay time τ. Such process is viewed as phase space reconstruction, which is utilized to get the potential complexity information of the vibration signals. The purpose of the phase space reconstruction is to determine m and τ. Here, we introduce mutual information method to determine τ and Cao method to determine m.

   Fraser and Swinney (1986) first introduced the concept of mutual information to measure the inherent dependence of one-dimensional signal. For the vibration signal x_i, the mutual information can be defined as

$$I_\tau = \sum_{i=1}^{n} P(x_i, x_{i+\tau}) \log \left\{ \frac{P(x_i, x_{i+\tau})}{P(x_i) P(x_{i+\tau})} \right\} \tag{3.78}$$

   where $x_{i+\tau}$ is the delayed vibration signal, $P(x_i)$ is the probability of the measure of x_i, $P(x_{i+\tau})$ is the probability of the measure of $x_{i+\tau}$, and $P(x_i, x_{i+\tau})$ is the joint probability distribution. When I_τ equals 0, it denotes that x_i and $x_{i+\tau}$ are completely uncorrelated. When I_τ reaches the minimal value, x_i and $x_{i+\tau}$ are as uncorrelated as possible and the redundant information of the two sequences is minimal. Therefore, τ is selected when I_τ reaches the first local minimal value.

   Cao method was put forward as the improved method of false neighbors (Cao, 1997). The advantage of Cao method is that the embedding dimension of stochastic signals can be effectively calculated after τ has been calculated. Its main idea is to compare the change value of distance between two adjacent points from dimension m to dimension $m + 1$. If this value is small enough, the m is the suitable embedding dimension. There are two judgment criterions, one for deterministic signals and the other for stochastic signals. The signals of mechanical systems are stochastic, thus we choose the judgment criterion for stochastic signals.

$$E'_m = \frac{1}{n - m\tau} \sum_{i=1}^{n-m\tau} \left| x_{i+m\tau} - x^N_{i+m\tau} \right| \tag{3.79}$$

   where $x^N_{i+m\tau}$ is the nearest neighbor of $x_{i+m\tau}$. The judgment criterion is calculated as

$$E(m) = E'_{m+1} / E'_m \tag{3.80}$$

   For the vibration signal accompanied by noise, the correlation between the data is changing with the change of the embedding dimension, and an appropriate embedding

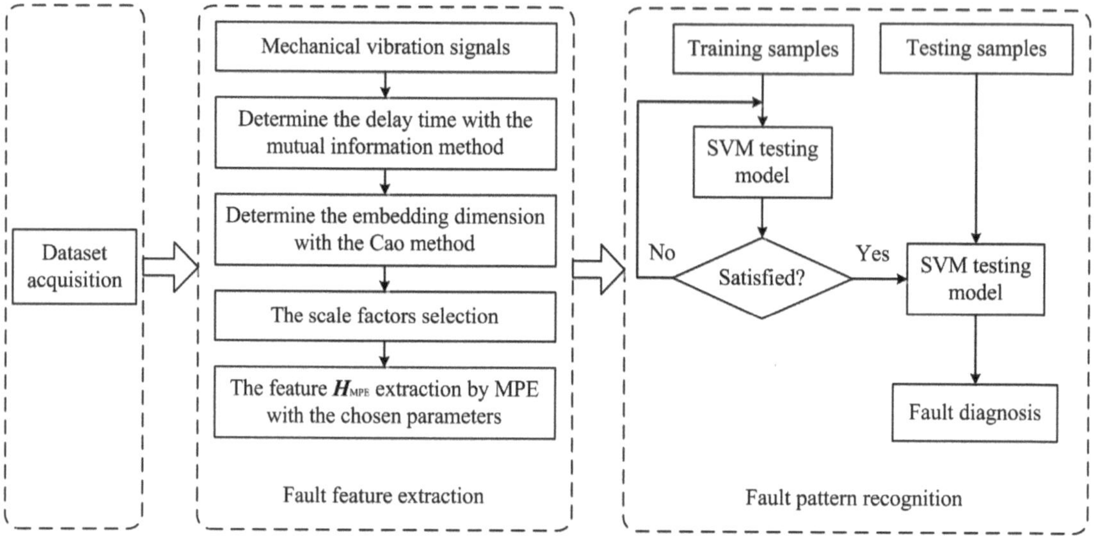

Figure 3.35. Flowchart of the method based on SVM.

dimension will be determined until the value of $E(m)$ does not change.

3. The intelligent diagnosis method based on SVM. An intelligent fault diagnosis strategy is proposed based on SVM, as shown in Fig. 3.35. The scale factors of MPE are selected as $1 \sim 5$ and the delay time and embedding dimension of MPE are decided by mutual information method and Cao method. The RBF kernel is chosen as the kernel function of SVM and grid search method is adopted to obtain the best values of C and γ as the optimal parameters of SVM. The proposed bearing fault diagnosis method is shown in Fig. 3.35 and the process is described as follows:

The feature vector of the signal calculated by MPE is defined as

$$\boldsymbol{H}_{MPE} = \left[H_p^{s_1} \; H_p^{s_2} \cdots H_p^{s_n} \right] \tag{3.81}$$

With the feature vector, the fault diagnosis method is shown in Fig. 3.35 and the process is described as follows:

a. Acquire vibration signals of rotating machinery.
b. Determinate the delay time and embedding dimension using phase space reconstruction.
c. Extract the feature vectors $\boldsymbol{H}_{MPE}$ of the data using MPE.
d. Partition the feature vectors $\boldsymbol{H}_{MPE}$ into a training set and a testing set.
e. Use the training set as the input of SVM with the RBF kernel and obtain the optimal parameters of SVM with grid search method.
f. Predict the health conditions of the testing set by the trained SVM.

3.3.2.5 Intelligent Diagnosis of Bearing Faults: A Motor Bearing Case Study

1. Data description: All the data used in this section are obtained from the Case Western Reverse Lab (Loparo, 2003). The data were collected from a motor-driven mechanical system, which consists of a 2 hp, three-phase induction motor, a torque transducer, a dynamometer, and several control electronics. The dynamometer is used to achieve desired torque load levels. The vibration data were acquired with accelerometers placed at the drive end of the motor. Inner race fault (IF), Outer race fault (OF), and roller fault (RF) were machined to the bearings using the EDM technique. The fault diameters were 0.18, 0.36 and 0.53 mm respectively and the depth of the point fault was 0.28 mm. The drive end bearing type is 6205-2RS JME SKF, deep groove ball bearing. A DAQ recorder collected the vibration data and the sampling frequency is 48 kHz. The details of the bearing dataset are shown in Table 3.9. The experiments were designed under the load of 1 hp. The bearing dataset used contains ten health conditions and each condition has 480,000 data points. The data points of each condition are divided into 100 samples and each sample has 4800 data points. The performance of the proposed method is assessed by using 20%

Table 3.9 Description of the Bearing Dataset

Datasets	Fault Type	Fault Diameter (mm)	Samples	Training Samples (%)	Testing Samples (%)	SVM Label
Bearing dataset	RF	0.18	100	20	80	1
	RF	0.36	100	20	80	2
	RF	0. 53	100	20	80	3
	IF	0. 18	100	20	80	4
	IF	0. 36	100	20	80	5
	IF	0. 53	100	20	80	6
	OF	0. 18	100	20	80	7
	OF	0. 36	100	20	80	8
	OF	0.53	100	20	80	9
	Normal	0	100	20	80	10

of samples as training samples and the remaining samples are used for testing.

2. Diagnosis results: We first investigate the parameter selection of MPE. In the proposed method, the mutual information method is used to determine the delay time τ, and the Cao method is used to determine the embedding dimension m.

   The mutual information is used to estimate the joint probability distribution $P(x_i, x_{i+\tau})$ of the data samples with many partitions in which the size of each element is tailored to the local situation. We used 128 partitions to calculate the mutual information I_τ under each health condition of the bearing dataset. Table 3.10 shows the dependence of I_τ on τ. The chosen τ for each health condition by the first minimum of I_τ is in bold. It can be seen that the most of the delay time for the samples is 4 when I_τ reaches its first minimum. So we choose 4 as the delay time for the bearing dataset.

   After the delay time τ is determined, we use the Cao method to numerically determine embedding dimension. In Fig. 3.36, the value of $E(m)$ is plotted as a function of m for the samples under each health condition. It shows that $E(m)$ becomes stable

Table 3.10 The Dependence of I_τ on τ of the Bearing Dataset

Health Condition	τ					
	1	2	3	4	5	6
RF 0.18	1.4252	0.5554	0.1659	**0.0285**	0.0765	0.2886
RF 0.36	1.6858	0.7983	0.3718	0.1531	**0.0932**	0.1247
RF 0.53	1.5506	0.6624	0.2446	**0.0682**	0.0705	0.2382
IF 0.18	1.4530	0.6055	0.2442	**0.1142**	0.1212	0.2076
IF 0.36	1.1112	0.3748	0.0953	**0.0446**	0.0813	0.1113
IF 0.53	1.6008	0.7028	0.2691	0.0717	**0.0484**	0.1860
OF 0.18	1.4743	0.6181	0.2667	**0.2019**	0.3864	0.8599
OF 0.36	1.4359	0.5643	0.1720	**0.0306**	0.0743	0.2862
OF 0.53	1.7595	0.8770	0.4651	0.2852	**0.2587**	0.3467
Normal	0.8513	0.2027	0.0576	**0.0466**	0.0709	0.0704

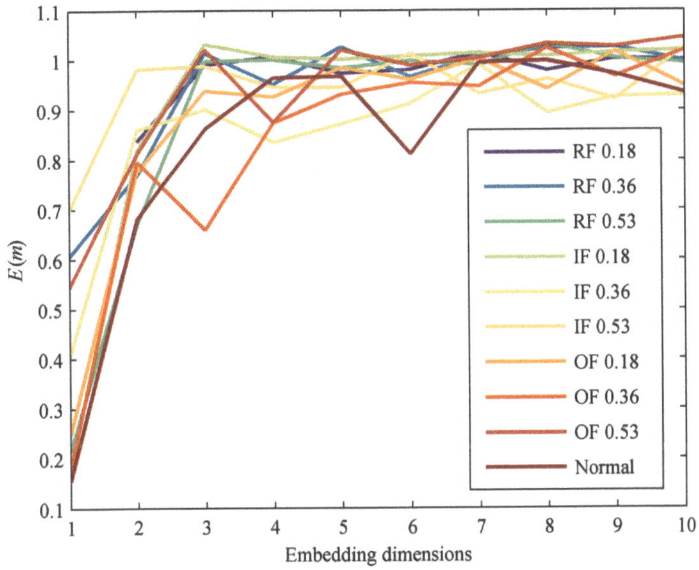

Figure 3.36. Values of $E(m)$ for the bearing dataset.

when m equals 4. After $m = 6$, the $E(m)$ values of samples for all health conditions vary in a stable interval, which means the values are stable enough for regarding them as no longer changing. Therefore, we choose that $m = 6$ for the bearing dataset.

After we determine the parameters of MPE, the MPE features extracted from the data samples are divided into two sets, that is, the training set and the testing set. The training set is used to train SVM, and the testing one is employed to predict the performance of the trained SVM. The diagnosis accuracy of the proposed method is 99.50%.

In addition, we classify the bearing dataset based on MPE with different delay times and embedding dimensions so as to investigate the influence of the MPE parameters on the diagnosis accuracies. The results are shown in Table 3.11. The accuracies of prediction beyond 99% are in bold. It shows that the peak accuracy of the bearing dataset is 99.50% when τ equals 4 and m equals 5 and 6, corresponding to the parameters selected. Therefore, the mutual information method and the Cao method in the proposed method are suitable for the parameter selecting of the MPE. Table 3.11 also shows that τ has more significant effects on diagnosis accuracies than m. Most accuracies above 99% are obtained by the proposed method when $\tau = 4$, but m can be chosen from 3 to 7. Therefore,

Table 3.11 Diagnosis Accuracy With Different Delay Times and Embedding Dimensions

	m (%)								
τ	**2**	**3**	**4**	**5**	**6**	**7**	**8**	**9**	**10**
1	13.62	91.50	94.00	94.62	96.25	96.37	95.25	94.25	94.62
2	16.25	94.12	97.12	97.87	98.12	98.00	95.87	94.00	86.62
3	13.37	96.50	98.50	98.75	98.50	97.62	95.12	90.12	84.50
4	11.62	**99.25**	**99.25**	**99.50**	**99.50**	**99.25**	98.12	97.75	90.87
5	13.37	**99.25**	98.50	**99.00**	98.87	98.37	95.00	85.12	79.50
6	13.50	97.62	96.00	98.12	98.62	98.75	97.62	92.62	71.87
7	13.62	93.50	97.75	96.87	95.50	91.87	86.75	81.25	77.87
8	16.87	92.12	97.25	97.50	97.87	95.00	85.75	85.00	78.87
9	14.12	92.12	98.12	98.12	98.50	96.75	95.75	89.50	78.25
10	14.87	95.00	96.37	97.50	96.37	94.87	85.00	73.75	64.25

in the proposed method, the time delay parameter τ should be chosen reasonably and precisely but the embedding dimension m can be chosen flexibility.

3.3.2.6 Epilog

Here, we present an intelligent fault diagnosis method based on SVM. First, the MPE is employed to evaluate various fault types and severities of the rolling element bearing. Second, in order that MPE performs better, the mutual information method and Cao method are applied to determine the delay time and embedding dimension of MPE. Finally, SVM is employed to identify the health conditions of the bearings based on features extracted by MPE. The diagnosis results show that the parameters of MPE are crucial to the diagnosis accuracy and the proposed method is superior in classifying ten health conditions of bearings. Furthermore, only five features are used and the accuracy over 99% is obtained in the case study. It reveals that MPE can quantitatively and effectively describe the complexity of various fault types and severities of the bearings. So this method does not need the sensitive fault feature selection, which could reduce the complexity of diagnosis system and may operate faster.

3.3.3 Relevant Vector Machine-Based Intelligent Fault Diagnosis

3.3.3.1 Introduction

From the diagnosis results in Section 3.3.2, it is verified that SVM performs well in intelligent fault diagnosis of rotating machinery. Nevertheless, SVM suffers from several limitations as follows. (1) The outputs of SVM represent decisions rather than posterior probabilities. (2) SVM is originally developed for the binary classification problems. Although several strategies could extend a set of SVMs to their multiclass version, the development of a single SVM used for a multiclass problem is still problematic. (3) There are several parameters, for example, penalty parameter C and kernel parameter γ, that must be found using a optimization method. (4) The kernel functions centered on training samples should satisfy Mercer's conditions.

RVM (Tipping, 2001), a new kernel technique for classification based on Bayesian framework, shares lots of characteristics of SVM and avoids the aforementioned limitations of SVM. The special advantages of RVM in solving classification problems can be summarized as follows. (1) RVM is able to build much sparser models than SVM, resulting in faster performance on test samples and maintaining comparable generalization ability. (2) RVM provides the membership probability of output members. (3) The kernel functions of the RVM algorithm are no longer constrained by the Mercer theorem and can be constructed freely. (4) The parameters of RVM are assigned automatically during the training process, instead of subjectively. Due to the advantages, RVM has been used in various research fields, like biomedicine, image processing, as well as mechanical fault diagnosis. For instance, Widodo et al. (2009) applied RVM to diagnose faults of rolling element bearings and obtained good results. Wang et al. (2014) developed an RVM-based monitoring system to classify tool wear conditions in the milling process. However, the study of RVM is still limited in intelligent fault diagnosis.

Here, a new method based on multiclass relevance vector machine (MRVM) is developed for classifying the health conditions of planetary gearboxes, where MRVM is an extended version of RVM to handle the multiclass tasks (Psorakis et al., 2010). This method uses two features, that is accumulative amplitudes of carrier orders (AACO) and energy ratio based on difference spectra (ERDS), to characterize the health conditions of planetary gearboxes, and adopts MRVM as the classifier. The two features are specifically designed for the planetary gearboxes based on the observation of their vibration characteristics and the investigation of their fault mechanism. The developed method is expected to classify different health conditions of multistage planetary gearboxes, such as a chipped tooth on the first-stage sun gear, a flaked needle on the first-stage planet bearing, a cracked tooth on the second-stage planet gear, and so on.

3.3.3.2 Relevance Vector Machine

RVM is originally derived for binary classification under a probabilistic Bayesian learning framework. It is able to predict the probability of the class that a sample belongs to (Tipping, 2001).

Given a set of samples $\{\boldsymbol{x}_i, t_i\}_{i=1}^{N}$, where $\boldsymbol{x}_i \in R^D$ is the ith D-dimensional feature and $\boldsymbol{t} \in \{1,2,3,\cdots,C\}$ are the class labels, the functional relationship between $\boldsymbol{x}_i$ and t_i can be modeled by RVM as

$$t_i = \sum_{i=1}^{N} w_i K(\boldsymbol{x}, \boldsymbol{x}_i) + w_0 \tag{3.82}$$

Considering the scalar output functions, it assumes that the outputs of RVM are class labels from the samples with additive noise, as shown in the following standard probabilistic formulation.

$$t_i' = t_i + \varepsilon_i = \sum_{i=1}^{N} w_i K(\boldsymbol{x}, \boldsymbol{x}_i) + w_0 + \varepsilon_i \tag{3.83}$$

where ε_i is the ith independent sample from a noise process that subjects to mean-zero Gaussian distribution with variance σ^2, $K(\boldsymbol{x}, \boldsymbol{x}_i)$ is a kernel function, and w_i is the ith weight of linear combination of the kernel functions.

The target of the RVM model can only take the value 0 or 1 for the binary classification problems. So the logistic sigmoid function $\sigma(y) = 1/(1 + e^{-y})$ is applied to generalize the model of RVM. Its likelihood is expressed as:

$$P(\boldsymbol{t} \mid \boldsymbol{w}) = \prod_{i=1}^{N} \sigma\left[y(\boldsymbol{x}_i; \boldsymbol{w})\right]^{t_i} \left\{1 - y(\boldsymbol{x}_i; \boldsymbol{w})\right\}^{1-t_i} \tag{3.84}$$

Since $P(\boldsymbol{t}|\boldsymbol{w})$ is a nonnormal distribution, the closed-form solutions of the weight posterior $p(\boldsymbol{w}|\boldsymbol{t}, \boldsymbol{\alpha})$ and the marginal likelihood $P(\boldsymbol{t}|\boldsymbol{w})$ cannot be analytically obtained, as well as the weight vector $\boldsymbol{w}$. So an approximation procedure based on the Laplace method is applied for solving the optimization problem in RVM.

1. Because $p(\boldsymbol{w}|\boldsymbol{t}, \boldsymbol{\alpha}) \propto P(\boldsymbol{t}|\boldsymbol{w})\, p(\boldsymbol{w}|\boldsymbol{\alpha})$, we could find the maximum posteriori weight with a fixed value of $\boldsymbol{\alpha}$:

$$\log\{P(t\,|\,w)p(w\,|\,\alpha)\} = \sum_{i=1}^{n}\left[t_i \log y_i + (1-t_i)\log(1-y_i)\right] - \frac{1}{2}w^{\mathrm{T}}Aw \quad (3.85)$$

where $A = \mathrm{diag}(\alpha_0, \alpha_1, \cdots, \alpha_N)$ and $y_i = \sigma\{y(x_i; w)\}$.

2. The Laplace method used is a quadratic approximation method for the log-posterior around its mode. We differentiate the maximum posteriori weight in Eq. (3.85) by twice and give the results as:

$$\nabla_w \nabla_w \log p(w\,|\,t, \alpha)\big|_{w_{\mathrm{MP}}} = -(\boldsymbol{\Phi}^T \boldsymbol{B}\boldsymbol{\Phi} + A) \quad (3.86)$$

where ∇ denotes the gradient operator and $\boldsymbol{B} = \mathrm{diag}(\beta_0, \beta_1, \cdots, \beta_N)$ is a diagonal matrix with $\beta_n = \sigma\{y(x_n)\}\left[1-\sigma(x_n)\right]$.

3. With the statistics $\sum = (\boldsymbol{\Phi}^T \boldsymbol{B}\boldsymbol{\Phi} + A)^{-1}$ and $\boldsymbol{w}_{\mathrm{MP}} = \sum \boldsymbol{\Phi}^T \boldsymbol{B}\boldsymbol{t}$ of the Gaussian approximation, the parameters α_i are able to be updated by:

$$\alpha_i^{\mathrm{new}} = \frac{1-\alpha_i \sum_{ii}}{w_{\mathrm{MP}i}^2} \quad (3.87)$$

where $w_{\mathrm{MP}i}$ is the ith element of the most possible maximum posteriori weight $\boldsymbol{w}_{\mathrm{MP}}$, and $\sum_{ii}$ is the ith diagonal element of the posterior weight covariance.

3.3.3.3 Multiclass Relevance Vector Machine

MRVM employs a hierarchical Bayesian framework to solve the multiclass problems (Psorakis et al., 2010). By introducing multinomial probit likelihood, MRVM is able to predict the probability of the sample belonging to a class, which is similar to RVM.

For the dataset $\{\boldsymbol{x}_i, t_i\}_{i=1}^{N}$, its kernel function is $\boldsymbol{K} \in R^{N\times N}$. Each row $\boldsymbol{k}_n$ of $\boldsymbol{K}$ denotes how a sample $\boldsymbol{x}_i$ is related to other samples.

The regression targets of MRVM, $\boldsymbol{Y} \in R^{C\times N}$, are the auxiliary variables, following a standardized noise model.

$$y_{cn}\,|\,w_c, \boldsymbol{k}_n \sim N_{y_{cn}}\left(w_c^T \boldsymbol{k}_n, 1\right) \quad (3.88)$$

Using the multinomial probit link, the regression targets of MRVM can be converted into the labels:

$$t_n = i, \quad y_{ni} > y_{nj}, \quad \forall i \neq j \quad (3.89)$$

So the probabilistic output of a sample is obtained by the resultant multinomial probit likelihood function:

$$P\left(t_n = i \mid w, k_n\right) = \varepsilon_{p(u)}\left\{\prod_{j \neq i} \Phi\left(u + \left(w_i - w_j\right)^T k_n\right)\right\} \tag{3.90}$$

where $u \sim N(0, 1)$, and $\boldsymbol{\Phi}$ is the Gaussian cumulative distribution function.

To ensure the sparsity of MRVM, the weight w_{nc} is subject to a standard normal distribution with zero-mean and variance α_{nc}^{-1}. And α_{nc} is the element of the scale matrix $A \in R^{N \times C}$ and is subject to a Gamma distribution. When the parameters a and b are small enough, most elements in $\boldsymbol{w}$ are restricted to zero, leading to a sparse solution.

Since the parameters in MRVM have certain dependency, it results in a hierarchical Bayesian framework. So the closed-form solution of weight $\boldsymbol{w}$ is derived as:

$$P(\boldsymbol{w} \mid \boldsymbol{Y}) \propto P(\boldsymbol{Y} \mid \boldsymbol{w}) P(\boldsymbol{w} \mid A) \propto \prod_{c=1}^{C} N\left(\left(\boldsymbol{K}\boldsymbol{K}^T + \boldsymbol{A}_c\right)^{-1} \boldsymbol{K} \boldsymbol{y}_c^T, \left(\boldsymbol{K}\boldsymbol{K}^T + \boldsymbol{A}_c\right)\right)^{-1} \tag{3.91}$$

where $\boldsymbol{A}_c$ is derived from the cth column of $\boldsymbol{A}$. So the maximum of a posterior estimator is:

$$\hat{\boldsymbol{w}} = \arg\max{}_{w} P(\boldsymbol{w} \mid \boldsymbol{Y}, \boldsymbol{A}, \boldsymbol{K}) \tag{3.92}$$

Then, the parameters can be updated by maximizing a posterior estimator value:

$$\hat{\boldsymbol{w}}_c = (\boldsymbol{K}\boldsymbol{K}^T + \boldsymbol{A}_c)^{-1} \boldsymbol{K} \boldsymbol{y}_c^T \tag{3.93}$$

Finally, the posteriori probability distribution of priori parameters of the weight vector is:

$$p(A \mid \boldsymbol{w}) \propto p(\boldsymbol{w} \mid A) p(A \mid a, b) \propto p \prod_{c=1}^{C} \prod_{n=1}^{N} G\left(a + \frac{1}{2}, \frac{w_{nc}^2 + 2b}{2}\right) \tag{3.94}$$

3.3.3.4 Intelligent Diagnosis Based on MRVM

The diagram of the intelligent diagnosis method using MRVM is shown in Fig. 3.37, and its process is detailed as follows:

1. Capture the vibration signals by accelerometers installed on different stages of planetary gearboxes under different operating conditions.
2. Extract two features, that is, AACO and ERDS, from the vibration signals. The extracted features combined with their labels can be comprised into a dataset.

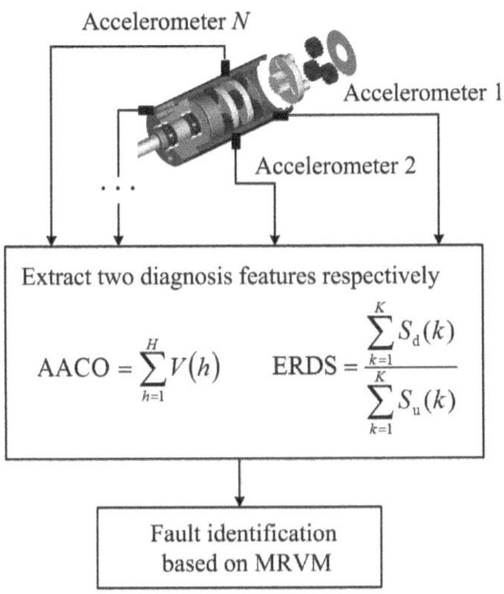

Figure 3.37. Flowchart of the proposed method based on MRVM.

3. Divide the dataset into a training set and a testing set.
4. Employ the training set to train the MRVM-based classifier.
5. Utilize the trained MRVM to classify the testing samples.

In this method, AACO and ERDS are specifically designed for fault diagnosis of planetary gearboxes, and they will be described as follows:

1. AACO: The characteristic frequencies of components in a planetary gearbox, like gears and bearings, are typically the multiples of the carrier rotating frequency. Thus, the amplitudes of the orders of the carrier rotating frequency will increase when a fault occurs in these components. To identify the change of the amplitudes of the orders, AACO is computed using the order spectrum of the rotating frequency of the carrier, and it is expressed as

$$\mathrm{AACO}=\sum_{h=1}^{H}V(h) \tag{3.95}$$

where $V(h)$ $(h=1,2,\cdots,H)$ reflects the amplitude at the hth order of the carrier rotating frequency in the order spectrum, and H is the number of orders in the order spectrum. The feature is computed by the following steps. (1) Calculate the carrier rotating frequency based on the transmission ratio and the shaft speed of the planetary gearbox. (2) Implement Fourier transform on the collected signals of the planetary gearbox and get their frequency spectrum. (3) Obtain the order spectrum according to the carrier rotating frequency. (4) Sum up the amplitudes of all the orders.

2. ERDS: A common strategy in fault diagnosis is to compare the signal under an unknown condition with that under the healthy condition so as to find the difference between them. If the health condition of a planetary gearbox changes, its vibration signals and the corresponding frequency spectra will also change. So in order to characterize the variation of the frequency spectrum, ERDS is developed using the difference spectrum of the signal under the healthy condition and that under the unknown condition of the planetary gearboxes. ERDS is calculated by normalizing the summation of the absolute amplitudes of the difference spectrum.

$$\mathrm{ERDS} = \frac{\sum_{k=1}^{K} s_d(k)}{\sum_{k=1}^{K} s_u(k)} \tag{3.96}$$

where $s_d(k) = |s_u(k) - s_h(k)|$ $(k = 1, \cdots, K)$ is the absolute amplitude at the kth spectrum line in the difference spectrum, K is the number of spectrum lines, and $s_u(k)$ and $s_h(k)$ reflect the amplitudes at the kth spectrum line of the frequency spectrum under the unknown health condition and that under the healthy condition, respectively. Five steps are needed to obtain ERDS. (1) Implement Fourier transform on the acquired signal of the planetary gearbox. (2) Choose a signal of a healthy planetary gearbox from the history dataset. (3) Get the frequency spectrum of the signal under the healthy condition with the Fourier transform. (4) Get the difference spectrum by subtracting the frequency spectrum under the healthy condition from that under the unknown one. (5) Sum up the absolute amplitudes of the difference spectrum and normalize the amplitudes of the spectrum under the unknown condition by the summation.

Since AACO and ERDS are designed based on the investigations of vibration characteristics and fault mechanism of planetary gearboxes, they are supposed to perform well in classifying the health conditions of the planetary gearboxes.

3.3.3.5 Intelligent Diagnosis of Planetary Gearboxes: A Case Study

1. Introduction of the experimental system and data collection. In Fig. 3.25, it shows the test rig containing a two-stage planetary gearbox. The experiments are conducted on this rig and vibration signals are collected to verify the proposed method. The motor speed is determined by a speed controller, and it allows

the test rig to operate at various speeds. Load is offered by the magnetic brake that is connected to the output shaft of the test rig, and this torque load can be adjusted using the brake controller. In the test rig, the health conditions of the two-stage planetary gearbox are what we concern. The inner sun gear in each stage of the planetary gearbox is surrounded by a stationary outer ring gear and three or four rotating planet gears. The torque is transmitted from the sun gear to the planets fixed on a planetary carrier, and the carrier transmits torque to the output shaft.

Since gears and bearings in planetary gearboxes easily suffer from damage, different types of faults are introduced to the bearings and the gears of the two-stage planetary gearbox. Six faults are simulated in the experiment, containing a cracked tooth on the first-stage sun gear, a pitted tooth on the first-stage sun gear, a cracked tooth on a planet gear in the first stage, a flaked needle on a planet bearing in the first stage, a chipped tooth on the second-stage sun gear, and a missing tooth on the second-stage sun gear. These faulty gears and bearings are displayed in Fig. 3.38. Thus, seven

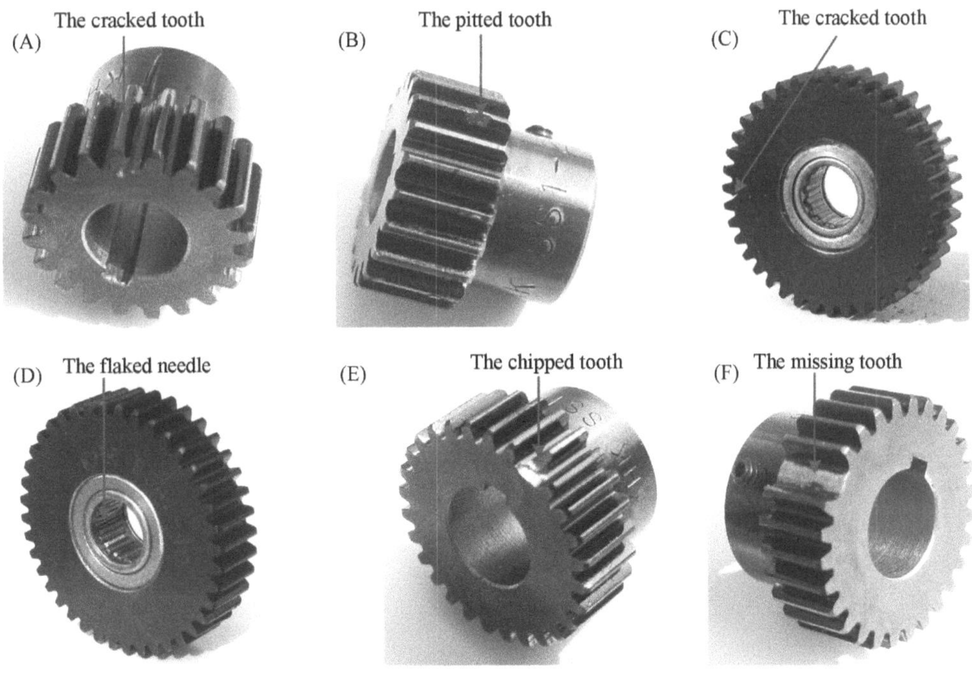

Figure 3.38. (A) A cracked sun gear in the first stage, (B) a pitted sun gear in the first stage, (C) a cracked planet gear in the first stage, (D) a flaked planet bearing in the first stage, (E) a chipped sun gear in the second stage, and (F) a missing tooth sun gear in the second stage.

health conditions are totally considered including the normal condition of the two-stage planetary gearbox and their information is listed in Table 3.12.

To collect the vibration signals, we install two accelerometers, that is, Accelerometer 1 and Accelerometer 2, on the first- and second-stage housings of the planetary gearbox, respectively. With a sampling frequency of 5120 Hz, the vibration signals for each health condition are collected at four motor speeds: 2100, 2400, 2700, and 3000 rpm, and under two loading conditions: no load and the load equaling 13.5 Nm. In the experiments, a vibration signal containing 20,480 data points composes a data sample. For each health condition at each motor speed and under each loading condition, 30 data samples are acquired with each accelerometer. We treat the same health condition at various speeds and under various loading conditions as one class. For instance, the samples collected from the pitted tooth on the first-stage sun gear at four different speeds and under two different loading conditions share the same label. So for each health condition, 480 data samples are obtained with the two accelerometers. Therefore, we totally get 3360 data samples for the seven health conditions of the two-stage planetary gearbox. Fig. 3.39 shows the examples of data samples for these health conditions, respectively.

In the vibration signals under the normal condition of the planetary gearbox, no impulses can be observed, while obvious

Table 3.12 Seven Health Conditions of the Two-Stage Planetary Gearbox

	Health Conditions	
Label	**First Stage**	**Second Stage**
1	Normal	Normal
2	The sun gear having a cracked tooth	Normal
3	The sun gear having a pitted tooth	Normal
4	A planet gear having a cracked tooth	Normal
5	A planet bearing having a flaked needle	Normal
6	Normal	The sun gear having a chipped tooth
7	Normal	The sun gear having a missing tooth

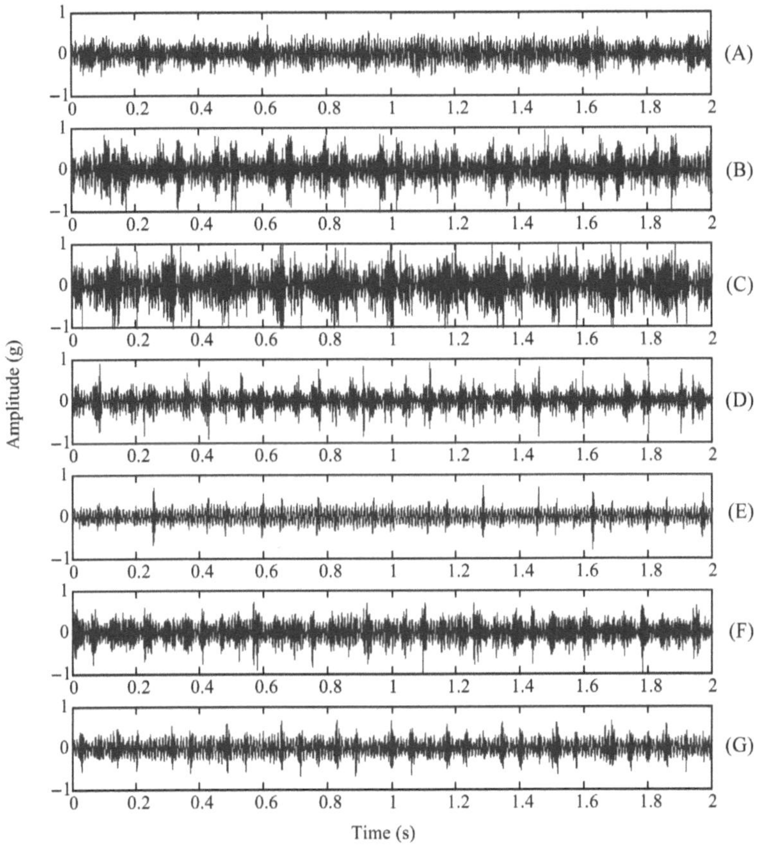

Figure 3.39. Typical data samples. (A) Normal, (B) a cracked sun gear in the first stage, (C) a pitted sun gear in the first stage, (D) a cracked planet gear in the first stage, (E) a flaked planet bearing in the first stage, (F) a chipped sun gear in the second stage, and (G) a missing tooth sun gear in the second stage.

impulses can be observed under the six faulty conditions. Although the vibration signals of the six faulty conditions show different characteristics, it is difficult to directly distinguish the health conditions with these vibration signals. Therefore, two features are used to characterize the health conditions of the planetary gearboxes, and MRVM is employed to classify the health conditions.

To thoroughly verify the proposed method, we implement three tests (Test 1, Test 2, and Test 3) using three different data subsets (Data subset 1, Data subset 2, and Data subset 3), which are selected from the whole data samples collected from the planetary gearbox test rig.

2. Test 1: Condition classification with the data collected from the first-stage housing. Data subset 1 is composed of 1680 data samples collected by Accelerometer 1 installed on the first-stage housing of the planetary gearbox. This data subset refers to seven different health conditions at four motor speeds and under two loading conditions. In Data subset 1, 50% of samples are used as the training samples and the remaining 50% are used for testing.

   Following the flowchart of the proposed method in Fig. 3.37, the two features, namely AACO and ERDS, are extracted from each training sample. The extracted features are displayed in Fig. 3.40A and B, respectively. Fig. 3.40C shows the cluster results with the two features. As we can see, the samples with the same labels are clustered well and the samples with different labels are separated. These extracted features combined with the corresponding health condition labels are employed to train the MRVM. After the training process, the AACO and ERDS features are extracted from the testing samples. We use the trained MRVM to classify the features so as to obtain the health condition that each testing sample belongs to. According to

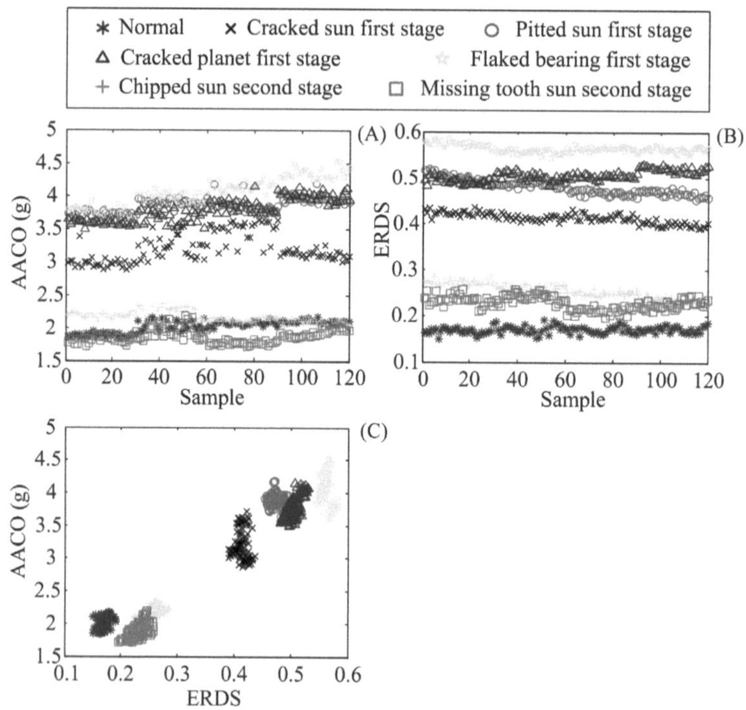

Figure 3.40. (A) AACO, (B) ERDS, and (C) AACO and ERDS.

Table 3.13 Identification Accuracy Using the Proposed Method for the Three Tests

Method	Test 1	Test 2	Test 3	
The proposed method	81.79%	81.43%	100%	95.95%

the number of the correctly identified testing samples, we can evaluate the classification accuracy of the proposed method. As is displayed in Table 3.13, the accuracy for Data subset 1 is 81.79%.

3. Test 2: Condition identification using the data obtained on the second-stage housing. The difference between Data subset 2 and Data subset 1 is that the former subset is collected by Accelerometer 2, which is installed on the second-stage housing. Data subset 2 also includes 1680 data samples, where 50% of samples are for training and the rest are for testing. For this data subset, the classification accuracy is 81.43%, which is very close to the accuracy in Test 1. So in Tests 1 and 2, the classification accuracies of the seven health conditions, less than 85%, may not be satisfied for application. The reason is that the adopted data are obtained by only one accelerometer. As is discussed in Section 3.2.4, the individual accelerometer may provide limited characteristic information for intelligent fault diagnosis. So we fuse the data measured by two accelerometers and apply the proposed method to achieve higher accuracies in classifying the health conditions of the planetary gearbox.
4. Test 3: Condition Identification using the data collected on both stages. We conduct an experiment on Data subset 3 to verify the improvement of the classification accuracy by fusing the data of two accelerometers. Data subset 3 is composed of the whole data samples, that is, 3360 data samples acquired by Accelerometer 1 and Accelerometer 2. Same as Test 1 and Test 2, half of the samples are utilized for training and the others for testing.

   Table 3.13 shows that the classification accuracy of Data subset 3 using the proposed method is 100%, which means that the proposed method correctly distinguishes all the samples of the seven health conditions with the data fusing strategy. Compared with Test 1 and Test 2 using the data collected from individual accelerometers, Test 3 takes full use of data from both accelerometers and enhances the classification accuracy largely.

Furthermore, to investigate the robustness of the proposed method at various motor speeds, another study is carried out. In this study, the training samples are composed of the data acquired at the two low motor speeds, that is, 2100 and 2400 rpm, and the testing samples are the data samples at two higher speeds, that is, 2700 and 3000 rpm. With this organization of the data samples, the classification accuracy of the proposed method changes from 100% of the previous study to 95.95% of this study. Although it decreases by 4.05%, the classification accuracy is acceptable for applications. It demonstrates that the proposed method is robust at various motor speeds.

3.3.3.6 Epilog

The health conditions of the planetary gearbox involve damage on both bearings and gears occurring in different stages, and the data are collected under various operating motor speeds and loads. To well classify these health conditions, we present an intelligent diagnosis method based on MRVM. In the method, the two features, that is, AACO and ERDS, are extracted from the collected data and used as the input of the MRVM. To validate the performance of the method, three tests are carried out. Through the results of these tests, the superiority of data fusing strategy is verified again. These results also illustrate the effectiveness and robustness of the proposed method in distinguishing the health conditions of planetary gearboxes.

It is a fact that intelligent fault diagnosis of planetary gearboxes largely depends on the extracted features input to the classifiers. Although many features have been proposed based on the fault mechanism of planetary gearboxes, there are still lots of issues to be investigated. As long as it is clearly revealed, the health conditions of the planetary gearboxes would be classified more correctly.

3.4 Deep Learning

In the previous sections, the effectiveness of the artificial intelligence technologies, such as neural networks and SVM, in intelligent fault diagnosis of rotating machinery is verified by plenty of diagnosis cases. They are definitely powerful tools. However, these intelligence techniques are limited by their ability to process mechanical signals in their original form. So we have to extract features from signals in time and frequency domain using signal processing techniques. So lots of actual effort in intelligent diagnosis methods goes into the design of feature extraction algorithms so as to obtain features from the signals. Such features take advantage of human ingenuity and special knowledge in fault diagnosis of rotating machinery, so we consider these extracted features as manual features. The manual features combined with the artificial intelligence technologies, whose advantage is automatic decision making, result in the current intelligent fault diagnosis methods. However, these methods suffer from the following shortcomings.

1. The features input into classifiers are extracted and selected by diagnosticians from the measured signals, significantly depending on prior knowledge about signal processing techniques and diagnostic expertise. It is not easy to master lots of signal processing techniques and gather much diagnostic experience.
2. The features are ordinarily extracted and selected according to a specific diagnosis issue and probably unsuitable for other issues. When handling a new diagnosis task, the feature extraction algorithms may need to be redesigned. For example, the features like AACO and ERDS are designed especially for the faults of gears. When dealing with bearing signals, one needs to use other features.
3. Nowadays, massive signals are acquired for monitoring the health conditions of rotating machinery, which promotes machinery health monitoring to enter the age of Big Data. Nevertheless, lacking a comprehensive understanding of mechanical big data, it is often difficult to ensure the extracted features carrying optimal information to classify the mechanical faults. So diagnosticians need to spend a lot of time in analyzing massive signals and grasping their properties, which is time-consuming and labor-intensive.

To deal with these weaknesses, it would be better to learn the features from original signals using advanced artificial intelligent techniques, instead of manually designing the features. It would make intelligent fault diagnosis methods less dependent on the prior knowledge or human labor so as to make mechanical fault diagnosis toward real artificial intelligence. The idea of deep learning holds the potential to overcome the weaknesses aforementioned in the intelligent diagnosis methods. In the following sections, we will introduce the concept of deep learning in detail.

3.4.1 Introduction to Deep Learning

Deep learning has been a new direction in machine learning research. In recent years, the techniques based on deep learning research have presented great impact on a wide range of information-processing work, such as image processing, object detection, speech recognition, and so on.

Deep learning refers to a class of machine learning techniques, where many layers of information processing stages in deep architectures are exploited for pattern classification and other tasks (Deng, 2014). So deep learning is obtained by multiple levels of representation, each of which is a simple layer that transforms the previous representation (starting with the original data) into a nonlinear representation at a higher representation. Based on the composition of enough nonlinear transformations, deep learning techniques have the ability in approximating complex nonlinear functions with a small error. Due to this advantage, deep learning techniques could adaptively capture the representative information from the signals of rotating machinery in time or frequency domain. For the tasks of intelligent fault diagnosis, higher layers of representation amplify discriminative information of the input signals and suppress their irrelevant variations. Therefore, the highlight of deep learning is that the features are learned from the original signals using a general-purpose learning procedure, instead of being designed by diagnosticians.

Deep learning isn't a new brand. Actually, it is derived from the concept of neural networks. The main model used in deep learning is called deep neural networks (DNNs). Thus, neural network is the precursor of deep learning. In 1980s, Hinton and other researchers helped to revive the interest in neural networks that made better use of many layers of neurons, and they believed that additional levels of layers would yield increased modeling power (Rumelhart et al., 1986). However, in practice, the neural networks of a deep architecture still required human involvement heavily, complex recognition tasks required much more computer power than we could provide and there was insufficient training sets for tuning the parameters of the neural networks. So the neural networks were given up gradually. The idea of the specific techniques for specific tasks becomes popular, namely features are extracted for specific problem and then fed to the classifiers for recognition. So the well-known classifiers, such as ANFIS, SVM and random forest are widely used for the classification tasks in a variety of application fields.

Things have changed in the last decade. Hinton, LeCun, and other researchers made some fundamental breakthroughs and these breakthroughs formed the concept of deep learning (LeCun et al., 2015). In 2006, Hinton proposed an efficient way to build a DNN and some theoretical studies of deep learning were conducted by Bengio and LeCun (Bengio, 2009; Jarrett et al., 2009; Hinton and Salakhutdinov, 2006). These studies illustrate the following important things. (1) It is needed to use deep architectures to efficiently learn the complex functions, which is able to achieve better generalization performance for the classification tasks. (2) A deep model is more able to reveal the characteristics of the input data. (3) The DNN can be effectively built by layer-by-layer pretraining, namely, pretraining one hidden layer at a time. Based on these points, several successful approaches of training DNNs are developed, such as deep belief networks and stacked autoencoders.

This success is extremely owed to use an unsupervised training criterion to perform a layer-by-layer initialization (Bengio et al., 2007). By optimizing the local unsupervised criterion, each layer can be trained using the output of the previous layer as its input. Since composing by multiple layers, DNNs can obtain the representation of the input data layer-by-layer and the representation of the higher level layer will be more abstract than that of the lower level layer. After this initialization, we perform a global fine-tuning to adjust the parameters of the deep networks in a supervised manner, which is used to refine the representation or features in each layer for the final classification. In other words, unsupervised learning promotes the DNN to produce good intermediate features of the input data, but it is unknown which feature is used for the classification tasks. So the final supervised learning may guide the DNN to amplify the useful features and suppress the irrelative features for the classification tasks.

Nowadays, deep learning has made major advances in solving numerous problems that are hard to overcome by the previous artificial intelligence techniques. It yields breakthroughs in various applications, which also encourages the development of the deep learning at a gallop. For instance, Google built one of the largest neural networks with deep learning, which had more than one billion connections and was led by Andrew Ng and Jeff Dean (Le, 2012). The neural network was trained by 10 million images that were sampled randomly from the YouTube videos. After the training process, it is shown that one neuron of the network fixated on images of cats while other neurons focused on human faces or other objects. Thanks to the ability of deep learning, the neural network was able to recognize the objects even though humans had not labeled them ever. In the famous visual recognition challenge, ImageNet Challenge, Krizhevsky et al. (2012) developed a deep learning-based method for classifying millions of images and got the best accuracy of 83.6%. In contrast, the highest classification accuracy using the traditional method, that is, manual features combined with linear classifiers, is only 74.2%. Also, deep learning has been used to improve the voice search on smartphones. Dean and his team replaced part of the speech system of Google's Android software with the part based on deep learning. Since this multiple-layer network allows for more precise training on the variants of a sound, it can identify the sound more precisely and robustly, even in noisy environments. In addition, deep learning has also presented its superiority in natural language understanding, the activity of drug molecules predicting, particle accelerator data analyzing, brain circuits reconstructing, and the effects predicting of DNA on gene expression and disease.

Deep learning has been applied to many domains of science, business, and government. However, there is little study on the intelligent fault diagnosis. We study the deep learning in intelligent fault diagnosis of rotating machinery, and show its ability in discovering intricate structures in mechanical signals. In the near future, deep learning will achieve more success in the field of mechanical fault diagnosis, since it can easily make the best of a large amount of data and available computation.

3.4.2 Stacked Tied-Weight Autoencoder-Based Fault Diagnosis

3.4.2.1 Introduction

In condition monitoring of rotating machinery, real-time data are collected from the machines so as to fully inspect its health conditions. Therefore, big data are acquired after long time operation of the machines. As the data are generally collected faster than diagnosticians can analyze, intelligent fault diagnosis methods are used to analyze the big data and provide accurate diagnosis results. Based on the results, we could take appropriate maintenance for the machines and ensure their normal operation. However, the intelligent fault diagnosis methods mostly used have several obvious deficiencies that we have described previously. (1) The features input into classifiers are designed by diagnosticians depending on prior knowledge about signal processing techniques and diagnostic expertise. And the features are selected according to a specific diagnosis issue and probably unsuitable for other issues. (2) The classifiers adopted have shallow architectures, and such simple architectures limit the capacity of the classifiers to learn the complex nonlinear relationships in fault diagnosis issues. Therefore, we use DNNs to accomplish both fault feature extraction and intelligent diagnosis.

There are a series of techniques used as the building block to establish DNNs, for example, autoencoder, restricted Boltzmann machine, independent component analysis, and so on. These methods help to provide good features of the mechanical signals. However, other than autoencoder, they often require the tuning of various hyperparameters to perform well, which is a great challenge for appropriately using them. For instance, the hyperparameters of the restricted Boltzmann machines that need to be set are: number of features, sparsity penalty, weight decay, learning rate, and momentum. Once these hyperparameters are set improperly, the learned features may result in poor diagnosis accuracy. Since autoencoder is simple and its tunable hyperparameter is only the number of features, we use a tied-weight autoencoder to build a DNN for the diagnosis purpose.

Based on the concept of deep learning, this section presents a DNN-based novel intelligent diagnosis method in fault diagnosis of rotating machinery. The training process of DNNs has two steps. The first step is pretraining the DNN using the tie-weight autoencoder layer-by-layer in an unsupervised way, and the second step is fine-tuning the pretrained DNN using the BP algorithm in a supervised manner. The pretraining process helps the fault feature mining and the fine-tuning process contributes to obtain the discriminative features for fault classification. Therefore, DNNs are utilized to implement both fault feature extraction and intelligent diagnosis in the presented method.

3.4.2.2 Tied-Weight Autoencoder

An autoencoder is a type of unsupervised neural networks that has three layers (Bengio et al., 2013). The output target of the autoencoder is its input data, as shown in Fig. 3.41. The autoencoder is composed of two following parts: encoder network and decoder network. The encoder network transforms the input data from a high-dimensional space into codes in a low-dimensional space and the decoder network reconstructs the inputs from the corresponding codes. So an autoencoder could be regarded as a nonlinear function that encodes the input data to a code vector, and the code vector has most of the information of the input data so as to reconstruct it.

The encoder network can be defined as an encoding function, which is denoted by f_θ. For each signal $\boldsymbol{x}^m$ of a dataset $\{\boldsymbol{x}^m\}_{m=1}^{M}$ collected from the rotating machinery, we define

$$\boldsymbol{h}^m = f_\theta(\boldsymbol{x}^m) \tag{3.97}$$

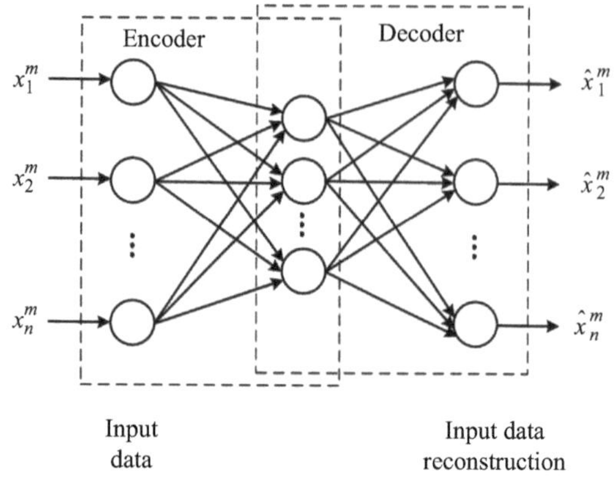

Figure 3.41. Architecture graph of an autoencoder.

where $\boldsymbol{h}^m$ is the encoded vector of $\boldsymbol{x}^m$.

The decoder network can be defined as a decoding function, which is denoted by $g_{\theta'}$. It maps the encoded vector $\boldsymbol{h}^m$ back into the original data space, producing the reconstruction as

$$\hat{\boldsymbol{x}}^m = g_{\theta'}(\boldsymbol{h}^m) \tag{3.98}$$

For the purpose of reconstructing the original input as well as possible, we use a cost function $L(\boldsymbol{x},\hat{\boldsymbol{x}})$ to measure the reconstruction discrepancy between $\boldsymbol{x}$ and $\hat{\boldsymbol{x}}$. By achieving the lowest $L(\boldsymbol{x},\hat{\boldsymbol{x}})$ over the M training examples, we can learn the encoding function and the decoding function simultaneously. Thus, the autoencoder training tries to learn the parameter sets $\boldsymbol{\theta}$ and $\boldsymbol{\theta}'$ through minimizing the reconstruction error:

$$\phi_{\mathrm{AE}}(\theta,\theta') = \frac{1}{M}\sum_{m=1}^{M} L\left(\boldsymbol{x}^m, g_{\theta'}\left(f_\theta(\boldsymbol{x}^m)\right)\right) \tag{3.99}$$

where the cost function $L(\boldsymbol{x},\hat{\boldsymbol{x}})$ is defined as

$$L(\boldsymbol{x},\hat{\boldsymbol{x}}) = \|\boldsymbol{x}-\hat{\boldsymbol{x}}\|^2 \tag{3.100}$$

And we use affine mappings as the forms of the encoding function and the decoding function:

$$f_\theta(\boldsymbol{x}) = s_f(\boldsymbol{W}_1\boldsymbol{x}+\boldsymbol{b}) \tag{3.101}$$

$$g_{\theta'}(\boldsymbol{x}) = s_g(\boldsymbol{W}_2\boldsymbol{x}+\boldsymbol{d}) \tag{3.102}$$

where s_f and s_g are the activation functions for encoding function and the decoding function, respectively. Commonly, we want to establish the relationship between s_f and s_g for better performance, so a tied-weight strategy is utilized, namely

$$\boldsymbol{W}_2 = \boldsymbol{W}_1^{\mathrm{T}} \tag{3.103}$$

Thus, the parameter sets of the tied-weight autoencoder are $\boldsymbol{\theta} = \{\boldsymbol{W}_1, \boldsymbol{b}\}$ and $\theta' = \{W_1^{\mathrm{T}}, \boldsymbol{d}\}$, where $\boldsymbol{b}$ and $\boldsymbol{d}$ are bias vectors, and $\boldsymbol{W}_1$ and $\boldsymbol{W}_1^{\mathrm{T}}$ are the weight matrices.

3.4.2.3 Deep Neural Network Using Stacked Tied-Weight Autoencoders

We can stack N tied-weight autoencoders to pretrain a DNN with N hidden layers. The process is displayed in Fig. 3.42. When given input signal $\boldsymbol{x}^m$, the input layer and the first hidden layer of the DNN are regarded as the encoder network of the first autoencoder, as shown in Fig. 3.42(A). We utilize $\boldsymbol{x}^m$ as input data and output target to train the first autoencoder. Through minimizing its reconstruction error, the trained parameters $\boldsymbol{\theta}_1 = \{\boldsymbol{W}_1, \boldsymbol{b}_1\}$ of the autoencoder are used to initialize the parameters of the first hidden layer of the DNN. The encoded vector $\boldsymbol{h}_1^m$ of the first autoencoder is computed as follows:

$$\boldsymbol{h}_1^m = f_{\theta_1}(\boldsymbol{x}^m) \tag{3.104}$$

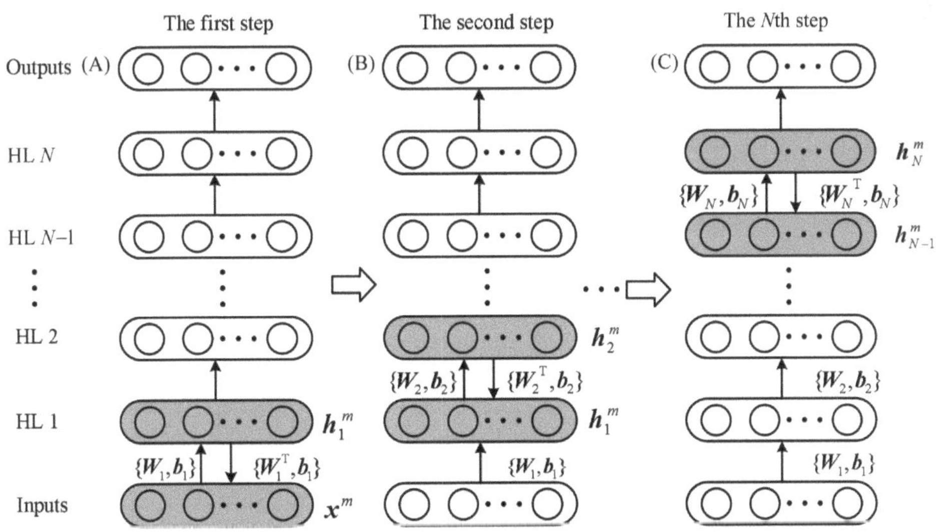

Figure 3.42. Diagram of illustrating the pretraining process (HL denotes hidden layer). (A) Train the first autoencoder of the DNN, (B) train the second autoencoder, and (C) train the Nth autoencoder.

Then, $\boldsymbol{h}_1^m$ is used as the input data and the output target to train the second autoencoder in Fig. 3.42(B). And the parameters $\theta_2 = \{\boldsymbol{W}_2, \boldsymbol{b}_2\}$ of the second hidden layer of the DNN are initialized after the training process. Finally, by continuing the training steps in the sequence, the Nth autoencoder is trained and the input signal $\boldsymbol{x}^m$ can be coded into $\boldsymbol{h}_N^m$ in Fig. 3.42C. In this way, all the hidden layers of the DNN are initialized. The encoded vector $\boldsymbol{h}_N^m$ of the $\boldsymbol{x}^m$ is computed as

$$\boldsymbol{h}_N^m = f_{\theta_N}\left(\boldsymbol{h}_{N-1}^m\right) \tag{3.105}$$

where $\boldsymbol{\theta}_N$ is the parameter set of the Nth autoencoder.

It can be seen that the hidden layers of the DNN can be pretrained after training the N tied-weight autoencoders. This pretraining process is proven to help achieve better generalization in various classification tasks, so it is also expected to perform well in fault diagnosis of rotating machinery.

After the DNN is pretrained, fine-tuning process is employed for the DNN training. The output layer of the DNN contains the output labels of the classification tasks, which is calculated as

$$\boldsymbol{y}^m = f_{\theta_{N+1}}(\boldsymbol{h}_N^m) \tag{3.106}$$

where $\boldsymbol{\theta}_{N+1}$ is the parameter set of the output layer. In order to approximate the output label precisely, BP algorithm is utilized to minimize an error criterion between the output and its real label. Supposing that the label of the $\boldsymbol{x}^m$ is $\boldsymbol{d}^m$, the error criterion is described as

$$\phi_{\mathrm{DNN}}(\Theta) = \frac{1}{M}\sum_m L(\boldsymbol{y}^m, \boldsymbol{d}^m) \tag{3.107}$$

where $\Theta = \{\theta_1, \theta_2, \cdots, \theta_{N+1}\}$. The parameter set Θ can be obtained by the backpropagation algorithm.

3.4.2.4 Fault Diagnosis Method Based on Deep Neural Networks

In this section, an intelligent diagnosis method based on DNNs is presented, which is used to not only automatically extract the features from the collected signals of rotating machinery but also classify its health conditions. Since the frequency spectra are able to provide comprehensive information for the health conditions of rotating machinery, we use them as the input data of the DNNs.

As shown in Fig. 3.43, the method contains four procedures. (1) Acquire the frequency spectra of rotating machinery, operating under different health conditions and operating conditions.

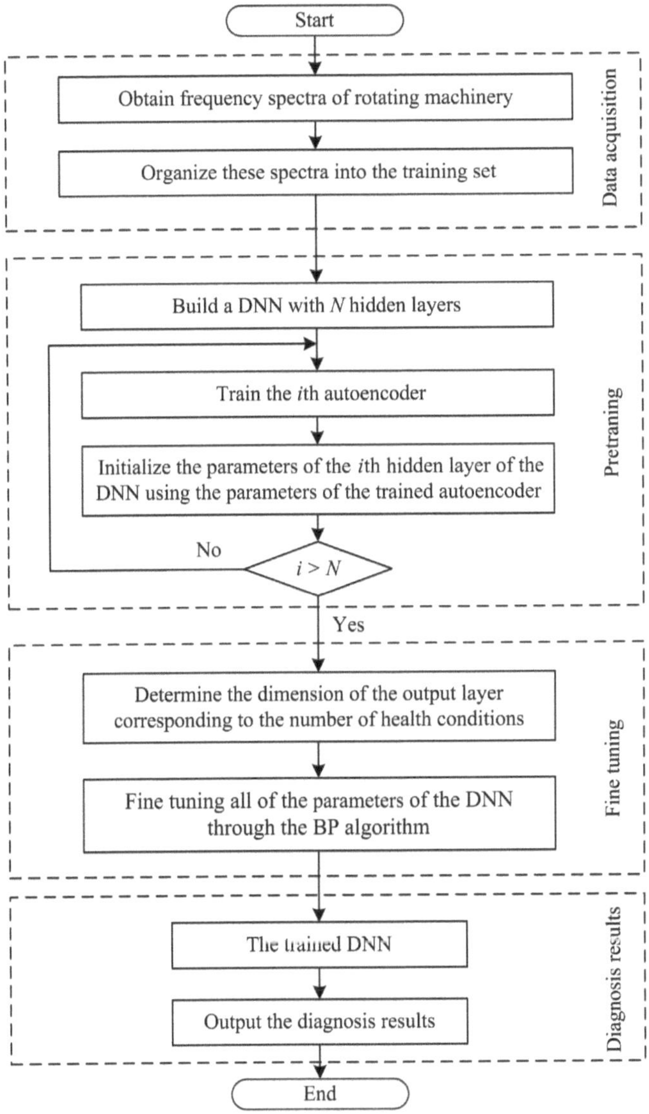

Figure 3.43. Flowchart of the DNN-based method.

The dataset $\{\boldsymbol{x}^i, \boldsymbol{d}^i\}_{i=1}^{M}$ is composed of these collected frequency spectra, where the sample $\boldsymbol{x}^i$ is the ith frequency spectrum, $\boldsymbol{d}^i$ is its health condition label and M denotes the number of the samples. (2) Divide the dataset into the training set and the testing set. And establish a DNN with multiple hidden layers, and utilize the training set (without its labels) to pretrain the DNN with a set of autoencoders. (3) Fine-tune the parameters of the DNN using the BP algorithm to minimize the error between the output labels produced by the training samples and the corresponding health condition labels. (4) Apply the trained DNN to verify its performance on the testing set.

In this method, the pretraining process of the DNN first helps learn multiple nonlinear transformations so as to capture the main information from the frequency spectra. Then the fine-tuning process enforces the pretrained DNN to discover the discriminative features. So the trained DNN is able to extract the essential features from the frequency spectra and establish the nonlinear mapping between the frequency spectra and its labels. Therefore, the method based on the DNN could implement fault characteristics mining and intelligent diagnosis of rotating machinery.

3.4.2.5 Intelligent Fault Diagnosis of Bearings Under Various Loads: A Case Study

1. Data description. The bearing data used come from a motor bearing test rig, which has been detailed in Section 3.3.2.5. Here, the test rig operated: (1) under normal condition (N); (2) with OF; (3) with IF; and (4) with RF. The faults were simulated on the drive-end bearings of the motor, and fault diameters are 0.18, 0.36, and 0.53 mm, respectively. The data were collected under three different loads, that is, 1, 2, and 3 hp. The sampling frequency was 48 kHz. So the bearing dataset involves ten health conditions, and its detailed description is shown in Table 3.14. There are 200 samples for each health condition under one load, where the same health condition under different loads is treated as one class. Each sample is actually a frequency spectrum of the bearings. Therefore, the bearing dataset contains 6000 samples, and it is used to verify the effectiveness of the proposed method.
2. Diagnosis results. In the proposed method, the DNN designed has five layers, where 600, 200, 100 are the node number of the first hidden layer, second hidden layer, and third hidden layer, respectively. We randomly select 50% of samples to pretrain the DNN in an unsupervised manner, and these selected samples are applied to fine-tune the parameters of the pretrained network in a supervised way. Then, the remaining samples are used to test the performance of the proposed method. It should be noticed that twenty trials are conducted for classifying the bearing dataset, and the final classification accuracy is averaged by the accuracies provided by these trials. Fig. 3.44A shows the accuracies of the twenty trials using the proposed

Table 3.14 Description of the Bearing Dataset

Load (hp)	The Number of Samples	Fault Type	Fault Diameter (mm)	Classification Label
	200/200/200/600	N	0	1
	200/200/200/600	RF	0.18	2
	200/200/200/600	RF	0.36	3
	200/200/200/600	RF	0.53	4
1~3	200/200/200/600	IF	0.18	5
	200/200/200/600	IF	0.36	6
	200/200/200/600	IF	0.53	7
	200/200/200/600	OF	0.18	8
	200/200/200/600	OF	0.36	9
	200/200/200/600	OF	0.53	10

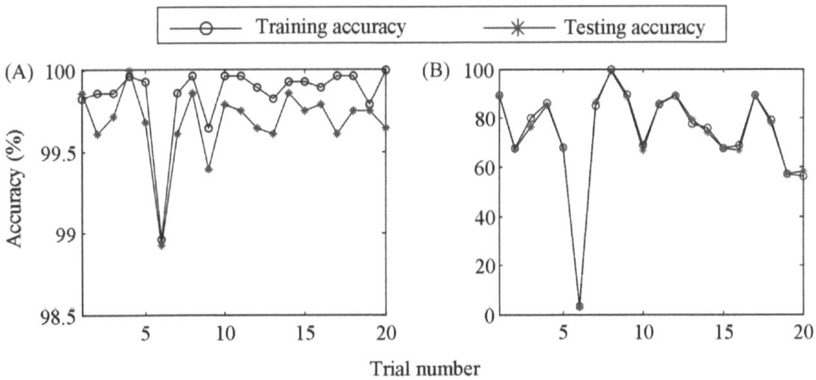

Figure 3.44. Diagnosis results of 20 trials of the bearing dataset using. (A) The proposed method and (B) the BPNN-based method.

method. In the trials, most of the training and testing accuracies are beyond 99% and some of accuracies are even 100%, which indicates that the proposed method can distinguish the ten health conditions of bearings with a high accuracy.

For comparison, we also process the bearing datasets using the method based on backpropagation neural network (BPNN), which is widely applied in intelligent diagnosis of rotating machinery. In this case study, the BPNN has the same architecture as the DNN, and the same training set is used

Table 3.15 Diagnosis Results of the Bearing Dataset

Methods	Training Accuracy (%)	Testing Accuracy (%)
The proposed method	99.85 ± 0.22	99.68 ± 0.22
The BPNN-based method	74.24 ± 20.31	73.74 ± 20.23

The format of the results is: average accuracy ± standard deviation.

to train the DNN and the BPNN, respectively. Fig. 3.44B displays the classification accuracies of the twenty trials using the BPNN-based method. It can be seen that these accuracies change greatly. Most of the accuracies vary around 74%, a few of accuracies are high, and the others are poor. To compare the diagnosis results between the proposed method and the BPNN-based method quantitatively, the average accuracies and the standard deviations in the twenty trials are computed for the bearing dataset, as given in Table 3.15. It shows that the average training and testing accuracies of the proposed method are 99.85 and 99.68%, and the corresponding standard deviations are both 0.22%. The results imply that the proposed method is able to effectively and stably classify not only the fault categories of bearings but also the fault severities. In contrast, the average accuracies of the BPNN-based method are much lower and the corresponding standard deviations are larger, which indicates the superiority of the proposed method to the BPNN-based method. It is worth being mentioned that the bearing dataset contains massive samples involving ten health conditions of bearings under different loads, and its classification accuracies demonstrate that the proposed method can diagnose these health conditions irrespective of the load fluctuation.

In Fig. 3.44B, an interesting phenomenon is observed that the 8th trial of the bearing dataset using the BPNN-based method provides similar accuracies with the same trial using the proposed method. So we plot the comparisons of training errors in Fig. 3.45. It can be seen that both the training errors are nearly 0 in the figure, which demonstrates that neural networks with deep architectures indeed have potential capability in extracting nonlinear characteristics from the frequency spectrum if the networks are well trained. However, the BPNN-based method gets stuck in a premature convergence at the

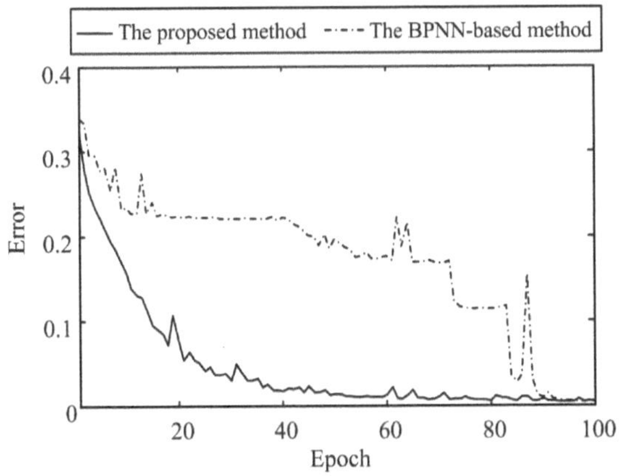

Figure 3.45. Training error comparison between the proposed method and the BPNN-based method in the 8th trial of the bearing dataset.

initial stage of the training process, while the training errors of the proposed method are updated smoothly into good solutions. It indicates that the proposed method is more robust than the BPNN-based method. Moreover, the BPNN-method happens to perform well due to random factors. Taking the 6th trial, for examples, the results using the BPNN-based method are extremely poor, and the training and testing accuracies are around 10%. It seems that in this trial, the trained BPNN has learned nothing useful for intelligent fault diagnosis. But the proposed method always performs well. Therefore, the results demonstrate the superiority of the proposed method to the BPNN-based method.

Furthermore, the traditional methods utilize signal processing techniques to extract features and employ a classifier to distinguish health conditions. However, the proposed method utilizes DNNs for both fault characteristic mining and intelligent fault diagnosis and is more intelligent than the traditional methods. Here, we also compare the proposed method with the traditional methods using the same bearing dataset published. A diagnosis method (Li et al., 2013) was proposed based on 19 time–domain and frequency–domain features and semisupervised distance-preserving self-organizing map. The method was used to classify three bearings with roller defects (0.18, 0.36, and 0.54 mm in diameter, respectively) and obtained the classification accuracy of 95.8%. However, the bearing dataset in this case study includes ten health conditions. The classification of this study is more complex, and the proposed method still gets the accuracy of 99.68%. A method using time–domain features and three-layer neural networks was proposed to classify the health conditions of bearings (de Almeida et al., 2015). The used dataset contained six health conditions under the load of 2 hp, and this method obtained 99.6% classification accuracy, whereas the classification accuracy of the proposed method is 99.68%. It shows that our results are slightly better, but more bearing conditions under different loads are classified in this case study. Jin et al. (2014) utilized six features about the percentage of energy corresponding to wavelet coefficients combined with nine time–domain statistical features to characterize ten health conditions of bearings under the load of 3 hp. Then these features were classified by the trace ratio linear discriminant analysis, and finally 100% diagnosis accuracy was achieved. The accuracy of the proposed method is a little lower than that of the method (Jin et al., 2014), but their method only distinguished the bearing health conditions under one load, which is simpler than the diagnosis issue in this study. Thus, the 0.32% difference between the two accuracies is acceptable. Through the compared results above, it indicates that the proposed method is effective in classifying bearing health conditions under varying loads.

To illustrate the ability of the proposed method in mining fault characteristics adaptively, PCA is used to visualize the features mined from the samples. In this study, the mined features are the inputs of the final layer of the DNN used since this layer is utilized to classify the health conditions in the architecture of the DNN. Thus, 100 features are extracted from each sample by the DNN. PCA is carried out on these 100 features and the first three PCs of the features are obtained, which is shown in Fig. 3.46. It can be seen that most features of the same health condition are gathered in the corresponding cluster and most features of the different health conditions are separated. Although RF with 0.36 mm defect and RF with 0.54 mm defect are not well separated by the first three PCs, the other PCs carry the discriminative information for classifying the samples of the two health conditions. Therefore, the proposed method still achieves the high diagnosis accuracies. It reveals that the proposed method is able to adaptively mine the fault characteristics from the bearing frequency spectra.

In conclusion, the results of the bearing diagnosis case show the effectiveness of the proposed method in mining fault

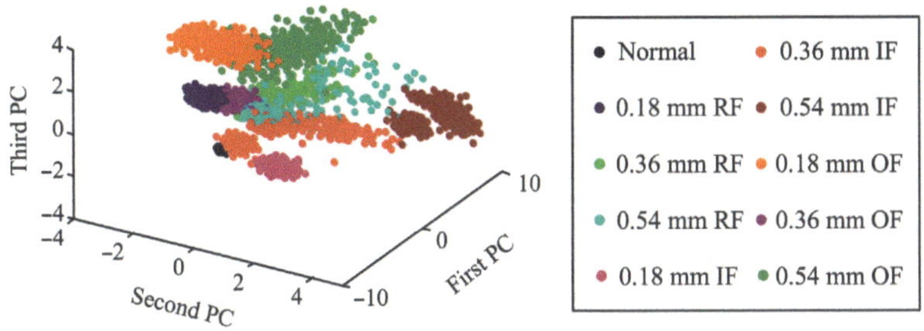

Figure 3.46. Scatter plots of principal components for the features of the bearing dataset.

characteristics and classifying mechanical health conditions. The main reason of its good performance lies in utilizing the autoencoders to pretrain DNNs. By using the code vector encoded by the previous autoencoder as the input to train the next autoencoder, the pretrained DNNs actually maintain the essential characteristics from the input data. Therefore, the pretraining process helps DNNs capture the characteristics of mechanical signals and facilitates the DNNs in refining the discriminative information of these signals in the fine-tuning process, leading to the high diagnosis accuracy.

3.4.2.6 Epilog

Aiming to automatic fault characteristic mining and fault classification, a DNN-based intelligent method is presented for recognizing the faults of rotating machinery. The proposed method is verified using a dataset collected from rolling element bearings. The dataset has massive samples and involves ten bearing health conditions under different loads. The classification results of the dataset show that the proposed method is able to adaptively mine fault characteristics from the frequency spectra irrespective of the load fluctuation and effectively classify the mechanical health conditions. Since mining fault characteristics automatically, the proposed method less relies on the prior knowledge, such as signal processing techniques and diagnostic expertise, than the traditional intelligent diagnosis methods. So by using the proposed method, new diagnosis applications can be carried out easily. There is some guidance to utilize the proposed method. First, hyperbolic tangent function is recommended as the activation function of DNNs since the sigmoid activation function suffers from the vanishing gradient problem badly. Second, half coefficients in the frequency spectra should be used as inputs because the coefficients of the spectra are symmetric. Finally, a deeper network can be tried for other applications, although the DNN with five layers has performed well in the case study.

We have not studied how to determine the architecture of the DNNs. When determining the architecture, we follow a simple strategy: The node number of the next layer is less than that of the previous layer. If so, the encoding process of the autoencoder can be regarded as a process of data compression or feature extraction. Moreover, DNNs are trained by the frequency spectra in this method. This may only work for rotating machinery and reciprocating machinery whose vibration signals measured are periodic. It is interesting to automatically mine the fault characteristics from time–domain signals so as to apply the method to other machines.

3.4.3 Sparse Coding-Based Fault Diagnosis

3.4.3.1 Introduction

DNNs are trained using frequency spectra in the above study. Although its performance is good in intelligent fault diagnosis, the frequency spectra need to be converted by fast Fourier transformation. Traditionally, the original signals collected from the machines are vibration signals, that is measured in the time–domain. So it is interesting to directly learn features from the vibration signals using artificial intelligence techniques.

Among these techniques, sparse coding is a promising one to effectively learn features from the vibration signals of rotating machinery. It emerges in theoretical neuroscience by the research of how the brain could form a useful representation from the signals obtained by receptor cells (Katkovnik et al., 2015). Földiák and Young (1995) found that each concept can be encoded by a few neurons firing strongly and other neurons do not fire. The brain computes compressed representations of these stimuli, leading to the improvement in both inference and energy efficiency. Viewing the neurons as representational elements and defining the cost of a representation by these neurons lead to the concept of sparse coding. By sparse coding, we want to encode the given image or signal as a combination of few elements out of a large possible set, and each element is regarded as a basis that helps discover representative patterns of the image or the signal. Olshausen and Field (1997) showed that basis functions learned by the natural images share the similar properties of the neurons in the primary visual cortex. Smith and Lewicki (2006) used the sparse coding to encode the auditory signal and verified that the learned basis is similar to cochlear filter and auditory nerve fibres. Sparse coding has been widely used in various fields and each of these has the same goal:

obtaining the sparse representations of signals and capturing its underlying structure (Zhang et al., 2013).

Here, we use sparse coding as a building block of a network and present a new intelligent diagnosis method. In the first learning stage, sparse coding, which is trained in an unsupervised manner, is used to learn representative features from the vibration signals of rotating machinery. Then in the second learning stage, softmax regression, which is a two-layer network, is trained to automatically classify the mechanical health conditions. Since the proposed method uses artificial intelligence techniques to learn the features, it does not depend on much prior knowledge and diagnostic expertise and may be more suitable for processing massive signals in the field of condition monitoring and fault diagnosis.

3.4.3.2 Sparse Coding

In signal processing, the earliest method attempts to find efficient linear approximations for the collected signals. Given a discrete signal $\boldsymbol{x} \in \Re^d$, and a set of basis vectors $\{\psi_i\}_{i=1}^d$, an approximation $\tilde{x} = \sum_{i=1}^{k} a_i \psi_i$ is calculated sujecting to $\|x - \tilde{x}\|_2 < \varepsilon$, where ε is a tolerable error (Huggins, 2005). This approximation is effective for disentangling the underlying pattern of the signal if it uses the best basis vectors and can be quickly calculated. Thus, the Fourier transform is widely used, where the bases are composed of sine functions and cosine functions at multiple frequencies. And researchers favor the orthogonal bases since the representation can be calculated by projecting the signal onto each basis vector independently. However, it is recognized that sines and cosines are not ideal bases that are suitable for representing the vibration signals of rotating machinery. Therefore, the issue of the linear approximations is determining the best bases used to represent signals. Recently, sparse coding is a natural approach to obtain the bases and the corresponding representation.

Also, sparse coding aims to decompose $\boldsymbol{x}$ into a linear combination of the basis vectors $\phi_i \in \Re^n$, for $1 \le i \le m$. The difference between sparse coding and the common linear approximation is that ϕ_i is unknown in sparse coding. So a sparse coding algorithm searches for the basis vectors ϕ_i and their coefficients $z_i \in \Re$ synchronously, which is expressed as

$$\boldsymbol{x} \approx \sum_{i=1}^{m} z_i \phi_i \tag{3.108}$$

where most z_i equal zero. The reason of enforcing most z_i to zero is analogous to the brain principle of "few neurons firing for any particular concept" (Földiák and Young, 1995). We define the dictionary $\boldsymbol{\Phi}$ as the matrix in which the ith column is the basis ϕ_i,

$$\boldsymbol{\Phi} = \left[\phi_1\, \phi_2 \cdots \phi_m\right]. \tag{3.109}$$

Thus, the task of sparse coding is rewritten as

$$\boldsymbol{x} \approx \boldsymbol{\Phi z} \quad \text{and} \quad \|z\|_0 \le s \tag{3.110}$$

where $\|\cdot\|_0$ is the ℓ_0 – norm, returning the number of nonzero coefficients for its vector input, and s is a small number. Particularly, if $s < n$, $\boldsymbol{z}$ is a compressed representation of the signal $\boldsymbol{x}$. If we have $s \ll m$ and obtain a good reconstruction of $\boldsymbol{x}$, the decomposition satisfies the brain principle.

The calculation of $\boldsymbol{\Phi z}$ is the reconstruction of the signal $\boldsymbol{x}$. However, in the field of fault diagnosis, we cannot perfectly reconstruct $\boldsymbol{x}$ since it is often overwhelmed by heavy noise. Instead, we allow a small error for the reconstruction.

$$\|\boldsymbol{x} - \boldsymbol{\Phi z}\|_2 \le \delta \quad \text{and} \quad \|z\|_0 \le s \tag{3.111}$$

Since the desired values of δ and s is unknown, we simultaneously minimize both formulas, achieving the balance between sparsity (low s) and reconstruction (low δ). Eq. (3.111) can be rewritten as

$$\arg\min \|\boldsymbol{x} - \boldsymbol{\Phi z}\|_2 + \lambda\|z\|_0 \tag{3.112}$$

where λ is the regularization parameter that controls the balance between sparsity and reconstruction.

Now, it is clear how Eq. (3.112) achieves the idea of "reconstruct $\boldsymbol{x}$ using few dictionary elements". Based on an appropriate value of λ, the term of $\|z\|_0$ enforces most elements of $\boldsymbol{z}$ to zero, and the term of $\|\boldsymbol{x} - \boldsymbol{\Phi z}\|_2$ drives $\boldsymbol{z}$ to obtain the essential information of $\boldsymbol{x}$ so as to reconstruct it as well as possible.

However, solving Eq. (3.112) is known as a NP-hard problem. So the common approach used is to relax the nonconvex ℓ_0 – norm to the convex ℓ_1 – norm. So Eq. (3.112) becomes

$$\arg\min \|\boldsymbol{x} - \boldsymbol{\Phi z}\|_2 + \lambda\|z\|_1 \tag{3.113}$$

In many cases, Donoho and Elad (2003) showed that by solving Eq. (3.113), we can yield the same solution that solves Eq. (3.112). The cost function in Eq. (3.113) is optimized by starting at initialized random values of ϕ_i and z_i, and then alternately optimizing the values. The first term optimizes coefficients z_i over the training samples by fixing the basis vectors ϕ_i, and the second term optimizes coefficients basis vectors ϕ_i by fixing the coefficients z_i.

3.4.3.3 Softmax Regression

Softmax regression is a generalized logistic regression technique for multiclass classification (Behley et al., 2013). It is useful for fault classification of rotating machinery, whose goal is to distinguish its multiple health conditions. Given a set of training samples $\{\boldsymbol{x}_1, \boldsymbol{x}_2, \cdots, \boldsymbol{x}_M\}$ with their label set $\{y_1, y_2, \cdots, y_M\}$, where $\boldsymbol{x}_i \in \Re^{N_\theta \times 1}$ and $y_i \in \{1, 2, \cdots, K\}$, the hypothesis of softmax regression tries to estimate the probability $p(y_i = j \mid \boldsymbol{x}_i)$ of each value of $j = 1, 2, \cdots, K$. So for each training sample $\boldsymbol{x}_i$, softmax regression outputs a K-dimensional vector that gives K estimated probabilities. Actually, this vector gives the K probabilities of an input sample belonging to each label. In detail, the hypothesis $h_\theta(\boldsymbol{x}_i)$ is expressed as

$$h_\theta(\boldsymbol{x}_i) = \begin{bmatrix} p(y_i = 1 \mid \boldsymbol{x}_i; \theta) \\ p(y_i = 2 \mid \boldsymbol{x}_i; \theta) \\ \vdots \\ p(y_i = K \mid \boldsymbol{x}_i; \theta) \end{bmatrix} = \frac{1}{\sum_{j=1}^{K} e^{\theta_j^{\mathrm{T}} x_i}} \begin{bmatrix} e^{\theta_1^{\mathrm{T}} x_i} \\ e^{\theta_2^{\mathrm{T}} x_i} \\ \vdots \\ e^{\theta_K^{\mathrm{T}} x_i} \end{bmatrix} \quad (3.114)$$

where $\theta = [\theta_1, \theta_2, \cdots, \theta_K]^{\mathrm{T}}$ is the set of parameters in softmax regression. It is noted that the term $\sum_{j=1}^{K} e^{\theta_j^{\mathrm{T}} x_i}$ normalizes the distribution of the hypothesis $h_\theta(\boldsymbol{x}_i)$, so that its elements sum to one. Therefore, the softmax regression model can be trained by minimizing the loss function as follows.

$$J(\theta) = -\frac{1}{M}\left[\sum_{i=1}^{M}\sum_{j=1}^{K} 1\{y_i = j\} \log \frac{e^{\theta_j^{\mathrm{T}} x_i}}{\sum_{l=1}^{K} e^{\theta_l^{\mathrm{T}} x_i}}\right] + \frac{\alpha}{2}\sum_{p=0}^{N_\theta}\sum_{j=1}^{K} \theta_{jp}^2 \quad (3.115)$$

where $1\{\cdot\}$ denotes the indicator function returning 1 if the condition is true, and 0 otherwise, and α is the weight decay term. The weight decay term forces the parameters of softmax regression into taking the values close to zero, and improves the generalization ability of the softmax regression. With the weight decay term (for any $\alpha > 0$), the function $J(\boldsymbol{\theta})$ is strictly convex so that the solution of the softmax regression model is unique (Behley et al., 2013).

3.4.3.4 Diagnosis Method Based on Sparse Coding

This section details the sparse coding-based method for intelligent fault diagnosis of rotating machinery, which utilizes sparse coding to learn features from the original vibration signals and

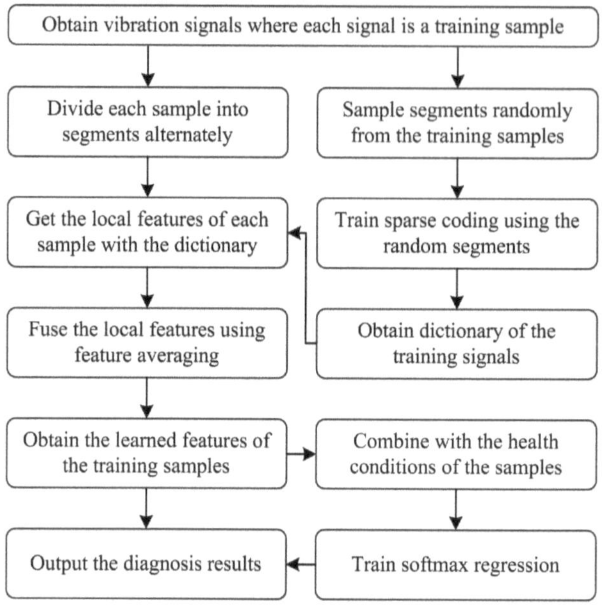

Figure 3.47. Flowchart of the proposed method.

uses softmax regression to classify the mechanical health conditions. As shown in Fig. 3.47, the method can be detailed in the following six steps:

1. For collect signals, we first collect the original vibration signals of rotating machinery under different health conditions. The collected signals compose the training set $\{x^i, y^i\}_{i=1}^{N}$, where each vibration signal is a training sample, $x^i \in \Re^{D\times 1}$ is the ith training sample containing D data points and y^i is its health condition label.
2. Train sparse coding and obtain the dictionary $\boldsymbol{\Phi}$. The basis number of the dictionary is N_b and the dimension of each basis is N_d, so the representation or feature vector $\boldsymbol{z}$ has N_b elements. We randomly sample N_{seg} segments from the training samples and each segment contains N_d data points. By training the sparse coding with the training samples, we can obtain the dictionary $\boldsymbol{\Phi}$.
3. Obtain the local features for each training sample $\boldsymbol{x}^i$. First, we divide each training sample into a set of segments $\left\{x_j^i\right\}_{j=1}^{J}$, where $J = D/N_d$. Then, under the trained dictionary $\boldsymbol{\Phi}$, each segment can be decomposed into the local feature vector $z_j^i \in \Re^{N_b\times 1}$ using sparse coding.

4. Calculate the learned feature for $\boldsymbol{x}^i$. We combine the local features into a feature vector $\boldsymbol{z}^i$ and the learned feature is calculated by

$$z^i = \left(\frac{1}{J}\sum_{j=1}^{J}\left|z_j^i\right|\right)^{\mathrm{T}} \tag{3.116}$$

5. Train softmax regression in a supervised manner. Once the learned feature set $\{z^i\}_{i=1}^N$ is obtained, we combine the set with corresponding label set $\{y^i\}_{i=1}^N$ to train softmax regression by minimizing Eq. (3.115).
6. Employ a set of test samples to verify the trained method. First, each test sample is alternately divided into segments. Then these segments are decomposed by the dictionary $\boldsymbol{\Phi}$ so as to obtain the local features. Next, the learned features of each test sample are obtained using Eq. (3.116). Finally, the health condition label of each test sample is decided by softmax regression based on the learned features.

3.4.3.5 Intelligent Fault Diagnosis of Bearings: A Locomotive Bearing Case Study

The data collected from the locomotive bearing test rig have been described in Section 3.2.3.5. The bearing dataset used involved seven different health conditions under three different loads. The seven health conditions are: normal condition (N), slight rub fault in the outer race (OF1), serious flaking fault in the outer race (OF2), slight rub fault in the inner race (IF), roller rub fault (RF), concurrent faults in the outer race and roller (ORF), and concurrent faults in the inner race and roller (IRF). There are 270 samples for each health condition under one load, and each sample is a vibration signal containing 1200 data points. Therefore, the dataset totally contains 5670 samples.

To distinguish the health conditions of the bearing dataset, the proposed method is trained by 30% of samples and the remaining samples are used to test the performance of the method. We randomly select 20,000 segments from the training samples and the segments are used to train sparse coding. The basis number of the dictionary is $N_\mathrm{b} = 50$ and the dimension of each basis is $N_\mathrm{d} = 100$.

1. Parameter Selection of the Method. There are several parameters needed to be determined before the training process of the presented method, that is, the regularization parameter λ of sparse coding in Eq. (3.112) and the weight decay term α of softmax regression in Eq. (3.115). We investigate the selection of the two parameters, respectively. It is noticed that 20 trials

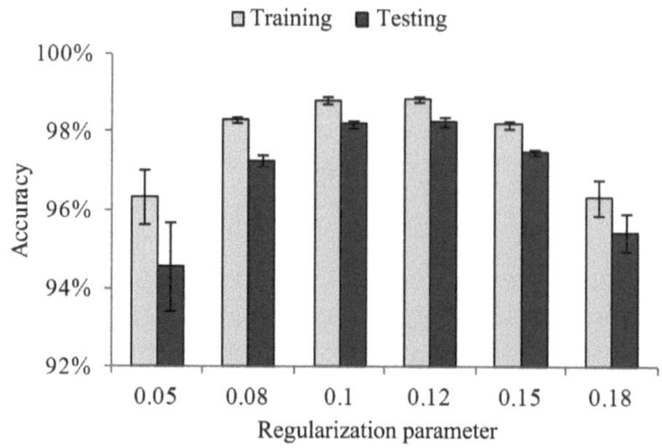

Figure 3.48. Influence of the regularization parameter λ.

are carried out in the following investigations so as to diminish the effect of the randomness.

First we investigate the selection of the regularization parameter λ. Traditionally, this parameter can be determined by the following equation (Mairal et al., 2010)

$$\lambda = 1.2 / N_{\mathrm{d}} \tag{3.117}$$

This empirical equation has been shown to obtain reasonable sparsity in the common applications. Here, we test the usability of the empirical equation in intelligent fault diagnosis. The weight decay term α equals 1E-5. The diagnosis results of using various values of λ are shown in Fig. 3.48. In the figure, the training accuracies and testing accuracies are averaged by 20 trials, and the error bars show the standard deviations. It can be seen that the training accuracy and the testing accuracy are the highest when $\lambda = 0.12$, corresponding to the value calculated by Eq. (3.117). The accuracies decrease with the values of λ far away from the value of 0.12 and the corresponding standard deviations increase. These results illustrate that Eq. (3.117) can be used to determine the values of the regularization parameter λ of the presented method. Although the equation may not be suitable for each case, it can provide a reference value for the classification problem. In this section, we choose 0.12 as the regularization parameter λ.

Then, we also investigate the weight decay term α in the softmax regression model. The diagnosis results are shown in Fig. 3.49. It can be seen that when α is too large, the training and testing accuracies are low and the corresponding standard deviations are high. The reason may be that a large α forces too many parameters of softmax regression to be zero. But a

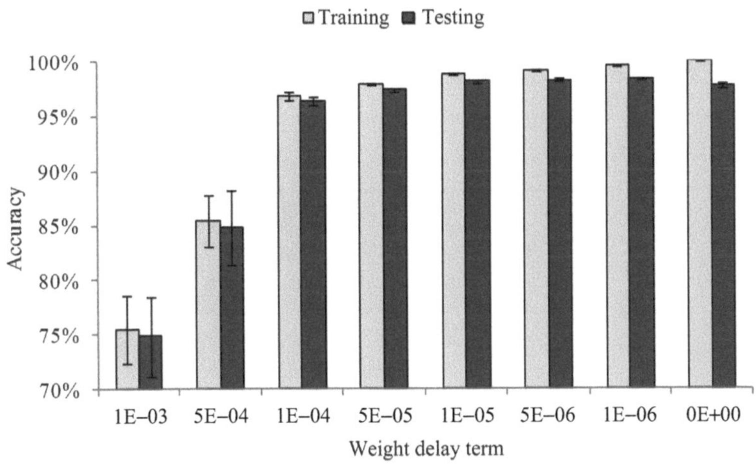

Figure 3.49. Influence of the weight decay term α.

serious overfitting phenomenon is observed when $\alpha = 0$. This indicates that the α is necessary for the softmax regression training. Since the accuracies and their standard deviations are steady after α = 1E-5, this value is chosen as the value of the weight decay term.

2. Influence of the Data Number Used for Training. Here, we investigate the influence of the data number used for training, that is, the percentage of samples used for training the presented method, and the number of the segments used for training sparse coding.

   Fig. 3.50 shows the diagnosis accuracies using various segment numbers when 30% of samples are randomly selected for training. It can be seen that the testing accuracy goes higher

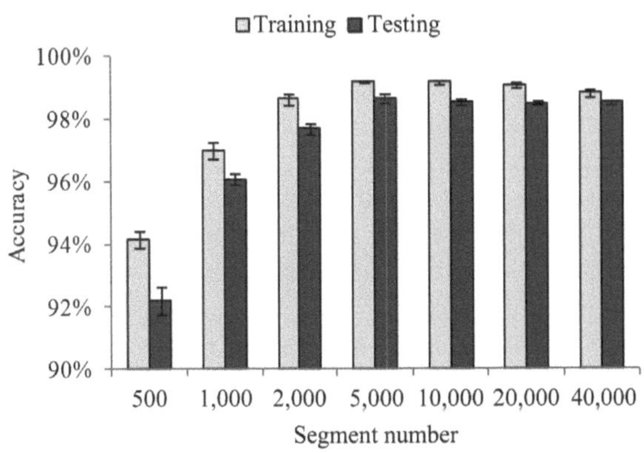

Figure 3.50. Diagnosis results using different segment numbers.

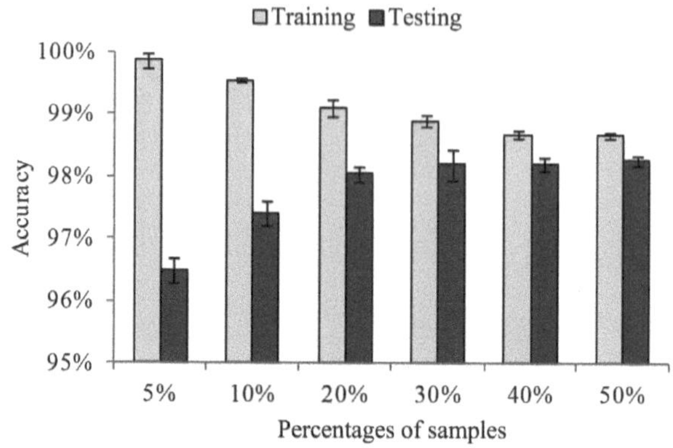

Figure 3.51. Diagnosis results using the proposed method trained by different percentages of samples.

and the standard deviation goes smaller when the segment number goes larger. It should be noticed that the segments are unlabeled data and is obtained much easier than the labeled data, which means that the proposed method is able to take advantage of unsupervised learning and improve its diagnosis accuracy. However, the average time increases exponentially as the segment number increases. To take the trade off between the time spent and diagnosis accuracy, we use 20000 segments for the motor bearing dataset.

We also investigate the diagnosis results of the proposed method trained by various percentages of samples, as shown in Fig. 3.51. It is certain that the testing accuracy will increase and its standard deviation will decrease with more samples used for training. It can be seen from the figure that an overfitting phenomenon occurs when the percentage of samples is low, and the phenomenon is alleviated when the percentage of samples increases. Although the overfitting phenomenon occurs when the percentage is 5%, the testing accuracy is still more than 96%, which indicates that the proposed method is able to perform well even in the case of lacking the labeled data. When the percentage increases to 30%, the testing accuracy is 98.2% with a small standard deviation of 0.26%, which is a satisfactory result. So in the studies, we use 30% of samples for training.

As shown in Fig. 3.47, sparse coding is used to obtain local features from original vibration signals, and the learned features are obtained by averaging these local features. To visualize the learned features, PCA is to analyze the learned features and the first three PCs are obtained, as displayed in

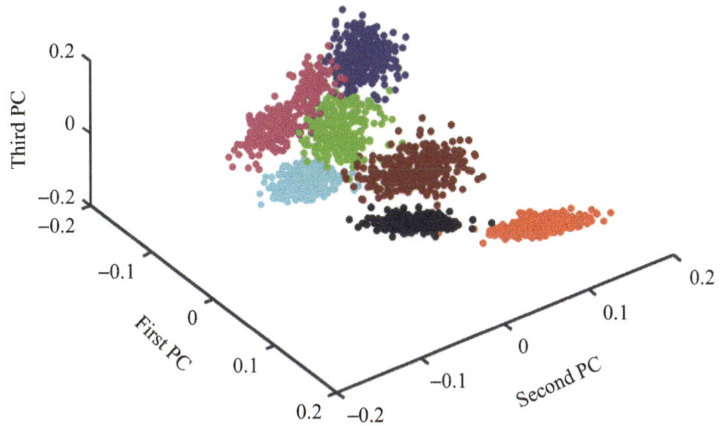

Figure 3.52. Scatter plots of principal components for the learned features of the bearing dataset.

Fig. 3.52. It can be seen that the features of the same health condition are clustered in the space, and the features of different health conditions are separated. It indicates that the proposed method is able to learn the features regardless of the varying loads.

3.4.3.6 Epilog

It would be highly desirable to make intelligent fault diagnosis not depend on human labor when processing mechanical big data, particularly in the step of feature extraction. So an intelligent diagnosis method is presented based on sparse coding. In the method, sparse coding can adaptively extract local features containing discriminative information from original vibration signals in an unsupervised manner, and then the features are computed by averaging the local features. Finally, softmax regression is used to classify the health conditions with the features in a supervised way. Through the case study of the bearing dataset, it is shown that the proposed method can automatically learn discriminative features from original vibration signals and effectively classify their health conditions. Additionally, the proposed method can take full use of unlabeled data and improve its classification accuracy as the data increase. In future work, more aspects of sparse coding in mechanical feature learning will be further studied, like the properties of the bases trained in the sparse coding and the diagnosis of other types of rotating machinery, so that we can bridge the gap between feature extraction based on signal processing techniques and feature learning based on advanced artificial intelligence techniques.

3.5 Conclusions

Fault diagnosis of rotating machinery has been of great practical significance in industry applications, in order to avoid economical losses and enhance machine availability. In traditional fault diagnosis methods, diagnosticians analyze the collected signals with signal processing techniques, such as envelope spectrum analysis and wavelet analysis, to extract characteristics that reflect the fault information. The main characteristics used refer to the fault characteristic frequency and its harmonics. Then the diagnosticians recognize the faults according to the existence of these characteristics in signals. These traditional methods are effective and used widely in fault diagnosis of rotating machinery. However, with the development of modern industry, condition monitoring systems are used to collect real-time data from machines to fully inspect the health conditions of rotating machinery, and massive data are acquired after long time operation of the machines. As the data are generally collected faster than diagnosticians can analyze, there is a demand for methods that can replace the diagnosticians to automatically make decision on the running health of the machines. Therefore, intelligent fault diagnosis becomes a hot topic and it includes the following three steps: signal acquisition, feature extraction, and selection based on the signal processing techniques, and fault recognition based on artificial intelligence techniques. In Sections 3.2 and 3.3, we have shown several intelligent diagnosis methods, and all of them perform efficiently in rapidly and efficiently processing collected signals and providing accurate diagnosis results. Based on the results provided by the methods, it is possible to promptly take appropriate maintenance actions and ensure healthy operation of the machines. Nevertheless, the main difference between intelligent fault diagnosis and traditional fault diagnosis lies in the fault recognition, and they both need the signal processing techniques. It means that intelligent fault diagnosis methods still need the human labor to design the feature extraction algorithms, so the intelligent diagnosis methods seem to be not as intelligent as they are named.

To make fault diagnosis towards real artificial intelligence, it would be much desirable to automatically learn the features from original signals using advanced artificial intelligent techniques, and deep learning may lift us to reach the goals. The main concept of deep learning has been detailed in Section 3.4, and the two presented methods automatically extract the features from frequency spectra and vibration signals, respectively. Since processing the original data intelligently, the presented methods do not depend on human labor and may be more suitable for processing massive signals, even mechanical big data in fault diagnosis of rotating machinery. Moreover, the main advantages of deep learning may be that it can easily extend the established model to a deeper one and is able to deal with more complex classification tasks in fault diagnosis. For instance, deep sparse coding could be tried for the applications in Section 3.4.3, namely we could input the features learned by the previous sparse coding to another sparse coding to obtain new features. Since the higher level features are more abstract and concentrate, the deep sparse coding may lead to better classification results.

In future work, deeper models are necessary to be studied and applied in fault diagnosis of rotating machinery as well as in processing mechanical big data. Furthermore, it may be interesting to interpret why deep learning methods could extract discriminative features, so that we can bridge the gap between feature learning using artificial intelligence techniques and manual feature extraction using signal processing techniques.

References

Aziz, W., Arif, M., 2005. Multiscale permutation entropy of physiological time series. 9th International Multitopic Conference, IEEE INMIC 2005. IEEE, 1–6.

Bandt, C., Pompe, B., 2002. Permutation entropy: a natural complexity measure for time series. Phys. Rev. Lett. 88, 174102.

Behley, J., Steinhage, V., Cremers, A., 2013. Laser-based segment classification using a mixture of bag-of-words. IEEE/RSJ International Conference on Intelligent Robots and Systems (IROS), 2013. IEEE, pp. 4195–4200.

Bengio, Y., 2009. Learning deep architectures for AI. Found. Trends Machine Learn. 2, 1–127.

Bengio, Y., Lamblin, P., Popovici, D., Larochelle, H., 2007. Greedy layer-wise training of deep networks. Adv. Neural Infor. Process. Syst. 19, 153.

Bengio, Y., Courville, A., Vincent, P., 2013. Representation learning: a review and new perspectives. IEEE Trans. Pattern Anal. Mach. Intell. 35, 1798–1828.

Burges, C.J., 1998. A tutorial on support vector machines for pattern recognition. Data Min.Knowl. Discov. 2, 121–167.

Cao, L., 1997. Practical method for determining the minimum embedding dimension of a scalar time series. Phys. D 110, 43–50.

Chen, S., Cowan, C.F., Grant, P.M., 1991. Orthogonal least squares learning algorithm for radial basis function networks. IEEE Trans. Neural Netw. 2, 302–309.

Cortes, C., Vapnik, V., 1995. Support-vector networks. Mach. Learn. 20, 273–297.

Cover, T.M., 1965. Geometrical and statistical properties of systems of linear inequalities with applications in pattern recognition. IEEE Trans. Electron. Comput., 326–334.

de Almeida, L.F., Bizarria, J.W., Bizarria, F.C., Mathias, M.H., 2015. Condition-based monitoring system for rolling element bearing using a generic multi-layer perceptron. J. Vib. Control 21 (16), 3456–3464.

Deng, L., 2014. A tutorial survey of architectures, algorithms, and applications for deep learning. APSIPA Trans. Signal Inform. Process. 3, e2.

Donoho, D.L., Elad, M., 2003. Optimally sparse representation in general (nonorthogonal) dictionaries via l1 minimization. Proc. Natl. Acad. Sci. 100, 2197–2202.

Földiák, P., Young, M.P., 1995. Sparse coding in the primate cortex. The Handbook of Brain Theory and Neural Networks, vol. 1. pp. 1064–1068, MIT press, Cambridge, MA.

Fraser, A.M., Swinney, H.L., 1986. Independent coordinates for strange attractors from mutual information. Phys. Rev. A 33, 1134.

Gunn, S.R., 1998. Support vector machines for classification and regression. ISIS Tech. Report, 15–48.

Hagan, M.T., Demuth, H.B., Beale, M.H., 1996. Neural network design. Pws Pub, Boston.

Haykin, S.S., Haykin, S.S., Haykin, S.S., Haykin, S.S., 2009. Neural networks and learning machines. Pearson Education Upper Saddle River.

Hinton, G.E., Salakhutdinov, R.R., 2006. Reducing the dimensionality of data with neural networks. Science 313, 504–507.

Huang, N.E., Shen, Z., Long, S.R., Wu, M.C., Shih, H.H., Zheng, Q., Yen, N.-C., Tung, C.C., Liu, H.H. The empirical mode decomposition and the Hilbert spectrum for nonlinear and non-stationary time series analysis. Proceedings of the Royal Society of London A: Mathematical, Physical and Engineering Sciences, 1998. The Royal Society, 903–995.

Huggins, P.S., 2005. Sparse coding via geometry. Yale University, New Haven.

Jang, J.-S.R., 1993. ANFIS: adaptive-network-based fuzzy inference system. IEEE Trans. Syst. Man Cybern. 23, 665–685.

Jarrett, K., Kavukcuoglu, K., Ranzato, M.A., LeCun, Y., 2009. What is the best multi-stage architecture for object recognition? IEEE 12th International Conference on Computer Vision, 2009. IEEE, pp. 2146–2153.

Jia, F., Lei, Y., Lin, J., Zhou, X., Lu, N., 2016. Deep neural networks: a promising tool for fault characteristic mining and intelligent diagnosis of rotating machinery with massive data. Mech. Syst. Signal Pr. 72, 303–315.

Jin, X., Zhao, M., Chow, T.W., Pecht, M., 2014. Motor bearing fault diagnosis using trace ratio linear discriminant analysis. IEEE Trans. Ind. Electron. 61, 2441–2451.

Katkovnik, V., Shevkunov, I., Petrov, N., Egiazarian, K., 2015. Wavefront reconstruction in digital off-axis holography via sparse coding of amplitude and absolute phase. Opt. Lett. 40, 2417–2420.

Krizhevsky, A., Sutskever, I., Hinton, G.E., 2012. Imagenet classification with deep convolutional neural networks. Adv. Neural Inf. Process. Syst., 1097–1105.

Le, Q., Ranzato, M., Monga, R., Devin, M., Chen, K., Corrado, G., Dean, J., Ng, A., 2012. Building high-level features using large scale unsupervised learning. In: Proceedings of the International Conference on Machine Learning.

LeCun, Y., Boser, B., Denker, J.S., Henderson, D., Howard, R.E., Hubbard, W., Jackel, L.D., 1989. Backpropagation applied to handwritten zip code recognition. Neural Comput. 1, 541–551.

LeCun, Y., Bengio, Y., Hinton, G., 2015. Deep learning. Nature 521, 436–444.

Lei, Y., He, Z., Zi, Y., Hu, Q., 2007. Fault diagnosis of rotating machinery based on multiple ANFIS combination with GAs. Mech. Syst. Signal Pr. 21, 2280–2294.

Lei, Y., He, Z., Zi, Y., 2008. A new approach to intelligent fault diagnosis of rotating machinery. Expert Syst. Appl. 35, 1593–1600.

Lei, Y., Li, N., Lin, J., He, Z., 2013. Two new features for condition monitoring and fault diagnosis of planetary gearboxes. J. Vib. Control, 1077546313486284.

Lei, Y., Lin, J., Zuo, M.J., He, Z., 2014. Condition monitoring and fault diagnosis of planetary gearboxes: a review. Measurement 48, 292–305.

Lei, Y., Liu, Z., Wu, X., Li, N., Chen, W., Lin, J., 2015. Health condition identification of multi-stage planetary gearboxes using a mRVM-based method. Mech. Syst. Signal Pr. 60, 289–300.

Lei, Y., Jia, F., Lin, J., Xing, S., Ding, S., 2016. An intelligent fault diagnosis method using unsupervised feature learning towards mechanical big data. IEEE Trans. Ind. Electron. 63, 3137–3147.

Li, W., Zhang, S., He, G., 2013. Semisupervised distance-preserving self-organizing map for machine-defect detection and classification. IEEE Trans. Instrum. Meas. 62, 869–879.

Loparo, K., 2003. Bearings vibration data set. Case Western Reserve University.

Loutridis, S., Douka, E., Hadjileontiadis, L., 2005. Forced vibration behaviour and crack detection of cracked beams using instantaneous frequency. NDT E Int. 38, 411–419.

Mairal, J., Bach, F., Ponce, J., Sapiro, G., 2010. Online learning for matrix factorization and sparse coding. J. Mach. Learn. Res. 11, 19–60.

McFadden, P., 1991. A technique for calculating the time domain averages of the vibration of the individual planet gears and the sun gear in an epicyclic gearbox. J. Sound Vibration 144, 163–172.

Micchelli, C.A., 1984. Interpolation of Scattered Data: Distance Matrices and Conditionally Positive Definite Functions. Springer, Berlin, Heidelberg.

Michalski, R.S., Carbonell, J.G., Mitchell, T.M., 2013. Machine Learning: An Artificial Intelligence Approach. Springer, Berlin, Heidelberg.

Olshausen, B.A., Field, D.J., 1997. Sparse coding with an over complete basis set: a strategy employed by V1? Vision Res. 37, 3311–3325.

Parker, D.B., 1985. Learning logic. Technical Report TR-47, Center for Computational Research in Economics and Management Science, Massachusetts Institute of Technology. Cambridge, MA.

Powell, M.J., 1987. Radial basis functions for multivariable interpolation: a review. Algorithms for approximation. Clarendon Press, 143–167.

Psorakis, I., Damoulas, T., Girolami, M., 2010. Multiclass relevance vector machines: sparsity and accuracy. IEEE Trans. Neural Netw. 21, 1588–1598.

Rumelhart, D.E., Hinton, G.E., Williams, R.J., 1986. Learning representations by back-propagating errors. Nature 323, 533–536.

Rumelhart, D.E., Mcclelland, J.L., Group, P.R., 1987. Parallel Distributed Processing. Massachusetts Institute of Technology, Cambridge, MA.

Smith, E.C., Lewicki, M.S., 2006. Efficient auditory coding. Nature 439, 978–982.

Tipping, M.E., 2001. Sparse Bayesian learning and the relevance vector machine. J. Mach. Learn. Res. 1, 211–244.

Vapnik, V.N., 1999. An overview of statistical learning theory. IEEE Trans. Neural Netw. 10, 988–999.

Villa, L.F., Reñones, A., Perán, J.R., de Miguel, L.J., 2012. Statistical fault diagnosis based on vibration analysis for gear test-bench under non-stationary conditions of speed and load. Mech. Syst. Signal Process. 29, 436–446.

Wang, G., Yang, Y., Xie, Q., Zhang, Y., 2014. Force based tool wear monitoring system for milling process based on relevance vector machine. Adv. Eng. Softw. 71, 46–51.

Werbos, P., 1974. Beyond regression: new tools for prediction and analysis in the behavioral sciences, Harvard University, Cambridge, MA.

Widodo, A., Yang, B.-S., 2007. Support vector machine in machine condition monitoring and fault diagnosis. Mech. Syst. Signal Process. 21, 2560–2574.

Widodo, A., Kim, E.Y., Son, J.-D., Yang, B.-S., Tan, A.C., Gu, D.-S., Choi, B.-K., Mathew, J., 2009. Fault diagnosis of low speed bearing based on relevance vector machine and support vector machine. Expert Syst. Appl. 36, 7252–7261.

Worden, K., Staszewski, W.J., Hensman, J.J., 2011. Natural computing for mechanical systems research: a tutorial overview. Mech. Syst. Signal Pr. 25, 4–111.

Wu, Z., Huang, N.E., 2009. Ensemble empirical mode decomposition: a noise-assisted data analysis method. Adv. Adapt. Data Anal. 1, 1–41.

Zanin, M., Zunino, L., Rosso, O.A., Papo, D., 2012. Permutation entropy and its main biomedical and econophysics applications: a review. Entropy 14, 1553–1577.

Zhang, X., 2010. Empirical Risk Minimization. Encyclopedia of Machine Learning. Springer, Berlin, Heidelberg.

Zhang, Q., Benveniste, A., 1992. Wavelet networks. IEEE Trans. Neural Netw. 3, 889–898.

Zhang, S., Yao, H., Sun, X., Lu, X., 2013. Sparse coding based visual tracking: review and experimental comparison. Pattern Recogn. 46, 1772–1788.

4

CLUSTERING ALGORITHM–BASED FAULT DIAGNOSIS

4.1 Introduction to Clustering Algorithm

In the previous chapter, we have introduced the concept of intelligent fault diagnosis using the individual classifiers, and have detailed several methods and their applications. The diagnosis results verify that intelligent fault diagnosis can indeed replace the diagnosticians to efficiently process the massive collected signals and automatically diagnose mechanical faults. But intelligent fault diagnosis cannot solve every diagnosis problem. Traditionally, training an artificial intelligence model needs a lot of labeled data, which greatly determines the diagnosis accuracy and generalization ability of the intelligent fault diagnosis methods. However, the labeled data are limited in some diagnosis cases, and such situation is even worse in the era of big data. So we need a technique that can automatically recognize the pattern inside the data so as to analyze the collected data without their labels (Lei et al., 2008a). Clustering may be a suitable technique for handling the diagnosis tasks without their labels.

Clustering aims to organize a set of samples into the corresponding clusters (or groups) based on similarity. Intuitively, samples within a right cluster are more similar to each other than they are to a sample that belongs to a different cluster (Jain et al., 1999). Clustering has been an indispensible task for exploratory data analysis in the fault diagnosis of rotating machinery. In the clustering problem, little prior information is available for the data collected, and the decision-maker should use assumptions about the data as few as possible (Xu and Wunsch, 2005). Therefore, clustering actually describes the methods for grouping unlabeled data. The main difference between the classification and clustering is whether the labeled data are used in the training process, and we also distinguish them as supervised learning and unsupervised learning. In classification (supervised), we are provided with a collection of labeled samples and the problem refers to labeling a new sample. So the given labeled samples are used to learn the descriptions of classes which in turn are used to label the new samples. In the case of clustering, the problem is to group a given collection of unlabeled patterns into meaningful clusters. In a sense, labels are associated with clusters but these labels are obtained solely from the data.

Clustering can be achieved by the algorithms that significantly differ in their notion of what constitutes a cluster and how the clusters are partitioned. Typical notions of clusters include the groups with small distances among the members of the clusters, dense areas of the data space, intervals, or particular statistical distributions (Berkhin, 2006). There are different techniques that use their assumptions for the components of the clustering process (Jain et al., 1999), involving K-means, K nearest neighbor, fuzzy clustering and so on, and we will detail the algorithms in the following sections.

Clustering used in a fault diagnosis task of rotating machinery is depicted in Fig. 4.1, including a feedback path where the clustering process can affect previous steps (Yiakopoulos et al., 2011; Soualhi et al., 2013; Lei et al., 2008b). It can be seen that the clustering mainly involves the following four steps.

1. *Data acquisition*. Mostly, the vibration signals are collected for inspecting the health conditions of rotating machinery, which is the same with the intelligent fault diagnosis methods.
2. *Data representation*. Data representation refers to the feature extraction, and the number and the scale of the features available to a clustering algorithm. Recall that feature extraction is the application of signal processing techniques to transform the vibration signals into new salient features, which can characterize the health conditions of the machinery. Feature selection is the process of searching the most effective features from the original features. Additionally, scaling the selected features (we can call it normalization) is an optional step to improve clustering performance.
3. *Definition of pattern proximity*. The pattern proximity is traditionally measured by a distance function based on pairs of patterns. A simple distance measure like Euclidean distance is often used to reflect dissimilarity between two patterns,

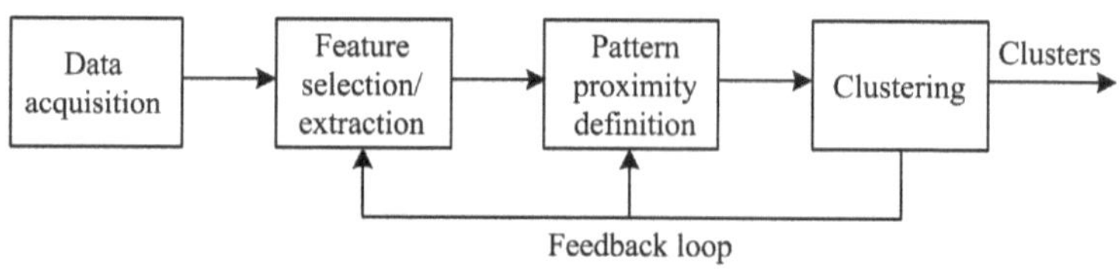

Figure 4.1. Stages of clustering in intelligent fault diagnosis.

whereas other similarity measures can be used to characterize the conceptual similarity between patterns (Liao, 2005). These distance measures are used in different techniques.

4. *Clustering*. The clustering step is able to be performed in a majority of ways. The output clustering can be hard, a partition of the samples into certain groups, or fuzzy, each sample has various degrees of membership belonging to each cluster. Then, the mechanical health conditions are determined based on the clustering results.

4.2 Weighted *K* Nearest Neighbor-Based Fault Diagnosis

4.2.1 Introduction

Inspecting the localized faults, such as the damage levels of gears, is essential in intelligent fault diagnosis of rotating machinery (Wang, 2008). The *K* nearest neighbor (*K*NN) algorithm (Beyer et al., 1999; Duda and Hart, 1973; Östh et al., 2015) may be a tool to deal with the problem of automatic fault level recognition. *K*NN has been proved to be a simple and stable technique, and performs well in plenty of real-world applications (Ghosh et al., 2005). In the *K*NN algorithm, a new sample is labeled by a majority vote of its neighbors. In other words, the sample is assigned to the most common label among its *K* nearest neighbors. However, unsatisfied performance of the *K*NN algorithm occurs when the label distribution is imbalanced and samples of different classes overlap in some regions in the feature space. One way to overcome this problem is to weight the *K*NN algorithm, taking into account the distance from the test sample to each of its *K* nearest neighbors. So a simple but reliable two-stage feature selection and weighting technique (TFSWT) based on compensation distance evaluation technique (CDET) is developed to improve the performance of the *K*NN algorithm, named the weighted *K* nearest neighbor (W*K*NN) algorithm.

In this algorithm, not only time- and frequency–domain statistical features but also feature parameters especially designed for gear damage detection are extracted to characterize the gear faults. Then, a few features obviously characterizing the gear conditions are selected from all the features using CDET. Meanwhile, the weights of the selected features are also obtained. The selected features are used as inputs of the W*K*NN algorithm and the obtained weights are used to measure the importance of the *K* nearest neighbors in label voting.

In view of the previous analysis, a new method using the W*K*NN algorithm to recognize the damage levels of gears is presented. Compared with the existing methods (Casimir et al., 2006; Lee et al., 2007; Lu et al., 2008), this method harnesses the merits that the computation of feature weights is simpler and the weights are easier to understand. To validate the performance of the proposed method, the experiments of different gear crack levels on a test rig were carried out. Vibration signals were measured from the gears under various load and speed conditions. The diagnosis results demonstrate that the method is able to recognize the gear crack levels effectively and automatically.

4.2.2 *K* Nearest Neighbor Algorithm

The *K* nearest neighbor (*K*NN) algorithm is a nonparametric method widely used for mechanical fault diagnosis (Rauber et al., 2015; Wang, 2008). In this algorithm, nearest neighbor distances serve as the foundation of clustering procedures. It assigns each unlabeled pattern to the cluster of its nearest labeled neighbors. A sample is assigned by the vote results of its neighbors, with the sample being assigned to the most common class among its *K* nearest neighbors (Everitt et al., 2011). If *K* equals 1, the sample is simply assigned to the label of that single nearest neighbor. While *K* is larger than 1, the sample is assigned to the label that most of the *K* nearest neighbors share. The process continues until all patterns are labeled. Fig. 4.2 shows a clustering example of the *K*NN algorithm. A test sample marked by solid circle should be assigned to either the first label marked by square or the second label marked by triangle. If *K* equals 3, the *K* nearest neighbors of the test sample are in the solid line circle,

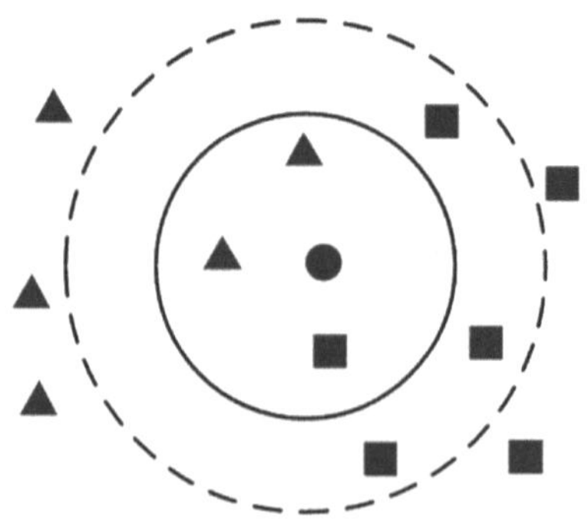

Figure 4.2. Illustration of the *K*NN algorithm.

and it will be assigned to the second label because there are two triangles and only one square inside the inner circle. If K equals 6 (dashed line circle), it is assigned to the first class (four squares and two triangles inside the dashed circle). Thus, *KNN* is a type of instance-based learning, or lazy learning, where the function is only approximated locally and all computation is deferred until labeled.

Given a set of samples $\{\boldsymbol{x}_m, y_m\}_{m=1}^{M}$, where $\boldsymbol{x}_m \subset \Re^D$ and y_m is the corresponding label. The *KNN* algorithm is based on the idea that the nearest samples to a target sample $\boldsymbol{x}'$ can deliver useful label information (Bishop, 2006). It assigns the class label of the majority of the K-nearest samples in data space. For this sake, we are able to define a similarity measure in data space. In $\Re^D$, it is reasonable to employ the Minkowski metric.

$$\|\boldsymbol{x}' - \boldsymbol{x}_m\|_p = \left(\sum_{d=1}^{D} \left| (x_d)' - (x_d)_m \right|^p \right)^{1/p} \tag{4.1}$$

where the Euclidean distance ($p = 2$) is commonly used. In the case of binary classification, the label set is employed, and *KNN* is defined as

$$f_{KNN}(\boldsymbol{x}') = \begin{cases} 1 & \text{if } \sum_{i \in N_K(\boldsymbol{x}')} y_i \geq 0 \\ -1 & \text{if } \sum_{i \in N_K(\boldsymbol{x}')} y_i < 0 \end{cases} \tag{4.2}$$

where K is neighborhood size and $N_K(\boldsymbol{x}')$ is the set of the K-nearest samples.

The *KNN* algorithm can also be extended to the multiclass classification version. For an unknown sample $\boldsymbol{x}'$, multiclass *KNN* predicts the label based on the K-nearest patterns in data space

$$f_{KNN}(\boldsymbol{x}') = \arg\max_{y} \sum_{i \in N_K(\boldsymbol{x}')} \Gamma(y_m = y) \tag{4.3}$$

where indicator function $\Gamma(\cdot)$ returns one, if its argument is true and zero otherwise.

The question arises that what size of neighborhood will achieve the best performance. The choice of K depends upon the data; generally, larger values of K reduce the effect of noise on the final results (Fan and Zuo, 2006), but make boundaries among classes less distinct. The value of K can be determined by various heuristic techniques. The special case where the class is predicted to be the class of the closest training sample, that is, when K equals 1, is called the nearest neighbor algorithm.

4.2.3 Weighted K Nearest Neighbor Algorithm

Although the KNN algorithm is easy to implement, it suffers from poor classification performance when samples of different classes overlap in some regions in the feature space. Due to the utilization of the Euclidean distance, the KNN algorithm is sensitive to scaling of the feature values. Therefore, a feature weighting technique is useful for overcoming the shortcomings of KNN. Here, we present an improved KNN algorithm, in which TFSWT via CDET is used. CDET is to select sensitive features and compute feature weights, and the feature weights are consequently used in the weighted K nearest neighbor algorithm.

Suppose that the feature set $\{p_{m,c,j}, m=1,2,\ldots,M_c; c=1,2,\ldots,C; j=1,2,\ldots,J\}$ is extracted from the data samples of gears. $p_{m,c,j}$ is the jth feature value of the mth sample under the cth health condition, M_c is the sample number of the cth health condition, C is the number of the health conditions, and J is the feature number of each sample.

1. Feature selection

   Although the J features may represent the health conditions of the gears from different aspects, they have different potentials in recognizing these conditions. So the sensitive features are selected from all the features based on CDET, and this process can be described as follows:

   a. Compute the distances between the samples that are under the same health condition

$$D_{c,j} = \sqrt{\frac{1}{M_c \times (M_c - 1)} \sum_{l,m=1}^{M_c} \left(p_{m,c,j} - p_{l,c,j}\right)^2}, l,m = 1,2,\ldots,M_c, l \neq m; \quad (4.4)$$

   and average these distances under C health conditions

$$D_j^{(w)} = \frac{1}{C} \sum_{c=1}^{C} D_{c,j}. \quad (4.5)$$

   b. Define the variance factor of the same health condition, which is calculated by

$$V_j^{(w)} = \frac{\max(D_{c,j})}{\min(D_{c,j})}. \quad (4.6)$$

c. Compute the average feature value for all the samples

$$a_{c,j} = \frac{1}{M_c} \sum_{m=1}^{M_c} p_{m,c,j}; \tag{4.7}$$

and then get the average distance for the samples under different health conditions

$$D_j^{(b)} = \sqrt{\frac{1}{C \times (C-1)} \sum_{c,e=1}^{C} \left(a_{e,j} - a_{c,j}\right)^2}, \quad c,e = 1,2,\ldots,C, c \neq e. \tag{4.8}$$

d. Define the variance factor of the different health conditions, which is computed as follows:

$$V_j^{(b)} = \frac{\max\left(\left|a_{e,j} - a_{c,j}\right|\right)}{\min\left(\left|a_{e,j} - a_{c,j}\right|\right)}, \quad c,e = 1,2,\ldots,C, c \neq e. \tag{4.9}$$

e. Define and calculate the variance factor as

$$\lambda_j = \left(\frac{V_j^{(w)}}{\max\left(V_j^{(w)}\right)} + \frac{V_j^{(b)}}{\max\left(V_j^{(b)}\right)}\right)^{-1}. \tag{4.10}$$

f. Compute the ratio between $D_j^{(b)}$ and $D_j^{(w)}$ by assigning the variance factor

$$E_j = \lambda_j \frac{D_j^{(b)}}{D_j^{(w)}}; \tag{4.11}$$

and then normalize $\boldsymbol{E}_j$ by using its maximum value so as to get the evaluation criteria

$$\overline{E_j} = \frac{E_j}{\max(E_j)}. \tag{4.12}$$

It can be easily seen that the larger $\overline{E_j}$ $(j = 1,2,\ldots,J)$ indicates that the features are better in distinguishing the gear health conditions. Therefore, the sensitive features can be selected from all features if their evaluation criteria $\overline{E_j} \geq \phi$, where ϕ is the preset threshold for the feature selection.

2. Feature weighting
The sensitive features can be selected in the stage of feature selection, but the selected features show different sensitivities

under different gear health conditions. Thus, feature weighting is needed to obtain more reliable diagnosis results. Generally, the feature weighting is a technique where each feature is multiplied by a number between 0 and 1, which reflects the ability of each feature in distinguishing different health conditions. In the Euclidean space, feature weighting is regarded to extend the axes according to the features related to the faults and shrink the axes corresponding to the insensitive features. First, the evaluation criteria $\overline{E_j}$ of all features are computed by using CDET. Second, a few sensitive features, whose evaluation criterion is larger than the threshold ϕ, are selected from the features in the original feature set. Third, CDET is employed again to evaluate these selected features. Finally, the evaluation criteria are computed for the selected sensitive features according to Eq. (4.12). The computed evaluation criteria are used as the feature weights $\boldsymbol{wf}$ ($\boldsymbol{wf} = (wf_1, \cdots, wf_d, \cdots, wf_D)$, where D is the number of the selected features, and these weights are assigned to the corresponding selected features so as to reflect their sensitivities in recognizing the health conditions of the gears.

3. Weighted K nearest neighbor
   Let TE_d denote the testing sample and TR_{md} denote the mth training sample. Based on Eq. (4.1), the weighted Euclidean distance between TE_d and TR_{md} can be expressed as:

$$D_m^{(wf)} = \left[\sum_{d=1}^{D} wf_d \left(TE_d - TR_{md} \right)^2 \right]^{1/2} \tag{4.13}$$

   where wf_d is the weight of the dth feature.
   By substituting $\boldsymbol{wf}$ calculated in the process of feature weighting into Eq. (4.13), the KNN algorithm based on the weighted Euclidean distance is presented, and it is named as the weighted K nearest neighbor (WKNN) algorithm.

4.2.4 Diagnosis Method Based on W*K*NN

Based on the WKNN algorithm, a diagnosis method for gear faults is proposed and displayed in Fig. 4.3. This method has the following four steps.

1. Vibration signals are collected from the gears under various health conditions. These signals are then preprocessed by Fourier transform and Hilbert transform in order to get the frequency spectra and the difference and residual signals.
2. The 25 features are extracted from the original vibration signals or the preprocessed signals. These features can be divided into

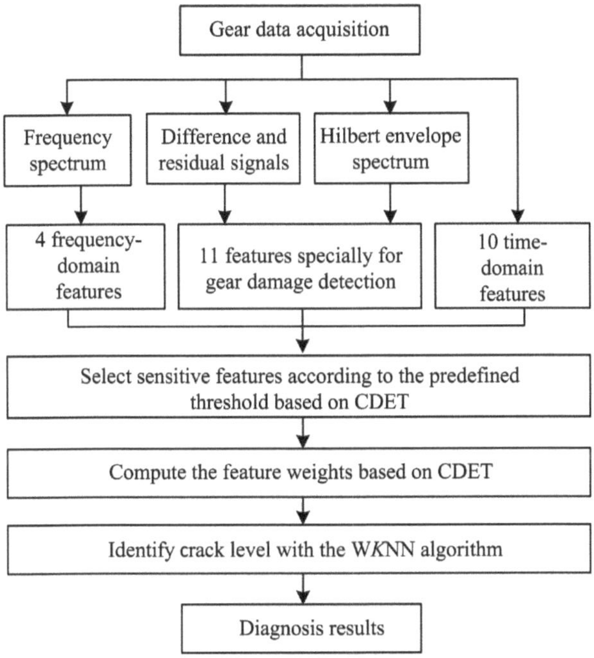

Figure 4.3. Flowchart of the WKNN-based diagnosis method.

three parts: time–domain features, features specially designed for gear damage detection, and frequency–domain features.

The first part contains 10 time–domain features that are widely used in the literature (Samanta, 2004; Melin et al., 2007; Zhou et al., 2016). They are mean, peak, standard deviation, root mean square, kurtosis, skewness, crest factor, shape factor, clearance factor, and impulse factor. The definitions of the 10 features have been described in Section 2.3.

The second part involves 11 statistical features that were originally presented in NASA technical reports and specially developed for gear damage detection (Decker, 2003; Decker and Lewicki, 2003). The definitions of these features are expressed as follows (Samuel and Pines, 2005).

*FM*0 is an indicator of major faults robustly used in the gear mesh

$$FM0 = \frac{PP_x}{\sum_{h=0}^{H} P_h} \tag{4.14}$$

where PP_x is the maximum peak-to-peak value of data point x_i $(i = 1,2,\cdots,n)$, P_h is the amplitude of the nth harmonic for the meshing frequency, and H is the total number of the harmonics. Traditionally, a complete data series is also named as a run ensemble. It can be further divided into M time records, and each of them contains n data points. For each time record, $FM0$ can be computed.

$FM4$ is developed using the difference signal. The kurtosis of the difference signal is used as $FM4$.

$$FM4=\frac{n\sum_{i=1}^{n}\left(d_i-\overline{d}\right)^4}{\left(\sum_{i=1}^{n}\left(d_i-\overline{d}\right)^2\right)^2} \tag{4.15}$$

where n is the number of data points in the time record x_i, d_i is the ith measurement of the difference signal of the time record, and $\overline{d}$ is the average value of the difference signal.

$FM4^*$ is designed for monitoring the development of a gear damage rather than the detection of the incipient faults. Among the M time records of the run ensemble, the first M' time records are regarded as the data collected under the normal condition. The remaining records include the information of the fault growth. $FM4^*$ can be expressed as

$$FM4^*=\frac{\frac{1}{n}\sum_{i=1}^{n}\left(d_i-\overline{d}\right)^4}{\left(\frac{1}{M'}\sum_{j=1}^{M'}\frac{1}{n}\sum_{k=1}^{n}\left(d_{jk}-\overline{d}_j\right)^2\right)^2} \tag{4.16}$$

where d_{jk} is the kth measurement in the ith time record collected under the normal condition.

$M6A$ is expressed as

$$M6A=\frac{\frac{1}{n}\sum_{i=1}^{n}\left(d_i-\overline{d}\right)^6}{\left(\frac{1}{n}\sum_{i=1}^{n}\left(d_i-\overline{d}\right)^2\right)^3} \tag{4.17}$$

Based on the $M6A$ parameter, $M6A^*$ is developed by differentiating the normal records from the fault records. It is defined as

$$M6A^*=\frac{\frac{1}{n}\sum_{i=1}^{n}\left(d_i-\overline{d}\right)^6}{\left(\frac{1}{M'}\sum_{j=1}^{M'}\left(\frac{1}{n}\sum_{k=1}^{n}\left(d'_{jk}-\overline{d'_j}\right)^2\right)\right)^3} \tag{4.18}$$

*NA*4 is based on the residual signal of a time record, and it is shown as

$$NA4 = \frac{\frac{1}{n}\sum_{i=1}^{n}\left(r_i - \bar{r}\right)^4}{\left(\frac{1}{M}\sum_{j=1}^{M}\left(\frac{1}{n}\sum_{k=1}^{n}\left(r_{jk} - \bar{r}_j\right)^2\right)\right)^2} \tag{4.19}$$

where r_i is the ith measurement of the residual signal of a time record, $\bar{r}$ is the average of the residual signal of the time record.

*NA*4* is viewed as an enhanced parameter of *NA*4. Its definition is similar to *FM*4* and is given by

$$NA4^* = \frac{\frac{1}{n}\sum_{i=1}^{n}\left(r_i - \bar{r}\right)^4}{\left(\frac{1}{M'}\sum_{j=1}^{M'}\left(\frac{1}{n}\sum_{k=1}^{n}\left(r_{jk} - \bar{r}_j\right)^2\right)\right)^2} \tag{4.20}$$

*NB*4 uses the envelope of the obtained band-pass filtered signal as its indicator of the localized damage in the gear tooth. *NB*4 is shown as

$$NB4 = \frac{\frac{1}{n}\sum_{i=1}^{n}\left(s_i - \bar{s}\right)^4}{\left(\frac{1}{M}\sum_{j=1}^{M}\left(\frac{1}{n}\sum_{k=1}^{n}\left(s_{jk} - \bar{s}_j\right)^2\right)\right)^2} \tag{4.21}$$

where $s(t)$ is the envelope calculated by $s(t) = |b(t) + i\{H[b(t)]\}|$, $b(t)$ is the signal band-pass filtered on the meshing frequency, and $H[b(t)]$ is the Hilbert transform of $b(t)$.

*NB*4* improves the performance of *NB*4 in tracking damage of the gear tooth, and it can be defined as

$$NB4^* = \frac{\frac{1}{n}\sum_{i=1}^{n}\left(s_i - \bar{s}\right)^4}{\left(\frac{1}{M'}\sum_{j=1}^{M'}\left(\frac{1}{n}\sum_{k=1}^{n}\left(s_{jk} - \bar{s}_j\right)^2\right)\right)^2} \tag{4.22}$$

Energy ratio (ER) is given as

$$ER = \sqrt{\frac{\frac{1}{n}\sum_{i=1}^{n}(d_i)^2}{\frac{1}{n}\sum_{i=1}^{n}(d'_i)^2}} \tag{4.23}$$

where d_i' denotes the ith measurement of the regular meshing components including the meshing frequencies and their harmonics, the shaft frequencies and their harmonics, and all first-order sidebands.

Energy operator (EOP) is proposed by first computing the value $x_i^2 - x_{i-1} * x_{i+1}$ for each data point x_i $(i=1,2,\cdots,n)$ of the signal collected. Then, the energy operator is calculated as the kurtosis of the resulting signal, and its definition is shown as follows

$$EOP = \frac{n\sum_{i=1}^{n}\left(re_i - \overline{re}\right)^4}{\left(\sum_{i=1}^{n}\left(re_i - \overline{re}\right)^2\right)^2} \quad (4.24)$$

where re_i is the ith measurement of the resulting signal, equaling $x_i^2 - x_{i-1} * x_{i+1}$, and $\overline{re}$ is the average value of the resulting signal.

The third part contains four statistical features using the frequency spectra of vibration signals. The four features may reflect some information that is not contained in the time–domain features. They are Mean Frequency (MF), Frequency Centre (FC), Root Mean Square Frequency (RMSF), and Standard Deviation Frequency (STDF), and all of them have been described in Section 2.4.

The 25 features are previously extracted from each of the data samples, where a data sample is one of the vibration signals collected from the gears. The 25 features of each vibration signal are a sample.

The presented W*K*NN algorithm is applied to recognize the health conditions of the gears. Concretely, the feature selection stage is applied to select the sensitive features, and the feature weighting stage is used to calculate the weights of the selected sensitive features. The sensitive features corresponding with their weights are input into the W*K*NN algorithm, and the diagnosis results are finally obtained.

4.2.5 Intelligent Diagnosis of Gear Crack Level: A Case Study

1. Experiment setup and data acquisition

   The experimental system covers a gearbox, a 3-hp motor to drive the gearbox, and a magnetic brake used for loading. A speed controller is applied to control the rotating speed of the motor, allowing the tested gear to operate at different speeds. The magnetic brake is connected to the output shaft of the gearbox, and its torque is adjusted by a brake controller.

Additionally, the gearbox is driven by a timing belt, and totally three shafts are in the gearbox. Gear 1 on shaft 1 which has 48 teeth meshes with gear 2 on shaft 2 having 16 teeth. Gear 3 has 24 teeth and meshes with gear 4 on the output shaft (shaft 3), and it has 40 teeth. In the experiments, gear 3 is used as the test gear.

Since a crack on gears is a common fault mode investigated in gear fault diagnosis, crack faults are simulated on the test gear. As shown in Fig. 4.4, we let α denote the crack angle, a denote half of the chordal tooth thickness and b denotes the face width. In the experiments, the different crack levels are introduced into the gear and the crack thickness is 0.4 mm, which are summarized in Table 4.1. As a consequence, three gears ($F0 \sim F2$) under different health conditions, that is, one normal gear and two faulty gears, are tested in the experiments.

The vibration signals were measured from the gearbox with each of the three gears using two accelerometers, which were produced by PCB Electronics. The accelerometers were mounted on the gearbox housing in the vertical and horizontal directions. A DSP Siglab analyzer and a laptop with the

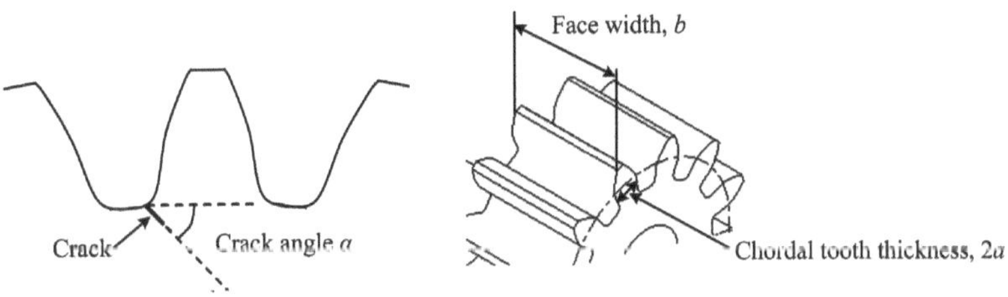

Figure 4.4. Crack angle, chordal tooth thickness, and face width of a gear.

Table 4.1 Geometry of the Crack Faults

Crack Fault Mode	Geometry of Fault			
	Crack Angle (Degree)	Depth (mm)	Width (mm)	Thickness (mm)
F0	—	0	0	0
F1	45	(1/4)a	(1/4)b	0.4
F2	45	(1/2)a	(1/2)b	0.4

data acquisition software were used to acquire the vibration signals. The motor provided 4 different driving speeds, that is, 1200, 1400, 1600, and 1800 rpm, and the loads of the magnetic brake involve loads 0, 1, and 2. Load 0 indicates that no load was applied to the gearbox. Load 1 indicates that half of the maximum load was applied to the gearbox, and load 2 indicates that maximum load was applied to the gearbox. The maximum load was computed theoretically so that the maximum stress on the gears was less than the allowable stress. The meshing frequencies and the maximum loads are summarized in Table 4.2. The sampling frequency is 5120 Hz and each vibration signal has 8192 points. Since the vibration signals in the vertical direction were more sensitive to the crack levels than those in the horizontal direction, they were utilized and further analyzed. There are two data samples collected under an identical operating condition. Therefore, 24 data samples are obtained for each crack level of the gears and there are totally 72 data samples for $F0 \sim F2$.

2. Diagnosis experiments and results
The vibration signals collected from the gear experimental system are employed to validate the effectiveness of the WKNN-based method in diagnosing gear faults. The evaluation criteria of the 25 features computed in the feature selection stage are displayed in Fig. 4.5A. Generally, it may be better to set the median of the range of the evaluation criteria as the threshold if there is no prior knowledge about how to set this threshold. Thus, we set the threshold to 0.5 for the feature selection. Correspondingly, the weights of the selected features computed in the feature weighting stage are shown in Fig. 4.5B. From

Table 4.2 Motor Speeds, Loads, and Characteristic Frequencies of the Gearbox.

Motor Speed (rpm)	Maximum Torque (N.m)	f1 (Hz)	f12 (Hz)	f2 (Hz)	f34 (Hz)	f3 (Hz)
1200.00	34.51	4.76	228.57	14.29	342.86	8.57
1400.00	29.58	5.56	266.67	16.67	400.00	10.00
1600.00	25.89	6.35	304.76	19.05	457.14	11.43
1800.00	23.01	7.14	342.86	21.43	514.29	12.86

Notes: f1, f2, and f3 denote the rotating frequencies of shaft 1, shaft 2, and shaft 3, respectively. f12 and f34 denote the meshing frequencies between gears 1 and 2, and gears 3 and 4, respectively.

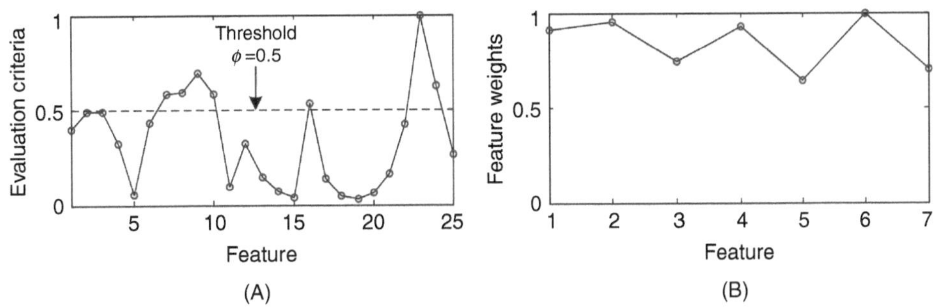

Figure 4.5. (A) Evaluation criteria of the 25 features and (B) feature weights of the selected 7 features.

Fig. 4.5A-B, it can be seen that seven features are selected as the sensitive features. They are FC, clearance factor, impulse factor, crest factor, shape factor, RMSF and *NA*4, and their weights are 1.000, 0.950, 0.930, 0.915 0.743, 0.700, and 0.640, respectively. Three different organizations of the collected data are used to carry out three diagnosis experiments, which will reflect the diagnosis ability of the W*K*NN-based method from different aspects. For comparison, the *K*NN without feature selection (Method 1), the *K*NN with random feature selection (Method 2) and the *K*NN with the feature selection without weighting (Method 3) are also applied to analyze the same data. For convenience, the W*K*NN-based method is regarded as Method 4 in the following experiments. For the four methods, the neighborhood parameter *K* changes from 1 to the number of the training samples so that we can select the optimal value of *K*.

3. Experiment 1

Since 24 samples are acquired for each of the three gear conditions *F*0, *F*1, and *F*2, the whole dataset totally contains 72 samples for the three conditions. So we use 36 data samples for training and the rest 36 samples are utilized to test. The information about the training and testing data in the first experiment is displayed in Table 4.3.

The W*K*NN-based method is used to recognize the three crack levels of the gears. The seven features selected in the feature selection stage are used as the input of W*K*NN. The evaluation criteria of the seven features in the feature weighting stage are adopted as the weights of the W*K*NN algorithm. The recognition accuracies of the W*K*NN-based method using various values of the neighborhood parameter *K* are displayed in Fig. 4.6. Table 4.4 displays the corresponding results of the identification accuracies.

Table 4.3 Data Description of the Three Experiments

Experiment	No. of Samples (train/test)	Fault Modes (train/test)	Motor Speeds (train/test)	Loads (train/test)	Label
1	12/12	F0/F0	1200~1800/1200~1800	0~2/0~2	1
	12/12	F1/F1	1200~1800/1200~1800	0~2/0~2	2
	12/12	F2/F2	1200~1800/1200~1800	0~2/0~2	3
2	12/12	F0/F0	1200, 1600/1400, 1800	0~2/0~2	1
	12/12	F1/F1	1200, 1600/1400, 1800	0~2/0~2	2
	12/12	F2/F2	1200, 1600/1400, 1800	0~2/0~2	3
3	8/16	F0/F0	1200~1800/1200~1800	0/1, 2	1
	8/16	F1/F1	1200~1800/1200~1800	0/1, 2	2
	8/16	F2/F2	1200~1800/1200~1800	0/1, 2	3

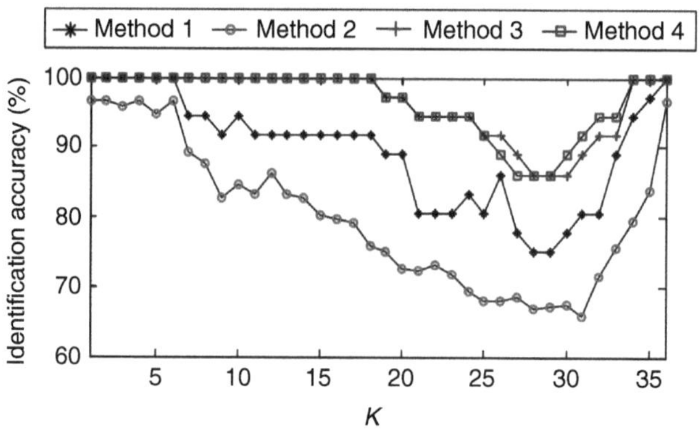

Figure 4.6. Accuracy comparison of the four methods for experiment 1.

In Method 1, all the 25 features are fed into the original *K*NN algorithm. In Method 2, seven features that randomly selected from the 25 features are used as the inputs and the original *K*NN algorithm is employed to diagnose the three gear cracks. To alleviate the effect of the randomness, the method is repeated fifteen times and the average accuracies are given. For Method 3, the selected seven sensitive features are fed to the *K*NN algorithm to distinguish the three gear cracks. The diagnosis results of the three methods are also displayed in Fig. 4.6 and Table 4.4, respectively.

Table 4.4 Diagnosis Results of the Four Methods of the Three Experiments

Experiment	Method 1 Accuracy (%)				Method 2 Accuracy (%)				Method 3 Accuracy (%)				Method 4 Accuracy (%)			
	Max	*Mean*	*Min*	**CPU** *Times (s)*	*Max*	*Mean*	*Min*	**CPU** *Times (s)*	*Max*	*Mean*	*Min*	**CPU** *Times (s)*	*Max*	*Mean*	*Min*	**CPU** *Times (s)*
1	100	89.6	75.0	0.23	96.7	80.2	65.9	0.17	100	96.5	86.1	0.25	100	96.7	86.1	0.27
2	97.2	87.4	77.8	0.23	83.3	77.3	66.3	0.17	100	99.2	97.2	0.25	100	99.6	97.2	0.27
3	97.9	87. 7	77.1	0.14	90.3	78.1	65.3	0.11	97.9	92.1	87.5	0.19	100	92.6	87.5	0.20

From the figure and table, it can be seen that using the feature selection, Methods 3 and 4 get higher recognition accuracies. The accuracies of these two methods both range from 86.1% to 100%. In Method 2, the seven features are randomly selected and the worst results (65.9~96.7%) are provided. Method 1 uses not only the seven sensitive features but also the other features unrelated to the faults in recognizing the crack levels, and obtains the moderate classification result ranging from 75.0% to 100%. The CPU times spent on the four methods are 0.23, 0.17, 0.25, and 0.27 s, respectively, and they are also displayed in Table 4.4.

4. Experiment 2

The training and testing data of this experiment are reorganized as shown in Table 4.3. The 36 samples collected at the motor speeds 1200 and 1600 rpm are used for training, while the 36 samples acquired at the motor speeds 1400 and 1800 rpm are employed for testing. This experiment aims to further investigate the generalization ability of the W*K*NN-based method when the method is tested by the samples collected at different motor speeds.

Applying the W*K*NN-based method to recognize the three levels of the gear cracks, the identification accuracies using different neighborhood parameter *K* are shown in Fig. 4.7 and Table 4.4, respectively. The diagnosis results of Methods 1~3 in this experiment are also displayed in Fig. 4.7 and Table 4.4 for comparison.

It can be seen that the similar accuracies are obtained by Methods 3 and 4, both of their correct rates ranging from 97.2% to 100%. The recognition accuracies of Method 1 range from 77.8% to 97.2%, which are lower than those of Methods 3 and 4.

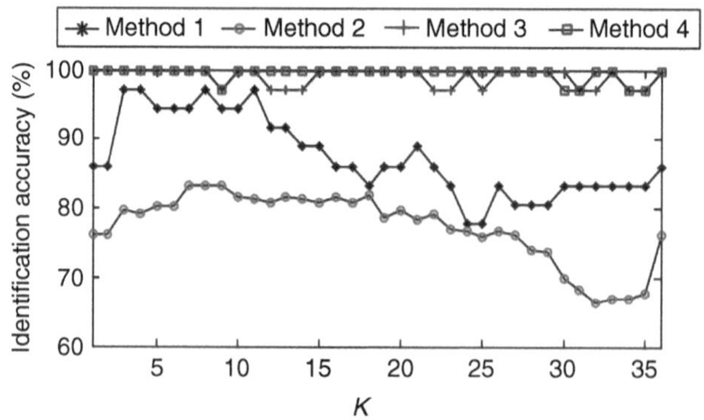

Figure 4.7. Accuracy comparison of the four methods for experiment 2.

Among the four methods, Method 2 gets the lowest accuracies (66.3~83.3%). The results of this experiment are similar to those of experiment 1, and they indicate that the generalization ability of Methods 3 and 4 is better than that of the others, Method 1 achieves the moderate accuracies and that Method 2 gets the worst ones. Since the computational complexity of experiment 2 is the same as that of experiment 1, the four methods spend the same CPU times in the two experiments. The CPU times in experiment 2 are 0.23, 0.17, 0.25, and 0.27 s for the four methods, respectively.

5. Experiment 3

In this experiment, we investigate the generalization ability of the WKNN-based method under the various loads. As described in Table 4.3, the 24 training samples collected under load 0, namely without torque, are used for training and the other 48 testing samples under loads 1 and 2 are applied for testing. Fig. 4.8 and Table 4.4 give the corresponding diagnosis results of the four methods. It is observed from these results that the identification accuracies of Methods 1, 2 and 3 range from 77.1% to 97.9%, 65.3% to 90.3%, and 87.5% to 97.9%, respectively. The highest accuracies are achieved by Method 4, namely the WKNN-based method, and they range from 87.5%~100%. As shown in Table 4.4, the CPU times in this experiment are 0.14, 0.11, 0.19, and 0.20 s for the four methods, respectively. They are shorter than CPU times in experiments 1 and 2 because of fewer training samples in experiment 3 and it has the lower computational complexity.

6. Discussion

First, the lowest diagnosis accuracies are obtained by Method 2, medium accuracies are achieved by Method 1 and the best

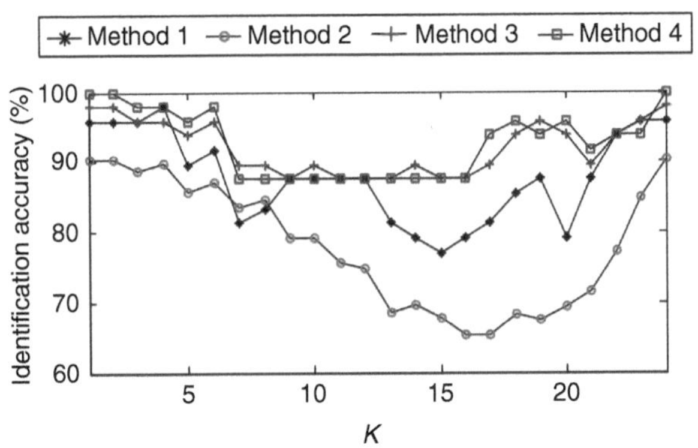

Figure 4.8. Accuracy comparison of the four methods for experiment 3.

diagnosis results are yielded by Methods 3 and 4 in these three experiments. The main difference is whether the sensitive feature selection is used or not. Since different features show varying potentials in recognizing crack levels of the gears, some features are sensitive to the crack levels but others are not. In Methods 3 and 4, the sensitive features are selected from the original feature set so as to distinguish the crack levels, and thus the accuracies are the highest. Although the feature number of Method 2 is the same as that of Methods 3 and 4, too much gear crack fault-unrelated information is contained in the features of Method 2 and the values of these features are overlapped in a high degree. So when the unrelated features are fed to W*K*NN to recognize the crack levels, they will disturb the diagnosis process and the diagnosis accuracies will decrease clearly, resulting in that Method 2 produces the worst diagnosis results. For Method 1, not only the sensitive features but also the fault-unrelated features are used to distinguish the crack levels, leading to the moderate diagnosis accuracies.

Second, by comparing the diagnosis accuracies of Methods 3 and 4, it is indicated that Method 4 is universally superior to Method 3. Though the ranges of the identification accuracies of the two methods are the same in experiments 1 and 2, the maximum accuracy (100%) of Method 4 in experiment 3 is higher than that of Method 3 (97.9%). It indicates that Method 4 performs better than Method 3 when the loads of gears change. Furthermore, in every experiment, Method 4 obtains higher mean accuracies (96.7% in experiment 1, 99.6% in experiment 2, and 92.6% in experiment 3) than Method 3 does (96.5% in experiment 1, 99.2% in experiment 2, and 92.1% in experiment 3). These comparisons confirm that the proposed W*K*NN method is able to not only select the sensitive features but also consider their different importance in identifying the crack levels of the gears. Thus, it produces the best diagnosis results among the four methods.

Third, based on the above result analysis, it is able to conclude that the proposed diagnosis method achieves higher diagnosis accuracies and better generalization capability than the other three methods. This may be attributed to three points. (1) The commonly used features in the time–domain, the features specially developed for gear damage detection and features in the frequency–domain are utilized to characterize the crack levels of the gears, providing much characteristic information from different aspects. (2) The sensitive features are selected, and other fault-unrelated features are discarded. (3) The selected sensitive features still show various potentials in identifying the crack levels of the gears. So the WKNN algorithm is used to conduct feature weighting so as to reflect the different importance of these sensitive features.

Next, from Table 4.4, it can be seen that Method 2 spends the shortest CPU times (0.17 s in experiment 1, 0.17 s in experiment 2, and 0.11 s in experiment 3) because of using seven features. In Method 1, all features are used, so it takes more CPU times in each experiment (0.23 s in experiment 1, 0.23 s in experiment 2, and 0.14 s in experiment 3). Since the feature selection process is carried out in Method 3 and both feature selection and weighting process are used in Method 4, the corresponding CPU times grow in Methods 3 and 4. Actually, in the two methods, sensitive features need to be selected and feature weights need to be calculated for only once. It is unnecessary to select the sensitive features and calculate the feature weights again in applications. Since the number of the sensitive features is also seven, the CPU times spent by the proposed method are the same as those spent by Method 2. Therefore, the proposed method is able to not only obtain the highest recognition accuracies but also decrease the computational burden.

Last, FC, clearance factor, impulse factor, crest factor, shape factor, RMSF, and *NA*4 are selected in the experiments. So the time–domain features are the most frequently used in the identification process of the gear cracks. Only *NA*4, among the 11 features specifically designed for gear damage detection, is used in the experiments. Theoretically, the 11 features are supposed to be effective in the gear crack diagnosis because they are related to the characteristic frequencies of the gear faults. But the 11 features are used only when the collected signals are stationary because of requiring the identification or removal of certain frequency elements from the signals. In our experiments, although the vibration signals were collected at the specified constant speed of the motor, the speed fluctuation exists. So it is impossible to extract features from such nonstationary signals using the 11 features specifically designed for gear damage detection.

4.2.6 Epilog

A diagnosis method is proposed for identifying the crack levels of gears based on a WKNN algorithm. In this method, features in time–domain, specifically developed for the damage detection of gears and in frequency–domain are extracted to characterize the gear cracks. To remove the fault-unrelated information, the feature selection and feature weighting via CDET are utilized.

With the selected sensitive features, the W*K*NN classification algorithm, which overcomes the shortcomings of *K*NN, is conducted to identify the crack levels of gears. To verify the effectiveness of the proposed method, gear crack experiments were carried out and the vibration signals were collected under different loads and at different motor speeds. The identification results show that the proposed method achieves higher identification accuracies and thus it can be used as a promising method for gear fault diagnosis.

4.3 Weighted Fuzzy *c*-Means–Based Fault Diagnosis

4.3.1 Introduction

In the previous section, the weighted *K*NN algorithm is used in fault diagnosis of gear faults, covering different gear crack levels under different load and speed conditions. The diagnosis results are satisfactory and thus demonstrate the effectiveness of the algorithm. However, a shortcoming of the *K*NN algorithm is that it is sensitive to the local structure of the data. The weighted *K*NN-based method may inherit this shortcoming, limiting its applications in fault diagnosis of rotating machinery. Among clustering algorithms, the *K*-means algorithm is not affected by the shortcoming, and it is also widely used owing to its simplicity. The *K*-means algorithm is a kind of hard clustering method, namely each sample belonging to a cluster or not. In the cases of mechanical fault diagnosis, however, it is expected to assign each sample to each cluster in a certain degree, for example, a likelihood of belonging to the cluster. As an extended version of the *K*-means algorithm, the fuzzy *c*-means (FCM) algorithm (Bezdek, 2013) is able to solve this problem.

Thanks to the superiority of the FCM algorithm in dealing with uncertainty and independence of labels, it has been widely studied and applied to fault diagnosis of rotating machinery (Zheng et al., 2015; Gao and Xie, 2000). According to the principle that similar samples gather in the same cluster and dissimilar samples are separated in different clusters, the algorithm utilizes a fuzzy mathematic strategy to partition the acquired samples into several homogenous groups or clusters. The main characteristic of the FCM algorithm is that each sample is subject to one cluster with a certain grade of membership (Xu and Wunsch, 2005). However, the FCM algorithm supposes that all the features make uniform contributions to the clustering, and does not take account of their different importance degrees. Therefore, the FCM algorithm may achieve inexact clustering results in intelligent fault diagnosis of machinery.

Several improved clustering algorithms have been reported in the literature, and they considered the importance of the features in the process of clustering. For instance, Chan et al. (2004) introduced an attribute-weighting clustering algorithm based on the development of a new approach to calculate the weight for each attribute. Wang et al. (2004) presented a weighted FCM algorithm using the weighted Euclidean distance. Frigui and Nasraoui (2004) developed a new procedure to simultaneously perform clustering and feature weighting, and applied it to segment images.

Here, aiming to effectively diagnose the mechanical faults, we present a new weighted fuzzy *c*-means (WFCM) algorithm based on a two-stage feature selection and weighting procedure. In this algorithm, compensation distance evaluation technique (CDET) is applied in sensitive feature selection and feature weighting. The evaluation criteria of the sensitive features are used as the weights, which are assigned to the corresponding features to reflect their different sensitivities. Consequently, the WFCM algorithm not only overcomes the weakness of the W*K*NN algorithm, but also presents merits that the computation of feature weights is simpler and the weights are easier to understand. To show the effectiveness of the WFCM algorithm, it is employed in fault diagnosis of locomotive roller bearings. The vibration signals are acquired from the bearings under different health conditions, including different fault types, fault severities, and compound faults. The diagnosis results imply the effectiveness and practicability of the new WFCM algorithm.

4.3.2 *K*-Means Algorithm

Suppose that we have a dataset $\{\boldsymbol{x}_m\}_{m=1}^{M}$ where $\boldsymbol{x}_m$ is a *D*-dimensional sample. The goal of the *K*-means algorithm is to partition the dataset into *K* clusters. Intuitively, a cluster may comprise a group of data points whose interpoint distances are smaller than the distances to the points of other clusters. We can formalize this intuition by using a set of *D*-dimensional vectors $\boldsymbol{\mu}_k$, where $k=1,\cdots,K$. $\boldsymbol{\mu}_k$ is called a prototype corresponding with the *k*th cluster and represents the center of the *k*th cluster. So we can find a set of vectors $\{\boldsymbol{\mu}_k\}$ and assign the samples to clusters so that the sum of the distances between each sample and its center $\boldsymbol{\mu}_k$ is minimized.

For each sample $\boldsymbol{x}_m$, we use a binary indicator variable $r_{mk} \in \{0,1\}$ to describe which cluster the sample $\boldsymbol{x}_m$ is assigned to. In other words, if sample $\boldsymbol{x}_m$ is assigned to the cluster *k*, then r_{mk}

equals 1 and r_{mk} equals 0 for other clusters. Based on the analysis, we can then define an objective function as

$$J = \sum_{m=1}^{M} \sum_{k=1}^{K} r_{mk} \left\| \boldsymbol{x}_m - \boldsymbol{\mu}_k \right\|^2 \tag{4.25}$$

which represents the sum of the squares of the distances between each sample and its assigned center $\boldsymbol{\mu}_k$. Therefore, the problem is transformed into finding values for the $\{r_{mk}\}$ and the $\{\boldsymbol{\mu}_k\}$ so that J can be minimized. The procedure we used involves two successive steps that refer to the successive optimizations with respect to the r_{mk} and the $\boldsymbol{\mu}_k$ in each iteration.

First, consider the optimization for r_{mk}. Since J is a linear function of r_{mk}, it is easy to obtain a closed form solution. Formally, this can be expressed as

$$r_{mk} = \begin{cases} 1 & \text{if } k = \arg\min_j \left\| \boldsymbol{x}_m - \boldsymbol{\mu}_j \right\|^2 \\ 0 & \text{otherwise.} \end{cases} \tag{4.26}$$

Then consider the optimization for $\boldsymbol{\mu}_k$ and meanwhile the r_{mk} is fixed. The objective function J can be minimized by letting its derivative with respect to $\boldsymbol{\mu}_k$ equal to zero

$$2\sum_{m=1}^{M} r_{mk} \left(\boldsymbol{x}_m - \boldsymbol{\mu}_k \right) = 0. \tag{4.27}$$

We can easily give a solution for $\boldsymbol{\mu}_k$,

$$\boldsymbol{\mu}_k = \frac{\sum_m r_{mk} \boldsymbol{x}_m}{\sum_m r_{mk}}. \tag{4.28}$$

In the K-means algorithm, the two steps of reassigning samples to clusters and recomputing the cluster means are repeated until the algorithm meets the stop criteria. Since each step reduces the value of J, convergence of the K-means algorithm is assured.

The K-means algorithm assigns each sample uniquely to one of the clusters. However, there may be some samples that lie roughly midway among cluster centers so the hard assignment may not be the most appropriate one. By adopting a probabilistic approach, the clustering algorithm can obtain "soft" assignments of data points to clusters, reflecting the level of uncertainty over the most appropriate assignment. This clustering algorithm is known as fuzzy clustering and it brings numerous benefits.

4.3.3 FCM Algorithm

Fuzzy c-means (FCM) is an extension of the K-means algorithm. Their major difference is that a sample is assigned to the clusters by a membership value r_{mk}, where $r_{mk} \in [0,1]$, instead of being assigned to one cluster. The goal of the FCM algorithm is to minimize the objective function expressed as:

$$J_{\text{FCM}} = \sum_{m=1}^{M} \sum_{k=1}^{K} (r_{mk})^{\lambda} \|\boldsymbol{x}_m - \boldsymbol{\mu}_k\|^2 \tag{4.29}$$

where λ is the weighting exponent of the partition matrix. All the memberships r_{mk} can be rewritten into the matrix form $\boldsymbol{R}$, which is called the fuzzy partition matrix.

To define the criteria for the membership values, we first define the sets:

$$L_m = \left\{ k \middle| 1 \le k \le K;\ \|\boldsymbol{x}_m - \mu_k\| = 0 \right\} \tag{4.30}$$

$$\tilde{L}_m = \{1, \cdots, K\} / L_m \tag{4.31}$$

Then we can compute the values of r_{mk} by

$$L_m = \varnothing \Rightarrow r_{mk} = \frac{1}{\sum_{l=1}^{K} \left(\dfrac{\boldsymbol{x}_m - \boldsymbol{\mu}_k}{\boldsymbol{x}_m - \boldsymbol{\mu}_l} \right)^{2/(\lambda-1)}} \tag{4.32}$$

The means of the cluster prototypes are calculated as

$$\boldsymbol{\mu}_k = \frac{\sum_{m=1}^{M} (r_{mk})^{\lambda} \boldsymbol{x}_m}{\sum_{m=1}^{M} (r_{mk})^{\lambda}} \tag{4.33}$$

In the FCM algorithm, the value for λ and the fuzzy partition membership matrix $\boldsymbol{R}$ should be initialized at the beginning. Typically, this is done by assigning values of the matrix randomly. The next step is to compute the means of each cluster and update the partition matrix alternately.

4.3.4 Weighted FCM Algorithm

Since the FCM algorithm supposes that all features make the uniform contributions to clustering and does not take their different importance degrees into account, the FCM algorithm frequently obtains inexact clustering results. Hence, a weighted fuzzy

c-means (WFCM) algorithm based on two-stage feature selection and weighting is presented. In this algorithm, features are first selected and weighted by CDET, and then the weighted features and their corresponding weights are used to enhance the performance of the FCM algorithm. So the enhanced algorithm contains the following three steps.

1. Feature selection
   Feature selection is performed by CDET, which has been described in Section 4.2.3. We calculate the evaluation criteria $\overline{E_j}$ for the jth feature and the bigger $\overline{E_j}$ indicates that the corresponding feature is more discriminative in separating the health conditions. Therefore, the sensitive features can be selected from the feature set when their distance evaluation criteria are bigger than the preset threshold ϕ.
2. Feature weighting
   The sensitive features selected from the original feature set may have different sensitivities in distinguishing the health conditions. Thus, feature weighting is necessary to improve the clustering algorithm. First, a few sensitive features are obtained from all the features. Then, CDET is employed again to evaluate the selected features. Next, the new distance evaluation criteria of the selected features are calculated. Finally, these new indexes act as feature weights $\boldsymbol{wf}$, where $\boldsymbol{wf} = \left(wf_1, \cdots, wf_d, \cdots, wf_D\right)$ and D is the number of the selected features. In fault diagnosis, we assign the feature weights to the corresponding features to indicate their sensitivities.
3. The WFCM algorithm
   Prior to clustering, the selected features are normalized. The value v_d of the dth feature is normalized by the following formula:

$$v'_d = \frac{v_d - \min(v_d)}{\max(v_d) - \min(v_d)}, d = 1,2,\ldots,D, \tag{4.34}$$

where D is the feature number.

The objective function J of the WFCM algorithm can be expressed as:

$$J_{\mathrm{WFCM}} = \sum_{k=1}^{K}\sum_{m=1}^{M}\left(r_{mk}\right)^{\lambda} d_{mk}^{(wf)}, \tag{4.35}$$

subject to

$$\sum_{k=1}^{K} r_{mk} = 1, \quad 0 \le r_{mk} \le 1, \tag{4.36}$$

where $d_{mk}^{(wf)}$ is the weighted Euclidean distance between the mth sample and the kth cluster center and it is defined as:

$$d_{mk}^{(wf)} = d^{(wf)}\left(\boldsymbol{x}_m, \boldsymbol{u}_k\right) = \left[\sum_{d=1}^{D} wf_d \left(x_{md} - u_{kd}\right)^2\right]^{1/2}, \tag{4.37}$$

Then we derive the computational formulae for the $\boldsymbol{\mu}_k$ and the r_{mk} as:

$$\boldsymbol{u}_k = \sum_{m=1}^{M} \left(u_{mk}\right)^{\lambda} \boldsymbol{x}_m / \sum_{m=1}^{M} \left(r_{mk}\right)^{\lambda} \tag{4.38}$$

and

$$r_{mk} = 1 / \sum_{a=1}^{K} \left(d_{mk}^{(wf)} / d_{ma}^{(wf)}\right)^{2/(\lambda-1)}, \tag{4.39}$$

where $d_{ma}^{(wf)}$ is the weighted Euclidean distance between the mth sample and the ath cluster center for $a = 1, \cdots, K$. The fuzzy clustering exponent λ is set to the middle value 2 in this algorithm (Pal and Bezdek, 1995).

4.3.5 Diagnosis Method Based on WFCM

Here, we detail the diagnosis method based on the proposed WFCM algorithm, whose flowchart is shown in Fig. 4.9.

First, similar to the diagnosis method using the WKNN algorithm, we extract features from the vibration signals involving various health conditions of rotating machinery. Both the time–domain and the frequency–domain features, listed in Table 4.5, are employed to recognize the mechanical health conditions. Most of these features have been defined in Sections 2.3 and 2.4. When a fault occurs in rotating machinery, its time–domain signal will change. The amplitude and distribution of the signal are different from those of the original time–domain signal under normal condition. At the same time, its distribution and frequency spectrum may also change so that new frequency components may generate and the corresponding convergence may be different. Among the time–domain features, x_m, x_{rms} and x_p may indicate the vibration amplitude and energy. x_{std}, x_{ske}, x_{kur}, CF, CLF, SF and IF may reflect the distribution of the vibration signal. Among the frequency–domain features, x_{mf} may show the energy of the frequency spectrum. x_{fc} and x_{rmsf} may describe the position change of the frequency elements. x_{stdf} may represent the convergence of the spectrum power. Thus, these 14 feature values are first obtained.

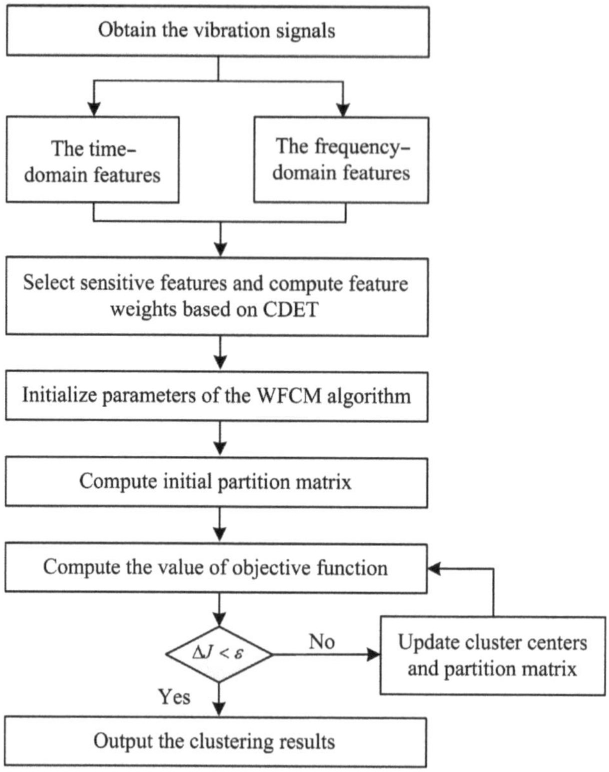

Figure 4.9. Flowchart of the WFCM-based diagnosis method.

Table 4.5 The Features Used in the Diagnosis Method Based on WFCM

Time–Domain Feature Parameters		Frequency–Domain Feature Parameters	
Feature	**Notation**	**Feature**	**Notation**
Mean	x_m	Mean frequency	x_{mf}
Standard deviation	x_{std}	Frequency centre	x_{fc}
Root mean square	x_{rms}	Root mean square frequency	x_{rmsf}
Peak	x_p	Standard deviation frequency	x_{stdf}
Skewness	x_{ske}	Spectrum peak ratio outer	$SPRO$
Kurtosis	x_{kur}	Spectrum peak ratio inner	$SPRI$
Crest factor	CF	Spectrum peak ratio roller	$SPRR$
Clearance factor	CLF		
Shape factor	SF		
Impulse factor	IF		

Second, the examination of the vibration signals shows the presence of low-frequency interference. Either high-pass or band-pass filtration should be used to remove the interference of the low-frequency components in the signals. So we adopt one high-pass filter and one band-pass filter. The cut-off frequency of the high-pass filter is 3.0 kHz, and the band-pass frequency of the band-pass filter ranges from 1.5 to 4.0 kHz. These frequencies are able to cover the signal components that contain the majority of the vibration energy. The 10 time–domain features are also extracted from each filtered signal. Thus, 10×2 time–domain features are obtained from these filtered signals.

Finally, the effects of the interfering signals within the filtered frequency band can be alleviated by demodulation. Demodulation detection makes the diagnosis independent of a particular machine because it focuses on the low-amplitude high-frequency broadband signals that characterize mechanical health conditions (Benko et al., 2005). So the Hilbert demodulation is conducted on the high-pass and band-pass filtered signals, respectively. Each Hilbert envelope spectrum obtained from the filtered signals is further used to extract the seven frequency–domain features, so 7×2 frequency–domain features are extracted. It should be noted that the number H of harmonics is 3 when *SPRO*, *SPRI* and *SPRR* are utilized. In a word, totally 48 feature values are extracted from a sample. These features are used as the inputs of the WFCM algorithm, and the health conditions are finally determined.

4.3.6 Intelligent Diagnosis of Incipient Faults: A Rolling Element Bearing Case Study

1. Data description

   Here, we use a bearing dataset to validate the WFCM-based diagnosis method. The dataset comes from the test bench of a locomotive roller bearing, and the details of the test bench can refer to Section 3.2.3.5. The bearing dataset totally contains 450 data samples under 9 bearing health conditions. So each bearing health condition includes 50 data samples, and each data sample is a vibration signal containing 8192 data points. Fig. 4.10 provides the typical data samples under each condition, their spectra and Hilbert envelope spectra. Fig. 4.10A–I respectively correspond to the nine conditions of the bearing: normal condition, slight rub fault in the outer race, serious flaking fault in the outer race, slight rub fault in the inner race, roller rub fault, compound faults in the outer and inner races, compound faults in the outer race and rollers, compound faults in the inner race

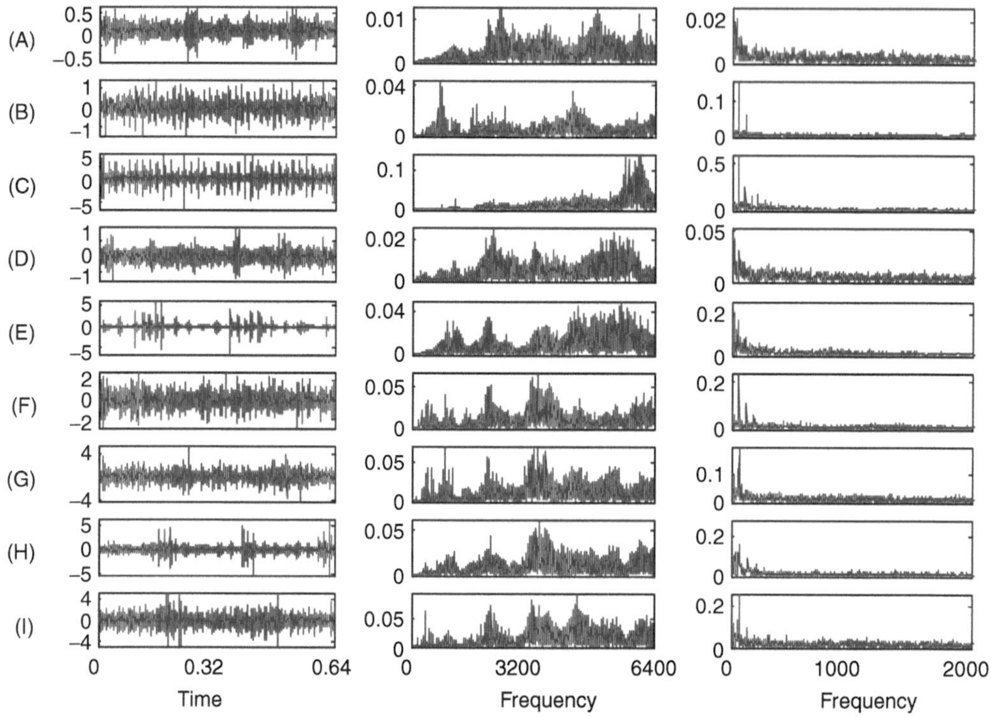

Figure 4.10. Typical data samples, their spectra and envelope spectra for nine conditions.

and rollers and compound faults in the outer and inner races and rollers. Among these conditions, the serious fault in the outer race is a serious flaking fault, and the other faults are all slight rub. It can be seen that the different faults cannot be distinguished by the original vibration signals and the frequency spectra. In the envelope spectra, the faults like serious flaking of outer race and roller rub faults may be recognized. However, it is not easy to distinguish all of the nine health conditions by using the vibration signals, frequency, and envelope spectra. Thus, the WFCM-based method is used to recognize the health conditions of locomotive roller bearings.

Three diagnosis experiments using three different datasets of locomotive roller bearings are carried out to demonstrate the effectiveness of the WFCM-based diagnosis method. For comparison, the diagnosis methods based on FCM algorithm (Method 1), and FCM algorithm with feature selection and no weighting (Method 2) are also applied to analyze the same datasets. For convenience, the WFCM-based diagnosis method is referred as Method 3.

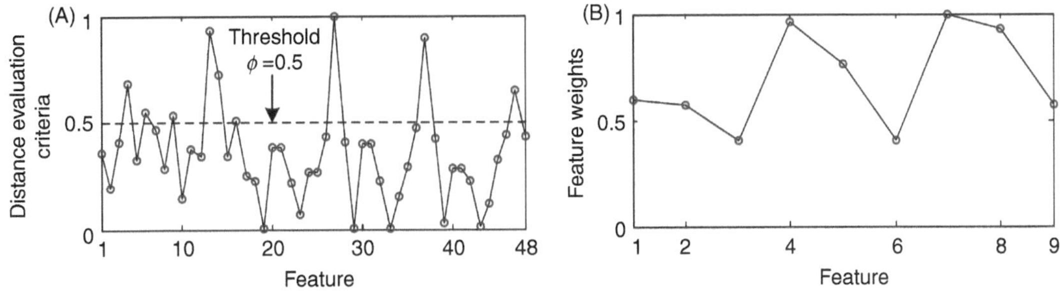

Figure 4.11. (A) Distance evaluation criteria of all features of experiment 1 and (B) feature weights of the sensitive features.

2. Experiment 1: clustering performance comparison of different fault categories
   In this experiment, the test dataset is composed of the samples under normal condition, serious fault in the outer race, inner race fault, and roller fault of bearings. The dataset is employed to test the performance of the proposed algorithm in recognizing the four single faults of locomotive roller bearings. As mentioned previously, the dataset for each health condition includes 50 samples, and therefore the dataset used here contains totally 200 samples.
   The distance evaluation criteria $\overline{E_j}$ $(j=1,2,\cdots,48)$ of the WFCM-based method are shown in Fig. 4.11A. When the threshold ϕ is set as 0.5, there are 9 sensitive features selected from all the 48 features using CDET. Simultaneously, Fig. 4.11B displays the feature weights of the nine sensitive features, which is also calculated by CDET.
   For visualization, principal component analysis (PCA) is applied to the clustering results obtained by the three methods. The plots of the first three principal components (PCs) of these results are given in Fig. 4.12. In Fig. 4.12A-B, it can be seen that many samples, which are circled by "o", are misclassified by Method 1 and Method 2. But in Fig. 4.12C, by using Method 3, most of the samples are correctly assigned to the corresponding clusters and only three samples of inner race fault, also circled by "o", are misclassified as the normal condition.
3. Experiment 2: clustering performance comparison of different fault severities
   To verify the effectiveness of the WFCM-based method in recognizing incipient faults that occur in the locomotive roller bearings, it is employed on a dataset including different fault severities. The data set totally includes 150 samples: 50 samples for normal condition, 50 samples belonging to slight fault in the outer race, and the rest 50 for serious fault in the outer race.

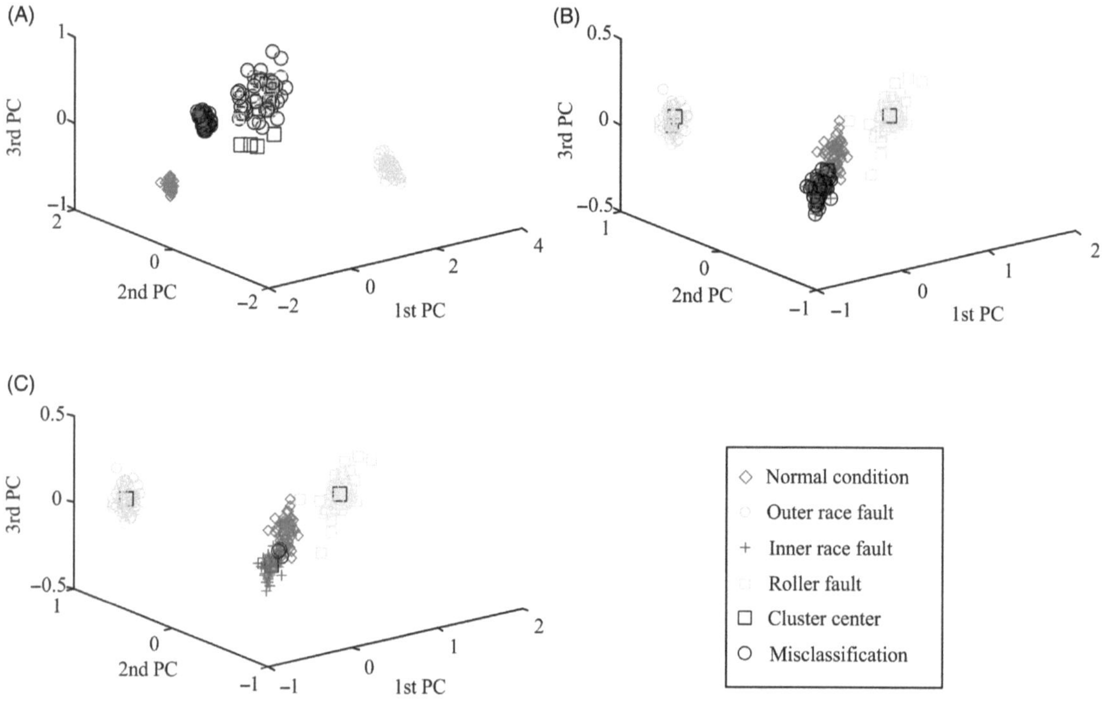

Figure 4.12. Scatter plots of the first three PCs for clustering results of experiment 1. (A) Method 1, (B) Method 2, and (C) Method 3.

Fig. 4.13A displays the distance evaluation criteria for all features. The feature weights of the sensitive features selected are shown in Fig. 4.13B, where the two-stage feature selection and weighting technique are adopted and the threshold ϕ is set to 0.5. PCA is used to compute the first three PCs of the experimental results of Method 1, Method 2 and Method 3, and their plots are shown in Fig. 4.14A–C, respectively. It can be seen from the

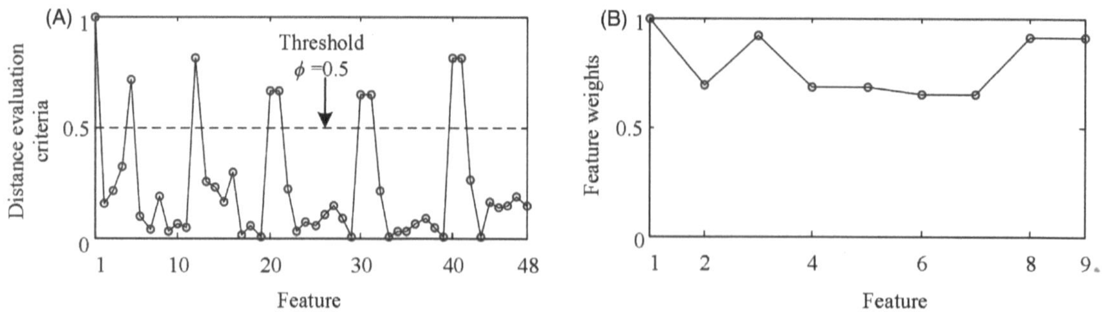

Figure 4.13. (A) Distance evaluation criteria of all features of experiment 2 and (B) feature weights of the sensitive features.

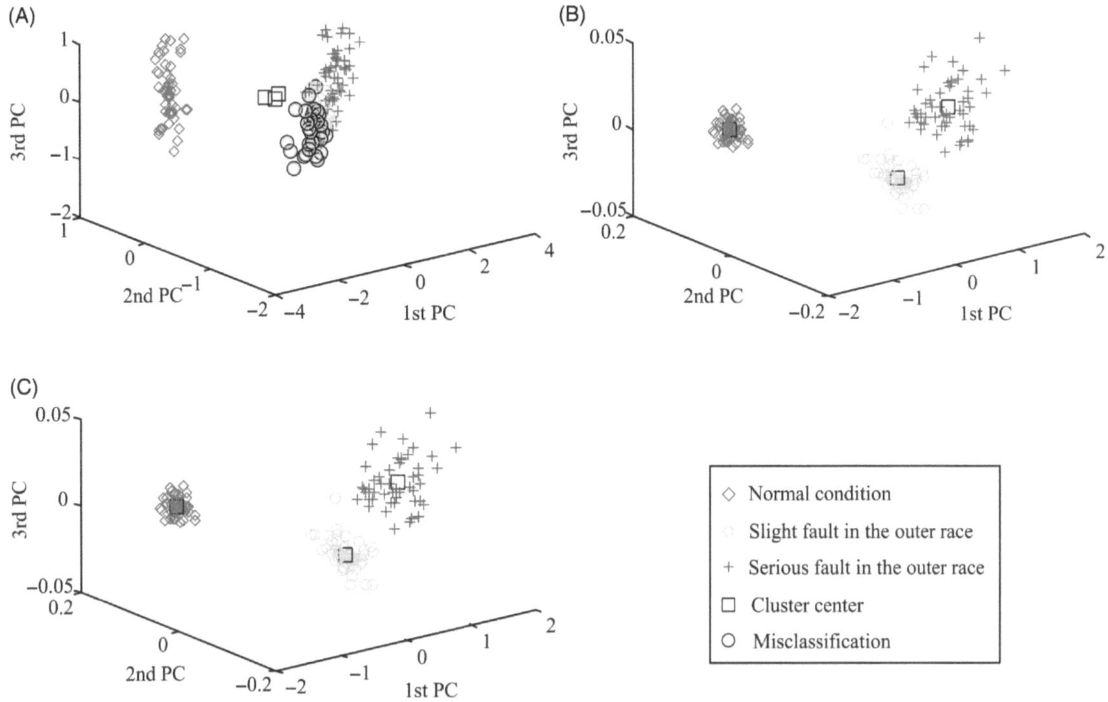

Figure 4.14. Scatter plots of the first three PCs for clustering results of experiment 2. (A) Method 1, (b) Method 2, and (c) Method 3.

three figures that a lot of samples with slight fault (circled by "o") are misclassified into the normal condition using Method 1. While using both Method 2 and Method 3, all samples are accurately assigned to the corresponding clusters.

4. Experiment 3: clustering performance comparison of the single and compound faults
Generally, besides the different types of a single fault and the different severities of a fault, compound faults also occur in rotating machinery. If the compound faults cannot be diagnosed as early as possible, they may lead to a serious consequence.
In this diagnosis experiment, a dataset including not only different fault types and severities but also compound faults is utilized to validate the practicability and robustness of the WFCM-based method. The dataset involves the nine health conditions of the locomotive roller bearings and thus is composed of 450 samples.
The distance evaluation criteria of the 48 features are displayed in Fig. 4.15A. When the threshold ϕ is set to 0.5, the nine sensitive features are selected based on CDET. Fig. 4.15B shows the

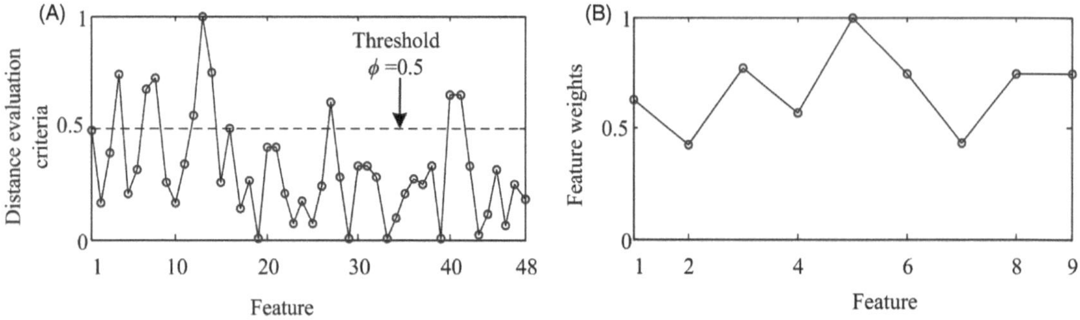

Figure 4.15. (A) Distance evaluation criteria of all features of experiment 3 and (B) feature weights of the sensitive features.

feature weights of the nine sensitive features obtained using the feature weighting.

Fig. 4.16 gives the clustering results of the bearings using the three different methods. As shown in Fig. 4.16A-B, plenty of samples (circled by "o") are still not recognized by Method 1,

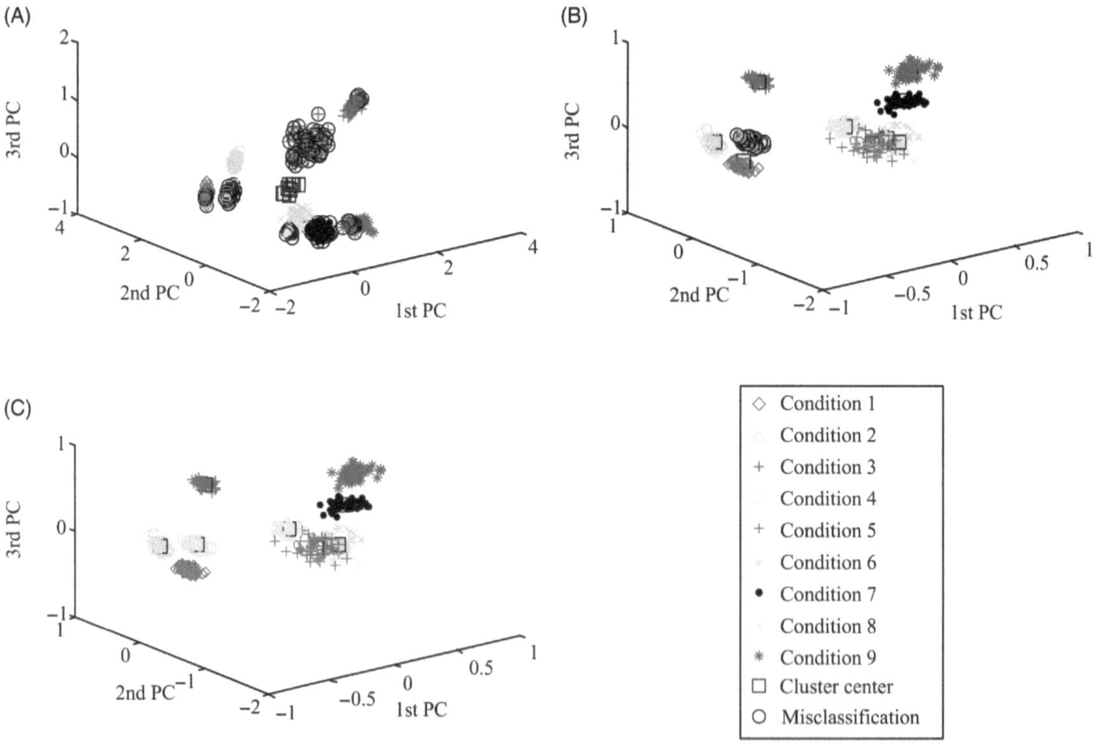

Figure 4.16. Scatter plots of the first three PCs for clustering results of experiment 3. (A) Method 1, (B) Method 2, and (C) Method 3.

Table 4.6 The Clustering Results Using the Diagnosis Method Based on WFCM

Algorithm	Item	Experiment 1	Experiment 2	Experiment 3
	Misclassification	100	34	185
Algorithm 1	Error (%)	50	22.67	41.11
	J	774.43	638.95	1184.80
	Misclassification	50	0	50
Algorithm 2	Error (%)	25	0	11.11
	J	23.14	1.34	63.19
	Misclassification	3	0	0
Algorithm 3	Error (%)	1.5	0	0
	J	8.25	0.91	24.82

and Method 2 misclassifies all the samples of inner race fault. Some of them are assigned as the normal condition and the rest are recognized as the condition of slight fault in the outer race. But utilizing Method 3, all samples are correctly assigned in Fig. 4.16C.

Table 4.6 gives the number of samples misclassified, the error rates and the objective function values produced by the three different methods in the three experiments. The table lists that the error rates of the three methods are respectively 50, 25 and 3% for experiment 1, 22.67%, 0, and 0 for experiment 2, and 41.11, 11.11%, and 0 for experiment 3. It can also be seen that the value J of the objective function from Method 1 to Method 3 obviously decreases in each of the three experiments.

5. Discussions

First, comparing Method 1 with Method 2, it can be found that the clustering accuracies (50% for experiment 1, 77.33% for experiment 2, and 58.89% for experiment 3) of the former method are obviously lower than the latter one (75% for experiment 1, 100% for experiment 2, and 88.89% for experiment 3). The main reason is that Method 1 lacks the feature selection process, leading to the poor results. For instance, a lot of samples are misclassified by Method 1 in Fig. 4.12. The reason is that different features may have different importance degrees in characterizing a fault. Some features are sensitive and closely related to the faults, but others are not. In Fig. 4.12A, some features contain too much fault-unrelated information and these features have a high degree of overlap between the clusters. Thus, when these features are used in clustering, its accuracies will decrease clearly.

Second, another comparison is between Method 3 and Method 2, and it indicates that Method 3 generally outperforms Method 2 according to the clustering accuracies. The accuracies of Method 3 (98.5% for experiment 1, 100% for experiment 2, and 100% for experiment 3) increase by 23.5%, 0, and 11.11% compared with Method 2 for each experiment, respectively. This confirms that the selection of sensitive features and their different importance degree in fault diagnosis are beneficial. Thus, the WFCM-based method obtains the best diagnosis results among the three methods.

Third, since experiment 2 is a simple three-class clustering problem, which is easy to solve, the best accuracies are obtained by both Methods 2 and 3. But the value of the objective function J of Method 3 is much smaller than that of Method 2 in this experiment, indicating that Method 3 is better. Therefore, Method 3 achieves more accurate results than the others and it helps reduce the vagueness and the uncertainty of clustering.

Fourth, in Fig. 4.12C, Fig. 4.14C, and Fig. 4.16C, it can be seen that each of the cluster centers is closer to the dense area of the corresponding cluster. This observation signifies that after feature selection and weighting, the WFCM-based method is able to remove the irrelevant features and highlight the importance of the sensitive features, so it improves the clustering performance.

Fifth, in the three experiments, the worst clustering results are provided by Method 1. Method 2 improves the clustering results by using feature selection and it even correctly recognizes the fault severities in experiment 2. But Method 2 presents poor performance when it is used to diagnose the compound faults. Contrarily, Method 3 not only reliably recognizes the different fault types and severities but also correctly assigned the compound faults. The main reason is that the WFCM-based method believes that the features have different importance degrees in distinguishing the mechanical health conditions. So it selects the sensitive features and assigns the feature weights so as to lift the clustering performance.

Last, in the feature selection of the three diagnosis experiments, it is found that two frequency–domain features x_{fc} (frequency centre) and x_{rmsf} (root mean square frequency) are most frequently selected. x_{mf} (mean frequency), x_{stdf} (standard deviation frequency), and SF (shape factor) are also selected sometimes. It is implied that these features are sensitive and should be used in fault diagnosis. The spectrum peak ratios of the outer race, the inner race, and the roller are seldom selected in the three experiments. It might show that the spectrum peak ratios of the envelope spectra are not always the good features for diagnosing bearing faults, especially for compound faults. Therefore, the feature selection is essential for clustering. It should be noted that CDET and PCA are used for different purposes, although they may both be used to reduce features. In this section, using CDET is to select sensitive features, while using PCA is to present the results and to see the results easily. Additionally, the selected features by CDET keep their physical meanings. While using PCA to reduce features, we cannot understand the physical meanings of the principal components. It indicates the superiority of CDET in feature selection.

4.3.7 Epilog

A fault diagnosis method based on a weight fuzzy *c*-means (WFCM) algorithm is developed for rotating machinery, which overcomes the existing shortcoming in the FCM algorithm by taking the importance degrees of features in clustering into account. In this method, in order to recognize the health conditions of rotating machinery, both time–domain and frequency–domain features are first extracted from the original and filtered vibration signals. Then, the two-stage feature selection and weighting technique using CDET are introduced to select sensitive features and compute the feature weights. Finally, they are incorporated into the FCM algorithm so the new clustering algorithm, WFCM, is developed. The WFCM-based method is applied to the incipient fault diagnosis and compound fault diagnosis of locomotive roller bearings. Compared with the other two diagnosis methods, the presented method presents better ability in identifying the different fault types, fault severities and the compound faults. Thus, the WFCM-based method may be a practicable and robust diagnosis method of rotating machinery.

4.4 Hybrid Clustering Algorithm–Based Fault Diagnosis

4.4.1 Introduction

In Section 4.3, a weighted fuzzy *c*-means (WFCM) algorithm is used to diagnose the mechanical faults, which utilizes a feature weight strategy to overcome the shortcoming of the FCM algorithm. WFCM supposes that all the features make different contributions to clustering. The diagnosis results of the bearing diagnosis cases demonstrate the effectiveness of the WFCM algorithm. However, the WFCM algorithm does not consider the following issues, which could be used for improving its performance. (1) It always treats each sample equally, and neglects their different influence for clustering. Since some samples may be outliers, they may affect the clustering accuracy. (2) It requires diagnosticians to previously set the number of clusters, which is commonly impossible to be known in advance. Therefore, the WFCM algorithm may provide the inexact clustering results in some diagnosis cases.

In Fig. 4.17, a simple example is displayed to illustrate that both feature weighting and sample weighting are needed in the clustering algorithm. Fig. 4.17A describes a data set that contains three clusters in a three-dimensional space, represented by the features f1, f2 and f3 respectively. Each cluster is made up of 12 samples, and these samples are located on two concentric spherical surfaces. The center of the concentric spherical surfaces is that of

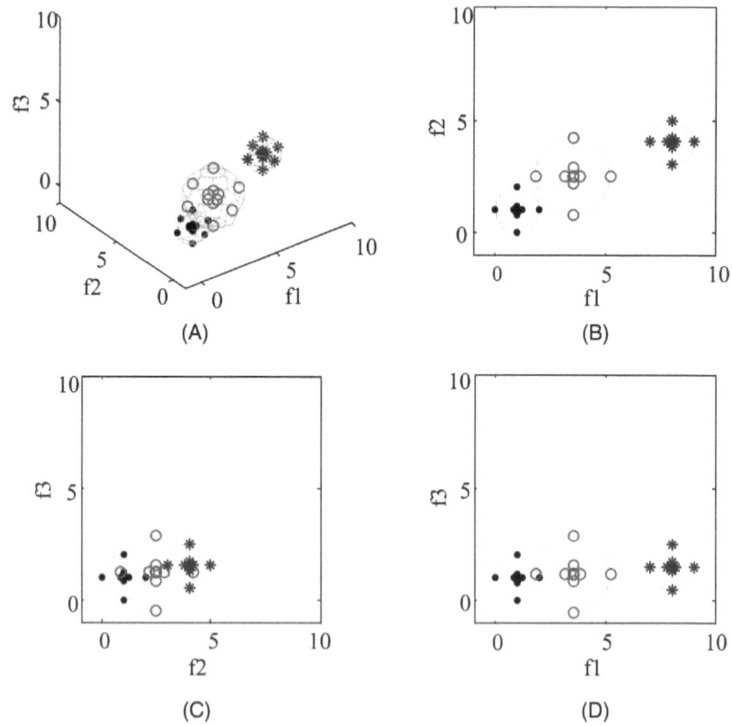

Figure 4.17. A data set containing three clusters and its projections on different planes. (A) Distribution in a three-dimensional space, (B) projection on f1–f2 plane, (C) projection on f2–f3 plane, and (D) projection on f1–f3 plane.

the corresponding cluster. Fig. 4.17B–D give the projections of the data set on different planes, respectively.

It can be seen from Fig. 4.17A that not all samples in their cluster have the uniform importance to clustering. The six samples located on the inner spherical surface are obviously more important for clustering than those located on the outer, namely the former is more representative for clustering. By comparing Fig. 4.17B–D, we can see that each feature has diverse importance to clustering. Concretely, f1 is superior to the others in discriminating these three clusters, f2 is inferior, and f3 is the worst. Therefore, through the analysis, not only feature weighting but also sample weighting are necessary to improve the performance of the clustering algorithm.

To overcome the shortcomings mentioned earlier, we developed a new hybrid clustering algorithm. This algorithm incorporates feature weighting, sample weighting and the validity index of a cluster into the original FCM algorithm, and is employed to incipient fault diagnosis of rotating machinery. The cluster number is determined by using the cluster validity index automatically. Feature weights are learned by a three-layer neural network (NN) in an unsupervised training manner, and the weights are then assigned to the relevant features to indicate their sensitivities to different faults. Sample weights are computed using a distribution density function of the samples and are assigned to the corresponding samples to reflect their importance to clustering. The highlights of the presented hybrid clustering algorithm are not only to find the optimal cluster number automatically, but also to take feature weighting and sample weighting into account simultaneously. Thus, it may present a better clustering performance than other algorithms.

4.4.2 Feature Weight Calculation Based on ANNs

Feature weighting can be viewed as a kind of feature selection. It assigns a number between 0 and 1 to a feature for indicating its sensitivity. In the Euclidean space, feature weighting aims to extend the axes according to the sensitive features and shrink the axes according to the features unrelated. Consequently, it enables the clustering algorithm to classify the nonspherical clusters. Fig. 4.18 shows the process of learning feature weights via a three-layer NN.

These weights are learned by minimizing the following objective function (Basak et al., 1998; Yeung and Wang, 2002):

$$E(wf) = \frac{2}{[N(N-1)]} \sum_{j<i} \frac{1}{2} [\rho_{ij}^{(wf)}(1-\rho_{ij}^{(1)}) + \rho_{ij}^{(1)}(1-\rho_{ij}^{(wf)})] \quad (4.40)$$

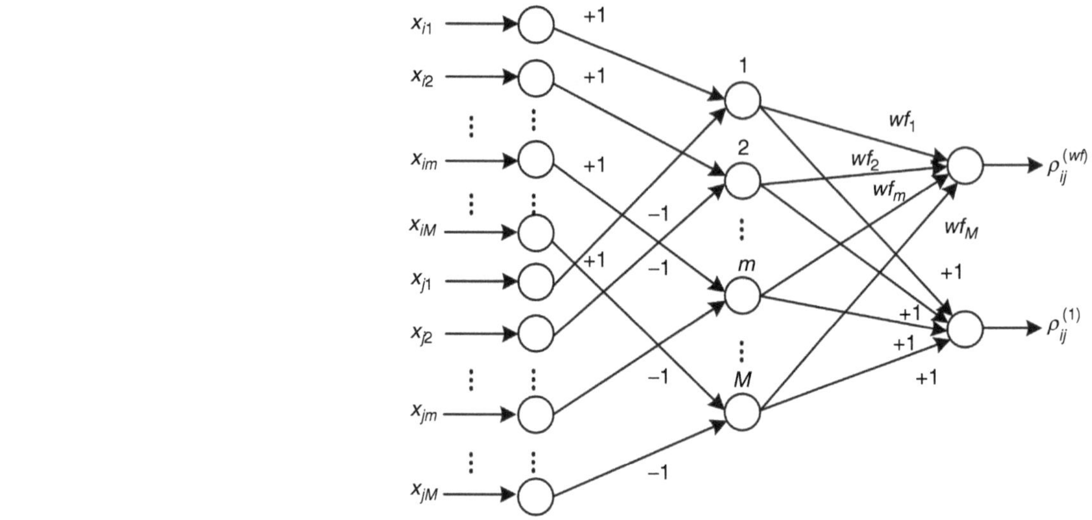

Figure 4.18. A three-layer NN used to learn feature weights.

where $\boldsymbol{wf} = \left(wf_1, \cdots, wf_m, \cdots, wf_M\right)$ is the feature weight vector. wf_m is the mth weight for the mth feature, and M is the number of features for each sample. N is the number of samples of a data set $\left\{\boldsymbol{x}_1, \cdots, \boldsymbol{x}_j, \cdots, \boldsymbol{x}_N\right\}$. $\boldsymbol{x}_j = \left(x_{j1}, \cdots, x_{jm}, \cdots, x_{jM}\right)$ is the jth sample, and x_{jm} is its mth feature. $\rho_{ij}^{(1)}$ denotes the similarity measure based on the standard Euclidean distance $d_{ij}^{(1)}$. $\rho_{ij}^{(wf)}$ represents the similarity measure based on the weighted Euclidean distance $d_{ij}^{(wf)}$.

The weighted Euclidean distance between the ith sample $\boldsymbol{x}_i$ and the jth sample $\boldsymbol{x}_j$ is defined as:

$$d_{ij}^{(wf)} = d^{(wf)}\left(\boldsymbol{x}_i, \boldsymbol{x}_j\right) = \left[\sum_{m=1}^{M} wf_m \left(x_{im} - x_{jm}\right)^2\right]^{1/2} \tag{4.41}$$

where wf_m denotes the sensitivity of the mth feature. If all the weights equal 1, the weighted Euclidean distance $d_{ij}^{(wf)}$ is back to the standard Euclidean distance $d_{ij}^{(1)}$.

The $\rho_{ij}^{(wf)}$ is defined as:

$$\rho_{ij}^{(wf)} = \frac{1}{\left(1 + \beta \cdot d_{ij}^{(wf)}\right)}, \tag{4.42}$$

where β is a positive parameter calculated by using the following equation:

$$\frac{2}{N(N-1)}\sum_{j<i}\rho_{ij}^{(1)} = \frac{2}{N(N-1)}\sum_{j<i}\frac{1}{1+\beta\cdot d_{ij}^{(1)}} = 0.5. \tag{4.43}$$

The reason of minimizing the objective function in Eq. (4.40) to obtain the feature weights is detailed as follows. This objective function will reach its minimum value if the similarity measures are close to 1 or 0. A good partition should have the property that the samples of one cluster are close to its center while those of different clusters are as separated as possible. It indicates that the samples of one cluster are similar $\left(\rho_{ij}^{(wf)} \to 1\right)$ and the samples of different clusters are separate $\left(\rho_{ij}^{(wf)} \to 0\right)$, which results in the low value of the objective function. Therefore, by optimizing the feature weights ***wf***, the similar samples in original space $\left(\rho_{ij}^{(1)} > 0.5\right)$ tend to be more similar $\left(\rho_{ij}^{(wf)} \to 1\right)$, while the dissimilar samples $\left(\rho_{ij}^{(1)} < 0.5\right)$ tend to be more separate $\left(\rho_{ij}^{(wf)} \to 0\right)$.

Fig. 4.18 shows that the FFNN architecture consists of three layers, in which the node numbers of the three layers are $2M$, M, and 2, respectively. In the input layer, a pair of samples are used as the inputs, namely the $2M$ features are the inputs for each node of this layer. Then the input layer transfers all the input features into the hidden layer. The mth input node of the hidden layer is the sum of the weighted outputs of the mth and the $(m+M)$th input nodes using the connection weights +1 and –1, respectively. In the output layer, $\rho_{ij}^{(wf)}$ is the summation of the weighted outputs of each hidden node and the corresponding weight wf_m. Contrarily, $\rho_{ij}^{(1)}$ is the sum of the weighted outputs of all the hidden nodes with the weights +1. In the training process, any two samples used as a pair are fed to the three-layer NN. It is noted that a dataset composed of N samples has $N(N-1)/2$ sample pairs.

We use $in_m^{(l)}$ and $out_m^{(l)}$ as notations that indicate the input and the output of the mth node $(m=1,\cdots,M)$ in the lth layer (l = 1, 2, 3), respectively. If the pairs fed to the NN are $(\boldsymbol{x}_i, \boldsymbol{x}_j)$, the inputs, the activation functions and the outputs of the three layers are given in Table 4.7.

We substitute the $\rho_{ij}^{(wf)}$ and $\rho_{ij}^{(1)}$ in Table 4.7 into the objective function $E(\boldsymbol{wf})$ in Eq. (4.40). By minimizing the objective function, the feature weights can be learned based on the back propagation algorithm. The weight change Δwf_m can be expressed as:

$$\Delta wf_m = -\eta\frac{\partial E(\boldsymbol{wf})}{\partial wf_m} = -\eta\frac{\partial E}{\partial out_1^{(3)}}\cdot\frac{\partial out_1^{(3)}}{\partial in_1^{(3)}}\cdot\frac{\partial in_1^{(3)}}{\partial wf_m}, \tag{4.44}$$

Table 4.7 Inputs, Activations, and Outputs of the Three-Layer NN

Layer	Input	Activation Function	Output
1	$in_m^{(1)} = x_{im}, in_{m+M}^{(1)} = x_{jm}$	$out = in$	$out_m^{(1)} = in_m^{(1)}, out_{m+M}^{(1)} = in_{m+M}^{(1)}$
2	$in_m^{(2)} = (+1)\cdot out_m^{(1)} + (-1)\cdot out_{m+M}^{(1)}$	$out = in^2$	$out_m^{(2)} = \left(in_m^{(2)}\right)^2$
3	$in_1^{(3)} = \sum_m \left(wf_m \cdot out_m^{(2)}\right)$, $in_2^{(3)} = \sum_m out_m^{(2)}$	$out = 1/(1 + \beta \cdot in^{1/2})$	$out_1^{(3)} = \rho_{ij}^{(wf)} = 1/\left[1+\beta\cdot\left(in_1^{(3)}\right)^{1/2}\right]$, $out_2^{(3)} = \rho_{ij}^{(1)} = 1/\left[1+\beta\cdot\left(in_2^{(3)}\right)^{1/2}\right]$

where

$$\frac{\partial E}{\partial out_1^{(3)}} = \frac{2}{N(N-1)} \cdot \sum_{j<i} \frac{1}{2}\left(1 - 2out_2^{(3)}\right), \tag{4.45}$$

$$\frac{\partial out_1^{(3)}}{\partial in_1^{(3)}} = \frac{\beta}{2\left(1+\beta\cdot\left(in_1^{(3)}\right)^{1/2}\right)^2 \cdot \left(in_1^{(3)}\right)^{1/2}} \tag{4.46}$$

$$\frac{\partial in_1^{(3)}}{\partial wf_m} = out_m^{(2)}, \tag{4.47}$$

In the equations, η is the learning rate and its optimal value for each training epoch is computed by golden section search method. This method finds the learning rate in the given range of learning rates. The learning rate attaching to the minimum of the objection function $E(\boldsymbol{wf})$ is selected as the learning rate in the current training epoch. For all sample pairs, FFNN first performs the forward calculations so as to get the value of the objective function. Then, the feature weight wf_m is updated iteratively by $wf_m + \Delta wf_m$ until the stopping criterions are satisfied.

Therefore, the NN used for feature weighting is different from the supervised learning NNs described in Section 3.2. It is trained in an unsupervised way, so the labels of training samples are not needed in the training. Actually, the outputs of the network are the similarity measures $\rho_{ij}^{(1)}$ and $\rho_{ij}^{(wf)}$ of the input samples. It is the same as supervised learning NNs in that both the training processes including the forward phase and backward phase. By using the two phases, the objection function is optimized and the weights of the features are finally obtained.

4.4.3 Sample Weight Computation With a Distribution Density Function

As we know, the distribution density of a sample goes larger as its surrounding samples increase. The larger the distribution density of the sample is, the more likely it will be a cluster center. Therefore, we utilize the distribution density function of a sample to calculate the sample weights.

For the ith sample $(i=1,\cdots,N)$, its distribution density function y_i can be defined as:

$$y_i = \sum_{j=1, j\neq i}^{N} \left(\frac{1}{d_{ij}^{(wf)}} \right), \tag{4.48}$$

subject to

$$d_{ij}^{(wf)} \leq r, \tag{4.49}$$

where r is the upper limit of the distribution density function. It will equal the value of weighted Euclidean distance if the probability distribution function of $d_{ij}^{(wf)}$ reaches its maximum value.

We can compute the sample weight vector $wp = (wp_1,\cdots,wp_i,\cdots,wp_N)$ by normalizing y_i according to Eq. (4.50). The element wp_i is the weight for the ith sample and indicates its typicality.

$$wp_i = \frac{y_i}{\sum_i^N y_i}. \tag{4.50}$$

4.4.4 Cluster Number Determination

Since the choice of the cluster number generally influences the clustering results, it needs to validate the cluster results. This validation is conducted by a cluster validity index, and it is used to select the optimal cluster number that can verify the correct and natural description of the data structure. Some numerical experiments for several artificial data sets and UCI databases have been used to compare the performance of various indexes including *PC* (Bezdek, 1975), *PE* (Bezdek†, 1973), *FS* (Fukuyama and Sugeno, 1989), *PMC* (Dave, 1996), *XB* (Xie and Beni, 1991), *FHV* (Gath and Geva, 1989), *K* (Kwon, 1998), *PCAES* (Wu and Yang, 2005), *OS* (Kim et al., 2004) and *PBMF* (Pakhira et al., 2004). The results demonstrate that the *PBMF* index performs better than the other indexes. Therefore, it is chosen as the cluster validity index, and the index is defined as (Pakhira et al., 2004):

$$PBMF(K)=\left[\frac{1}{K}\cdot\frac{E_1}{J_\lambda}\cdot D_K\right]^2, \tag{4.51}$$

where K is the cluster number, J_λ is a compactness measure of the K clusters, D_K is a separation measure of the K clusters, and E_1 is the value of J_λ when K equals 1.

The definition of the *PBMF* index reflects that the larger its value is, the better the clustering results are. Therefore, the cluster number maximizing the *PBMF* index is chosen as the optimal number of the clusters. And more details of the *PBMF* index can be found in the reference (Pakhira et al., 2004).

4.4.5 Diagnosis Method Using the Hybrid Clustering Algorithm

The hybrid clustering algorithm is developed by integrating the FCM algorithm with feature weighting by the three-layer NN, sample weighting by the distribution density function and the cluster validity index *PBMF*. The flowchart of the hybrid clustering algorithm is shown in Fig. 4.19. It can be seen that the algorithm is composed of an inner iterative loop and an outer iterative loop. In the inner iterative loop, clustering is performed by minimizing the objective function in Eq. (4.41), which is similar to the FCM algorithm. In the outer iterative loop, it computes the validity index of clusters provided by the inner iterative loop under various numbers of clusters.

The objective function J of the hybrid clustering algorithm is defined as:

$$J(\boldsymbol{U},\boldsymbol{Z},\boldsymbol{wp},\boldsymbol{wf};\boldsymbol{X})=\sum_{k=1}^{K}\sum_{i=1}^{N}wp_i\left(u_{ik}\right)^{\lambda}\left(d_{ik}^{(wf)}\right)^2, \tag{4.52}$$

subject to

$$\sum_{k=1}^{K}u_{ik}=1, \quad 0\le u_{ik}\le 1, \tag{4.53}$$

where $\boldsymbol{U}=[u_{ik}]$ is the fuzzy partition matrix for $i=1,\cdots,N$ and $k=1,\cdots,K$, and u_{ik} indicates the membership degree for the ith sample belonging to the kth cluster. λ denotes the fuzzy clustering exponent. $d_{ik}^{(wf)}$ is the weighted distance between the ith sample and the center of the kth cluster, and it is expressed as:

$$d_{ik}^{(wf)}=d^{(wf)}\left(\boldsymbol{x}_i,\boldsymbol{z}_k\right)=\left[\sum_{m=1}^{M}wf_m\left(x_{im}-z_{km}\right)^2\right]^{1/2}, \tag{4.54}$$

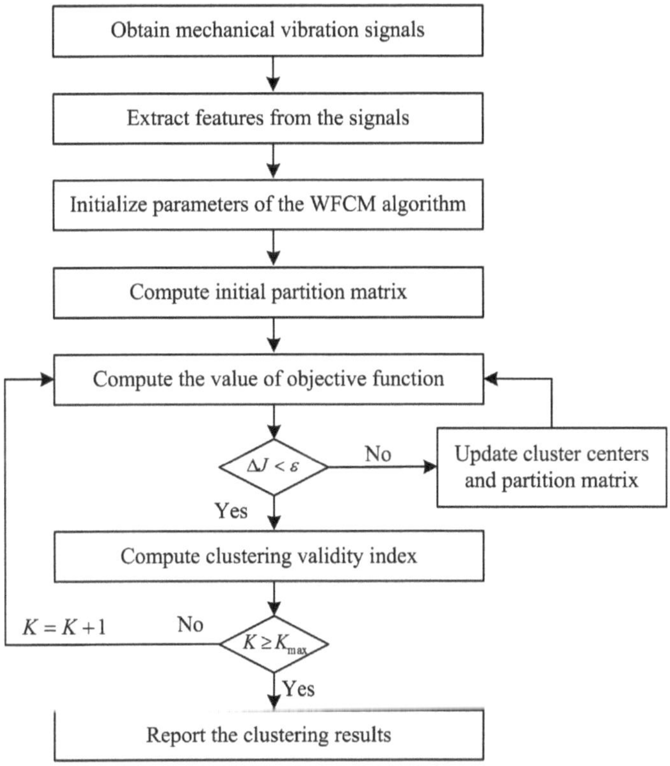

Figure 4.19. Process of the hybrid clustering algorithm.

where $Z = \{z_1, \cdots, z_k, \cdots, z_K\}$ is the set of cluster centers, $z_k = (z_{k1}, \cdots, z_{km}, \cdots, z_{kM})$ is the center of the kth cluster and z_{km} is its mth element.

Based on the feature weights $\boldsymbol{wf}$ learned by the three-layer NN and the sample weights $\boldsymbol{wp}$ computed by the distribution density function, we can obtain the computational formulae of $\boldsymbol{z}_k$ and u_{ik} as:

$$z_k = \frac{\sum_{i=1}^{N} wp_i (u_{ik})^{\lambda} \boldsymbol{x}_i}{\sum_{i=1}^{N} wp_i (u_{ik})^{\lambda}} \tag{4.55}$$

and

$$u_{ik} = \frac{1}{\sum_{a=1}^{K} \left(\frac{d_{ik}^{(wf)}}{d_{ia}^{(wf)}} \right)^{2/(\lambda-1)}} \tag{4.56}$$

where $d_{ia}^{(wf)}$ is the weighted distance between the ith sample for $i=1,\cdots,N$ and the center of the ath cluster for $a=1,\cdots,K$.

Based on the hybrid clustering algorithm, a fault diagnosis method of rotating machinery is presented here. As shown in Fig. 4.19, its procedural steps can be summarized as follows.

1. Obtain the signals of rotating machinery under various health conditions. These signals compose the dataset.
2. Extract features from the dataset. Features are used to characterize the different faults for the original vibration signals, and they compose the set of samples.
3. Use the hybrid clustering algorithm to cluster the dataset and assign the label for each sample. Concretely, we first obtain the feature weights of the samples using the three-layer NN trained by the back propagation algorithm, and compute their sample weights using the distribution density function, respectively. Second, the parameters of the hybrid clustering algorithm are initialized, including the number of clusters, the maximum number of clusters K_{max}, the fuzzy clustering exponent λ, fuzzy partition matrix $[u_{ik}]$, and a threshold value ε. Next, we update the fuzzy cluster centers $\boldsymbol{z}_k$ using Eq. (4.55) and the fuzzy partition matrix $[u_{ik}]$ using Eq. (4.56). If the difference ΔJ between the two interval values of the objective function J exceeds the set threshold ε, the algorithm will continually update the $\boldsymbol{z}_k$ and $[u_{ik}]$. Finally, we compute the *PBMF* index of clusters derived and find the maximum value of *PBMF*. If the value reaches its maximum value, the algorithm will use the corresponding value of K as the optimal number of clusters, and fuzzy partition matrix $\boldsymbol{U}$ is just the clustering results. Otherwise, the parameters of the number of clusters are reinitialized, and the algorithm updates the fuzzy cluster centers $\boldsymbol{z}_k$ and the fuzzy partition matrix $[u_{ik}]$ again, until the *PBMF* index reaches its maximum value or the algorithm satisfies stop criteria.
4. The final clustering results of the hybrid clustering algorithm are the diagnosis results for the dataset of rotating machinery.

It is noted that the initial number of clusters is 2. According to a rule of thumb (Bezdek, 1998), the K_{max} value should not exceed $\sqrt{N}$ where N is the number of samples. Following the principle of choosing the fuzzy clustering exponent λ in the FCM algorithm, the hybrid clustering algorithm also selects the same range of the exponent, that is, the optimum of λ is selected ranging from 1.5 to 2.5.

4.4.6 Intelligent Diagnosis of Compound Faults: A Locomotive Bearing Case Study

1. Case study description
   In Section 4.3, types of the locomotive roller bearing faults including outer race fault, inner race fault, roller fault and their compound faults are studied thoroughly. The incipient faults of the locomotive roller bearings traditionally behave as slight rub. If these faults cannot be recognized as early as possible, they will rapidly become serious flaking faults, which may result in the breakdown of the locomotives. Therefore, we study the fault diagnosis of incipient faults for the locomotive roller bearings. As same as the test rig used in Section 4.3, a bearing data set is obtained from locomotive roller bearings involving the following three health conditions: (1) normal conditions (NC); (2) with slight rub faults occurring in outer races (SRF); and (3) with serious flaking faults occurring in outer races (SFF). There are 50 data samples for each health condition, where each data sample contains 8192 data points.
2. Feature extraction
   Here, six of the features usually used for the fault diagnosis of roller bearings in time–domain are extracted from each sample. The six features include: root mean square (RMS), peak value (PV), kurtosis value (KV), crest factor (CF), impulse factor (IF) and clearance factor (CLF), and they have been described in Chapter 2. Based on the amplitude probability density function of the bearing vibration signals, these features can be computed and compared with those of normal bearing signals (Sun et al., 2004). In Reference (Howard, 1994), it has been shown that the peak value and root mean square are able to reflect the vibration amplitude and energy, and the kurtosis value, crest factor, impulse factor and clearance factor are used to characterize the impact existing in the bearings. Among them, the kurtosis value and crest factor are sensitive to incipient faults while are insensitive to the various operating conditions of bearings. The impulse factor and clearance factor can well reflect the spikiness of the sharp impulses, which are generated by the contact between a defect and the mating surface (Sun et al., 2004; Li and Pickering, 1992). The diagnosis cases aim to diagnose the local defects in outer race. Such defects typically occur in the locomotive roller bearings and generate a series of impact vibrations, causing the increase of the vibration amplitude. Therefore, we utilize these six features to represent the

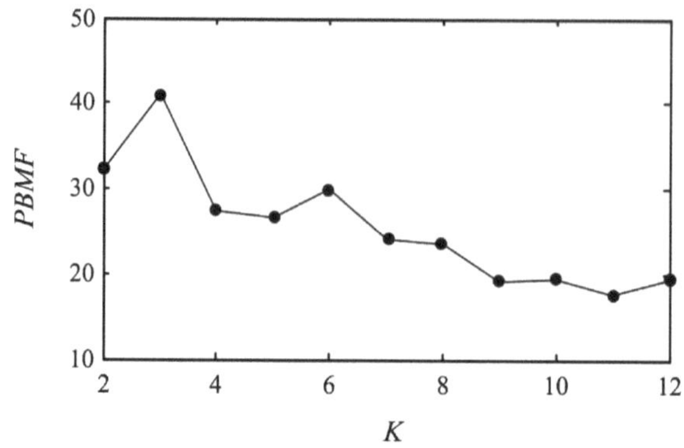

Figure 4.20. Values of the *PBMF* index using various numbers of clusters.

health conditions of the bearings. Since the six features show the different importance for characterizing each bearing condition, we assign the weights to the corresponding features so as to indicate their different sensitivities in the bearing fault diagnosis.

3. Diagnosis results

We employ the presented hybrid clustering algorithm for clustering on the bearing dataset and Fig. 4.20 shows the values of the *PBMF* index with various numbers of clusters ranging from 2 to 12. It is evidently seen that the *PBMF* index reaches its maximum value when K equals 3. This illustrates that the dataset is composed of three clusters, which is coincident with the fact.

To visualize clustering results, we apply PCA method to the results clustered by the presented algorithm for the dataset. The first two principal components (PCs) of the results are displayed in Fig. 4.21. We can see that more than 97% of variability of the features is accounted by the two PCs. To show the effectiveness of the hybrid clustering algorithm, Fig. 4.22 also displays the clustering results of the original FCM algorithm, in which the cluster number is set to 3. It can be seen from this figure that some samples, which is circled by "o", are misclassified by the FCM algorithm. On the contrary, by using the presented hybrid clustering algorithm, all samples are classified correctly to the corresponding clusters.

To quantitatively compare the presented algorithm and the FCM algorithm, Table 4.8 provides the error rates, the values of the objective function and the CPU-times of the two algorithms. It shows that the error rates of the proposed algorithm and the FCM algorithm are 0 and 10%, respectively. Their values of the objective function are 1.2 and 16, respectively. The proposed

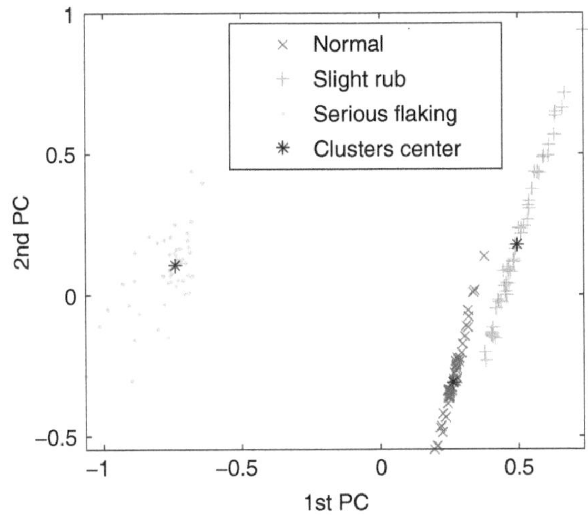

Figure 4.21. Scatter plots of PCs for clustering results using the hybrid clustering algorithm.

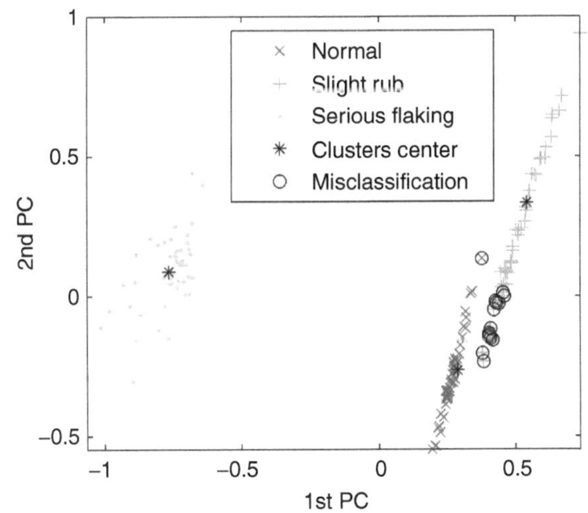

Figure 4.22. Scatter plots of PCs for clustering results using the FCM algorithm.

algorithm and the FCM algorithm spend the CPU-times of 2.765 and 0.047 s respectively to carry out their program.

By comparing these results, it can be found that the presented algorithm not only determines the number of clusters, but also gives the accurate rate of 100%, so it obviously outperforms the FCM algorithm. Although the FCM algorithm can well diagnose the serious flaking faults, it displays poor performance in

Table 4.8 A Comparison Between the Two Clustering Algorithm

	The Hybrid Clustering Algorithm			The FCM Algorithm		
	NC	SRF	SFF	NC	SRF	SFF
NC	50	0	0	49	14	0
SRF	0	50	0	1	36	0
SFF	0	0	50	0	0	50
Error (%)	0			10		
J	1.2			16.0		
Time (s)	2.765			0.047		

diagnosing the incipient slight rub faults. By using the FCM algorithm, 15 samples out of the incipient slight rub fault samples and the normal condition samples are not assigned to the right labels. However, the hybrid clustering algorithm can accurately diagnose the incipient slight rub faults, the normal conditions as well as the serious flaking faults. The reasons can be analyzed as follows. First, the presented algorithm introduces the effective cluster validity index into its training process. By using the *PBMF* index, the best number of clusters can be selected objectively and automatically from the possible cluster numbers. Second, in the hybrid clustering algorithm, it is supposed that the 6 time–domain features and the 50 samples for each cluster have various importance in recognizing these three health conditions of the bearings. So the algorithm is able to adaptively acquire the information from the dataset, and calculate the feature weights for the different features and the sample weights for the different samples, which improves the performance of the clustering. Finally, it can be seen from Table 4.8 that the objective function value J of the presented algorithm is obviously lower than that of the FCM algorithm, which demonstrates that the presented algorithm helps reduce the vagueness and the uncertainty for clustering and therefore performs better.

An interesting phenomenon observed is that each center of the three clusters in Fig. 4.21 is closer to the dense area of the samples than that in Fig. 4.22. It illustrates that the cluster centers are close to the representative samples in the presented clustering and are far away from the vague samples. This observation owes to the strategy of sample weighting. It is noted that the sample weighting emphasizes the influence of the representative samples and weakens the effect of vague samples.

In addition, it can be seen in Table 4.8 that the proposed algorithm spends more CPU-time than the FCM algorithm. The reason is that the hybrid clustering algorithm integrates several methods, so it is more complex than the FCM algorithm. However, by adopting feature weighting, sample weighting and the *PBMF* index, the hybrid clustering algorithm can overcome the shortcomings of the FCM algorithm so that the clustering accuracy can be improved greatly. Thus, it is actually a trade-off between computational complexity and performance improvement. Moreover, it is a fact that the hybrid clustering algorithm improves its performance at the price of learning feature weights, calculating sample weights and computing the *PBMF* index. Therefore, it is worth of the cost of computational complexity, and this hybrid clustering algorithm can be efficiently employed to fault diagnosis of rotating machinery.

4.4.7 Epilog

A hybrid clustering algorithm is presented to overcome the existing shortcomings in the FCM algorithm. In this algorithm, three techniques, that is, feature weighting based on a three-layer NN, sample weighting using a distribution density function and a cluster validity index named *PBMF*, are incorporated into the FCM algorithm. Unlike the existing algorithms that only solve one of the problems of the FCM algorithm, this hybrid algorithm simultaneously overcomes the three existing problems. To show the effectiveness of the presented algorithm, it is employed to diagnose the incipient faults of locomotive roller bearings. The diagnosis results imply that the algorithm can not only determine the optimal number of clusters but also improve the clustering accuracy. Therefore, the hybrid clustering algorithm is a promising approach for the incipient fault diagnosis of bearings, and of course, it can also be applied to diagnose faults of other rotating machinery.

4.5 Conclusions

In this chapter, we focus on the usage of clustering in the fault diagnosis of rotating machinery. Different from intelligent fault diagnosis, which utilizes features of the vibration signals combined with their labels to train individual classifiers and make decisions, clustering algorithm–based fault diagnosis does not need the work of labeling the collected signals, and it is able to automatically mine the pattern within the unlabeled samples and reliably recognize the health conditions of the machinery. Thus, clustering algorithm–based fault diagnosis takes full advantage of unlabeled data and reduces the human labor in fault decision. These properties make the clustering algorithm–based fault diagnosis play an increasingly important role in the era of big data.

The traditional clustering algorithms, however, suffer from the following three shortcomings. (1) These algorithms suppose that all features make the uniform contributions to clustering and do not take their different importance degrees into account. (2) Every sample is treated equally, and their different influence to clustering is neglected. (3) It is required to predefine the number of clusters, but it is not always possible to know the number in advance. Aiming to overcome these shortcomings, we proposed three improved clustering algorithms. The algorithm in Section 4.2 is the weighted K nearest neighbor algorithm, and it incorporates the feature selection and feature weighting into the K nearest neighbor algorithm by considering the different abilities of the features in characterizing mechanical health conditions and the various weights of the features in clustering. The algorithm in Section 4.3 also considers the effect of feature selection and feature weighting, but it is based on a more effective clustering algorithm, the fuzzy c-means algorithm, to recognize the mechanical health conditions. This new algorithm is called weighted fuzzy c-means algorithm, which deals with uncertainty and assigns a sample to a cluster by a membership value, leading to better performance. The algorithm in Section 4.4 extends the weighted fuzzy c-means algorithm to a hybrid clustering algorithm and simultaneously solves the three shortcomings mentioned earlier by the feature weighting, the sample weighing and automatic cluster number determination. The fault diagnosis methods based on the three proposed clustering algorithms are used to recognize the various health conditions of rotating machinery, and all of them achieve better clustering results than the diagnosis methods based on the traditional clustering algorithms. It illustrates that using feature selection, feature weighting and sample weighing described in this chapter is a good strategy for clustering. The readers can use these strategies in their own algorithms so as to improve the recognition performance of the algorithms.

Although clustering offers a solution to release human labor in the learning process, the feature extraction in the clustering-based diagnosis methods still depends on diagnosticians (Lei and Zuo, 2009). Thus, clustering algorithm is limited in its potential for completely automated data analysis. Some methods of automatic feature extraction have been shown in Chapter 3, and we are willing to combine the idea of the automatic feature extraction and the automatic recognition into an integrated method, so that we can step toward real intelligence in processing the mechanical big data. We will focus on this topic in future work.

References

Basak, J., De, R.K., Pal, S.K., 1998. Unsupervised feature selection using a neuro-fuzzy approach. Pattern Recognit. Lett. 19, 997–1006.

Benko, U., Petrovcčicč, J., Juričić, Đ., Tavčar, J., Rejec, J., 2005. An approach to fault diagnosis of vacuum cleaner motors based on sound analysis. Mech. Syst. Signal Process. 19, 427–445.

Berkhin, P., 2006. A survey of clustering data mining techniques. Grouping Multidimensional Data. Springer, Berlin, Heidelberg, pp. 25–71.

Beyer, K., Goldstein, J., Ramakrishnan, R., Shaft, U., 1999. When is "nearest neighbor" meaningful? In International conference on database theory. Springer, Berlin, Heidelberg, pp. 217–235.

Bezdek, J., 1998. Pattern Recognition in Handbook of Fuzzy Computation. IOP Publishing Ltd, Boston, NY.

Bezdek, J.C., 1975. Mathematical models for systematics and taxonomy. In: Proceedings of eighth international conference on numerical taxonomy, pp. 143–166.

Bezdek, J.C., 2013. Pattern Recognition With Fuzzy Objective Function Algorithms. Springer, New York.

Bezdek†, J.C., 1973. Cluster validity with fuzzy sets. J. Cybernetics 3, 58–73.

Bishop, C.M., 2006. Pattern Recognition and Machine Learning. Springer, New York.

Casimir, R., Boutleux, E., Clerc, G., Yahoui, A., 2006. The use of features selection and nearest neighbors rule for faults diagnostic in induction motors. Eng. Appl. Artif. Intell. 19, 169–177.

Chan, E.Y., Ching, W.K., Ng, M.K., Huang, J.Z., 2004. An optimization algorithm for clustering using weighted dissimilarity measures. Pattern Recognit. 37, 943–952.

Dave, R.N., 1996. Validating fuzzy partitions obtained through c-shells clustering. Pattern Recognit. Lett. 17, 613–623.

Decker, H.J., 2003. Effects on diagnostic parameters after removing additional synchronous gear meshes. DTIC Document.

Decker, H.J., Lewicki, D.G., 2003. Spiral bevel pinion crack detection in a helicopter gearbox. DTIC Document.

Duda, R.O., Hart, P.E., 1973. Pattern Classification and Scene Analysis. Wiley, New York.

Everitt, B.S., Landau, S., Leese, M., Stahl, D., 2011. Miscellaneous clustering methods. Cluster Analysis, fifth Edition, 215–255.

Fan, X., Zuo, M.J., 2006. Gearbox fault detection using Hilbert and wavelet packet transform. Mech. Syst. Signal Process. 20, 966–982.

Frigui, H., Nasraoui, O., 2004. Unsupervised learning of prototypes and attribute weights. Pattern Recognit. 37, 567–581.

Fukuyama, Y., Sugeno, M., 1989. A new method of choosing the number of clusters for the fuzzy c-means method. In: Proceedings of Fifth Fuzzy System Symposium, 247, 247–250.

Gao, X., Xie, W., 2000. Advances in theory and applications of fuzzy clustering. Chin. Scie. Bull. 45, 961–970.

Gath, I., Geva, A.B., 1989. Unsupervised optimal fuzzy clustering. IEEE Trans. Pattern Anal. Mach. Intell. 11, 773–780.
Ghosh, A.K., Chaudhuri, P., Murthy, C., 2005. On visualization and aggregation of nearest neighbor classifiers. IEEE Trans. Pattern Anal. Mach. Intell. 27, 1592–1602.
Howard, I., 1994. A Review of Rolling Element Bearing 'Vibration Detection, Diagnosis and Prognosis'. DTIC Document.
Jain, A.K., Murty, M.N., Flynn, P.J., 1999. Data clustering: a review. ACM Comput. Surv. 31, 264–323.
Kim, D.-W., Lee, K.H., Lee, D., 2004. On cluster validity index for estimation of the optimal number of fuzzy clusters. Pattern Recognit. 37, 2009–2025.
Kwon, S.H., 1998. Cluster validity index for fuzzy clustering. Electron. Lett. 34, 2176–2177.
Lee, H., Kim, E., Park, M., 2007. A genetic feature weighting scheme for pattern recognition. Integr. Comput. Aid. Eng. 14, 161–171.
Lei, Y., He, Z., Zi, Y., Chen, X., 2008a. New clustering algorithm–based fault diagnosis using compensation distance evaluation technique. Mech. Syst. Signal Process. 22, 419–435.
Lei, Y., He, Z., Zi, Y., Hu, Q., 2008b. Fault diagnosis of rotating machinery based on a new hybrid clustering algorithm. Int. J. Adv. Manuf. Technol. 35, 968–977.
Lei, Y., Zuo, M.J., 2009. Gear crack level identification based on weighted K nearest neighbor classification algorithm. Mech. Syst. Signal Process. 23, 1535–1547.
Li, C., Pickering, C., 1992. Robustness and sensitivity of non-dimensional amplitude parameters for diagnosis of fatigue spalling. Cond. Monit. Diagn. Technol. 2, 81–84.
Liao, T.W., 2005. Clustering of time series data—a survey. Pattern Recognit. 38, 1857–1874.
Lu, J., Zhao, T., Zhang, Y., 2008. Feature selection based-on genetic algorithm for image annotation. Knowledge Based Syst. 21, 887–891.
Melin, P., Castillo, O., Ram Rez, E.G., 2007. Analysis and Design of Intelligent Systems Using Soft Computing Techniques. Springer, Berlin, Heidelberg.
Östh, J., Clark, W.A., Malmberg, B., 2015. Measuring the scale of segregation using k-nearest neighbor aggregates. Geogr. Anal. 47, 34–49.
Pakhira, M.K., Bandyopadhyay, S., Maulik, U., 2004. Validity index for crisp and fuzzy clusters. Pattern Recognit. 37, 487–501.
Pal, N.R., Bezdek, J.C., 1995. On cluster validity for the fuzzy *c*-means model. IEEE Trans. Fuzzy Syst. 3, 370–379.
Rauber, T.W., De Assis Boldt, F., Varejao, F.M., 2015. Heterogeneous feature models and feature selection applied to bearing fault diagnosis. IEEE Trans. Ind. Electron. 62, 637–646.
Samanta, B., 2004. Gear fault detection using artificial neural networks and support vector machines with genetic algorithms. Mech. Syst. Signal Process. 18, 625–644.
Samuel, P.D., Pines, D.J., 2005. A review of vibration-based techniques for helicopter transmission diagnostics. J. Sound Vibration 282, 475–508.
Soualhi, A., Clerc, G., Razik, H., 2013. Detection and diagnosis of faults in induction motor using an improved artificial ant clustering technique. IEEE Tran. Ind. Electron. 60, 4053–4062.
Sun, Q., Chen, P., Zhang, D., Xi, F., 2004. Pattern recognition for automatic machinery fault diagnosis. J. Vibration Acoustics 126, 307–316.
Wang, W., 2008. An enhanced diagnostic system for gear system monitoring. IEEE Trans. Syst. Man Cybern. B Cybern. 38, 102–112.

Wang, X., Wang, Y., Wang, L., 2004. Improving fuzzy *c*-means clustering based on feature-weight learning. Pattern Recognit. Lett. 25, 1123–1132.

Wu, K.-L., Yang, M.-S., 2005. A cluster validity index for fuzzy clustering. Pattern Recognit. Lett. 26, 1275–1291.

Xie, X.L., Beni, G., 1991. A validity measure for fuzzy clustering. IEEE Trans. Pattern Anal. Mach. Intell., 841–847.

Xu, R., Wunsch, D., 2005. Survey of clustering algorithms. IEEE Trans. Neural Netw. 16, 645–678.

Yeung, D.S., Wang, X., 2002. Improving performance of similarity-based clustering by feature weight learning. IEEE Trans. Pattern Anal. Mach. Intell. 24, 556–561.

Yiakopoulos, C., Gryllias, K.C., Antoniadis, I.A., 2011. Rolling element bearing fault detection in industrial environments based on a K-means clustering approach. Expert Syst. Appl. 38, 2888–2911.

Zheng, Z., Jiang, W., Wang, Z., Zhu, Y., Yang, K., 2015. Gear fault diagnosis method based on local mean decomposition and generalized morphological fractal dimensions. Mech. Mach. Theory 91, 151–167.

Zhou, H., Chen, J., Dong, G., Wang, H., Yuan, H., 2016. Bearing fault recognition method based on neighbourhood component analysis and coupled hidden Markov model. Mech. Syst. Signal Process. 66, 568–581.

5

HYBRID INTELLIGENT FAULT DIAGNOSIS METHODS

5.1 Introduction

The health operation of machinery highly depends on bearings and other rotating elements which easily suffer from faults (Loutridis, 2008a; Halim et al., 2008; Jafarizadeh et al., 2008; Zuo et al., 2005). These faults may cause the decline of performance level and the breakdown of mechanical equipment. Therefore, it is significant for ensuring the reliable operation of machinery to conduct incipient fault diagnosis. Up to now, all kinds of fault diagnosis methods or techniques have been studied and applied extensively to extract incipient mechanical faults and further make an appropriate maintenance decision (Lei et al., 2011a, 2011b, 2013).

Artificial intelligent techniques (Lei et al., 2008b, 2008c; Loutridis, 2008b; Rafiee et al., 2007), such as *K* nearest neighbor (*K*NN) (Lei et al., 2009), adaptive neuro-fuzzy inference systems (ANFIS) (Lei et al., 2007; Boyacioglu and Avci, 2010; Jang, 1993), multilayer perceptron (MLP) neural network (Yilmaz and Özer, 2009; Riedmiller, 1994), and radial basis function (RBF) neural network (Lu et al., 1998; Chen et al., 1991), have been studied widely and applied effectively to automatic monitoring and diagnosis of mechanical health conditions. The reliability of fault monitoring and diagnosis systems is highly improved by introducing artificial intelligent technologies (Abu-Mahfouz, 2005; Yang et al., 2002; Hansen, 1998). It is a promising technique for mechanical fault diagnosis and also opens a window for intelligent diagnosis. Initially, the research mainly focuses on individual classifier to improve its accuracy and even produces a novel classifier. Thereby, lots of classifiers with different classification algorithms and structural characteristics have been developed. They greatly enrich the family of the classifiers. Therefore, the research on individual classifier has been well-established and the corresponding fundamental theory about all kinds of classifiers has also been built. However, the classification accuracy of single classifier is insufficient for complicated multifault problem to make a reliable diagnosis and decision for rotating machinery.

In order to increase the reliability, later the combination of multiple classifiers has been increasingly studied to overcome the limitations of individual classifier and achieve higher classification accuracy (Samanta, 2004; Rasheed et al., 2008; Bi et al., 2008; Huenupán et al., 2008; Kittler et al., 1998; Fumera and Roli, 2005; Zio et al., 2008). It is suggested that these classifiers differing in input feature set or classification algorithm usually display complementary classification behavior. Therefore, if the classification results of multiple classifiers can be combined by integration techniques to generate the final classification result, the final result may be superior to the best result of each individual classifier on one feature set (Lei et al., 2010, 2008a, 2009, 2007). Obviously, the combination of multiple classifiers can achieve more reliable diagnosis results than single classifier.

Based on this motivation of multiple classifier combination, there exist two important factors: multiple classifiers and input feature sets. It is well known that the classifier and feature set are two key portions during intelligent fault diagnosis. Therefore, a problem arises spontaneously whether different classifiers or input feature sets will produce complementary classification performance. In order to explore this problem, this chapter mainly studies multiple classifier combination from two different aspects: classifiers and input feature sets. Therefore, three kinds of combination schemes are investigated and analyzed in this chapter. They contain three categories: (1) the different classification engines but the same input feature sets (Section 5.2), (2) the same classification engines but different input feature sets (Section 5.3), and (3) the different classification engines and different input feature sets (Section 5.4). Note that different classification engines have two implications: the same classifier but different parameters, for example, multiple weighted *K*NN with different *K* and different classifiers with distinct classification algorithms. Dependent on three different combination strategies, multidimensional hybrid intelligent diagnosis method is investigated and complementary behaviors between different classifiers or input feature sets are deeply analyzed in this chapter.

5.2 Multiple W*K*NN Combination-Based Fault Diagnosis

5.2.1 Motivation

The *K*NN classifier, as one of the simplest and most attractive statistical classifiers, has been studied extensively and applied successfully in many pattern recognition fields. However, the *K*NN classifier has the following two main drawbacks one is that it faces severe challenges from difficult discrimination when the samples of different classes overlap in some regions of feature space and the other is how to seek an optimal value of the neighborhood parameter *K*. For the overlapping samples from different classes, Zhou and Chen (2006) developed a new cam weighted distance to eliminate the curse of dimensionality. In this method, the weighted distance is used to optimize the distance measurement dependent on the analysis of interprototype relationship. Wang et al. (2007) significantly increased the classification accuracy of *K*NN by introducing a new *K*NN rule. Up to now, many researchers have proposed different kinds of adaptive or discriminative metrics to increase the classification accuracy. In general, the aforementioned proposed methods, all focus on the improvement of *K*NN classifier. It is not easy for us to achieve the goal in engineering applications due to the complexity of algorithm itself. Moreover, it is important for improving *K*NN classifier to move a small step forward. In order to increase the classification accuracy more effectively, a feature weighting technology is applied to overcome the problem of overlapping samples. The improved *K*NN classifier is named as the weighted *K* nearest neighbor classifier (W*K*NN). The W*K*NN method can achieve the improvement of *K*NN classifier from the view of input feature sets rather than the classifier itself. Therefore, it is easier to be applied in practice.

For the selection of neighborhood parameter, usually *K* is estimated by the method of cross-validation. However, the optimal value of *K* in a classification process not only relies on the entire data set, but also on the specific observation to be classified. Therefore, a too large or small *K* value may cause the false discrimination of sample classes and further produce an incorrect maintenance decision. In order to achieve reliable diagnosis and maintenance, instead of using any single value of *K*, combing a finite sequence of *K*NN classifiers indexed by different *K* is studied in this section. Consequently, a *K* value approximate to ideal *K* value will be obtained to alleviate the effect of *K* on final classification results. The improvement of the classification accuracy is attributed largely to the complementary effect between multiple *K*NN classifiers with different *K* values.

Therefore, a classification system combining multiple W*K*NN classifiers with different *K* is more accurate and robust than a single classifier. Based on these improved schemes, the multiple W*K*NN combination is supposed to be a promising method in the field of pattern recognition and consequently may generate an effective approach to incipient fault diagnosis of mechanical equipment. Finally, this method is applied to fault diagnosis of the rolling element bearings and its effectiveness is further verified.

5.2.2 Multiple W*K*NN Combination

In the process of *K*NN classification, every training sample is generally represented in a *D*-dimensional space by virtue of each value of its *D* features. The testing sample is then also represented in the same space and its *K*NN is chosen. Then, we make the class of each of these *K* neighbors tallied and the class with the most "votes" is chosen as the classification of the testing sample. The *K*NNs are generally decided by calculating the Euclidean distance between the testing sample and each of the training samples. Since Euclidean distance is calculated in the process of *K*NN classification, *K*NN classifier is sensitive to scaling of the feature values. In order to solve this problem, therefore, the *K*NN is improved effectively by using a feature weighting technique. Feature weighting is a more common method than feature selection in which each feature is multiplied by a weight value proportional to the ability of the feature to discriminate different classes. Then, the weighted Euclidean distance between the testing sample Te_d and the mth training sample Tr_{md} can be written as:

$$d_m^{(wf)} = \left[\sum_{d=1}^{D} wf_d (Te_d - Tr_{md})^2 \right]^{1/2}, \ d = 1, 2, \ldots, D \text{ and } m = 1, 2, \ldots, M \quad (5.1)$$

where wf_d represents the weight of the dth feature and D and M are the numbers of features and training samples, respectively.

To calculate the weight wf_d, an compensation distance evaluation technique (CDET) referred to Section 4.2.3 in Chapter 4, is implemented to evaluate the D features, which are performed in the stage of feature selection. Consequently, the final sensitivity measurement of each feature can be obtained. The final sensitivity measurements are seen as the weights, wf_d ($d = 1, 2, \ldots, D$), of the corresponding features and assigned to each feature to improve the performance of the *K*NN classifier. As a result, the *K*NN classifier improved by the weighted Euclidean distance metric is called as the W*K*NN.

In addition, it is also noticed that the selection of neighbor parameter *K* is a traditional problem for the *K*NN classifier. Most of *K*NN classifiers with the fixed neighbor parameter *K* are used in the pattern recognition, which ignores the effects of data sets on final classification results. Moreover, this kind of fixed-value strategy not only depends on prior knowledge of specialist but also cannot match optimally with each data set. To make the selection of parameter *K* more reasonable and effective, it will be achieved by combining the results of the multiple *K*NN classifiers with different *K* instead of the fixed value of *K*. Multiple *K*NN combination may not only solve the sensitive problem of the selection of neighbor parameter K to data sets and specific observation but also increase the final classification accuracy. In order to effectively combine the classification results of multiple KNN classifiers, a weighted averaging technique is applied to combine the results of multiple classifiers to make a final decision. Considering that WKNN overcomes the shortcoming of the traditional KNN, the WKNN is adopted to implement the classification rather than KNN. Thus, the final result of multiple WKNN combination is described as:

$$\begin{cases} \overline{c_n} = \sum_{g=1}^{G} w_g \overline{c_{n,g}}, n=1,2,\ldots,N_a, g=1,2,\ldots,G \\ \sum_{g=1}^{G} w_g = 1, \quad w_g \geq 0 \end{cases} \tag{5.2}$$

where $\overline{c_n}$ and $\overline{c_{n,g}}$ denote the classification results of the nth sample by the WKNN combination and the gth WKNN, respectively; w_g is the weight related with the gth single WKNN classifier; N_a is the number of all samples and G is the number of the WKNN classifiers to be combined. Based on a rule of thumb that investigators usually apply, it is noticed that the maximum of neighbor parameter K should not exceed $\sqrt{N_{\mathrm{tr}}}$ where N_{tr} is the number of the training samples. For the finite sequence of WKNN, the K value is from 1 to $\lfloor \sqrt{N_{\mathrm{tr}}} \rfloor$ with an increment of 1, where $\lfloor \sqrt{N_{\mathrm{tr}}} \rfloor$ denotes the largest integer less than or equal to $\sqrt{N_{\mathrm{tr}}}$. Therefore, the number G of the WKNN to be combined is equal to $\lfloor \sqrt{N_{\mathrm{tr}}} \rfloor$.

The weight w_g can be evaluated through utilizing all kinds of technologies by optimizing an objective or fitness function. In this subsection, the genetic algorithms (GAs) can be applied to seek the optimal weights of multiple WKNN classifiers. The fitness or objective function is usually expressed as the following equation:

$$\begin{cases} f = \dfrac{1}{1+E_{\mathrm{tr}}} \\ E_{\mathrm{tr}} = \left[N_{\mathrm{tr}} \sum_{n=1}^{N_{\mathrm{tr}}} (c_n - \overline{c_n})^2 \right]^{1/2}, \ n=1,2,\ldots,N_{\mathrm{tr}} \end{cases} \tag{5.3}$$

where E_{tr} denotes the root mean square training errors and c_n stands for the real result of the nth training sample.

Therefore, a multiple WKNN combination scheme is implemented in this subsection. It has two advantages: one is that traditional Euclidean distance is improved to increase the performance of KNN classifier by the feature weighting technique which considers the intrinsic characteristics of each feature and the other is that the classification accuracy is improved by combining multiple W*K*NN classifiers with different neighborhood parameter K instead of single W*K*NN classifier with a fixed K. However, a complete fault diagnosis method not only includes effective intelligent classification algorithm but also contains effective feature extraction and selection. Thus, next subsection will exhibit a complete intelligent fault diagnosis method based on multiple W*K*NN classifier combination.

5.2.3 Diagnosis Method Based on Multiple W*K*NN Combination

Feature extraction and selection is a key procedure in the whole intelligent fault diagnosis process. Generally, if effective features, sensitive to mechanical faults, are extracted from the collected vibration signals, it will yield twice the result with half the effort. Well, first we should extract as much features as possible from vibration signals by virtue of various advanced signal processing methods or techniques. Then, the sensitive or effective features should be selected from feature sets using certain feature selection methods. However, according to signal processing tools in Chapter 2 we know that a signal can be analyzed in time, frequency, and time–frequency domains. In other words, features can also be extracted from the time, frequency, and time–frequency domains of vibration signals to uncover the operating conditions of machinery from different aspects.

To monitor the health conditions of machinery, therefore, 10 common time–domain statistical features are extracted from the time domains of original vibration signals. These time–domain statistical features include mean value (X_{m}), standard deviation (X_{sd}), root mean square (X_{rms}), peak value (X_{peak}), skeness (X_{skeness}), kurtosis (X_{kurtosis}), crest factor (X_{crest}), clearance factor ($X_{\mathrm{clearance}}$), shape factor (X_{shape}), and impulse factor (X_{impulse}), as shown in Table 2.1 in Chapter 2. Successively, four common frequency–domain statistical features, including mean frequency (X_{mf}), frequency center (X_{fc}), root mean square frequency (X_{rmsf}), and standard deviation frequency (X_{stdf}) as shown in Table 2.2 in Chapter 2, are also extracted from FFT spectra of original vibration signals. Therefore, 14 feature values are produced from both time and frequency domains of original vibration signals.

To acquire much more characteristic information from original vibration signals, many advanced signal processing techniques are also utilized to produce much more subsignals, thereby extracting more features from them. Moreover, these signal processing techniques may uncover valuable features from original vibration signals. Wavelet transform (WT) has been studied widely for condition monitoring and fault diagnosis in recent years (Sun and Tang, 2002; Mallat and Hwang, 1992). However, it is still difficult for us to select a suitable basis function by virtue of different mechanical faults because different basis functions have their own merits in the time and frequency domains. Therefore, the analysis results may largely depend on the selection of wavelet basis function. The lifting scheme is a new wavelet construction method, which implements the WT in the time domain (Sweldens, 1998). It can not only avoid the annoyance of basis function selection but also closely associate with original vibration signals. Consequently, it is appropriate for mechanical health monitoring to process the vibration signals. In this proposed method, wavelet packet transform based on lifting scheme (WPTLS) with level 3 is utilized to further decompose each of the original signals (Duan et al., 2007). Therefore, eight frequency-band signals are obtained using the WPTLS. Similar to the feature extraction from the original vibration signals, 10 statistical features in the time domain are extracted from each of the 8 frequency-band signals, respectively. Then, additional 10×8 time–domain statistical features are generated. Demodulation can not only weaken the effects of interfering components in each frequency-band signal of the WPTLS but also focus on the low-amplitude and high-frequency broadband signals characterizing operating condition of rolling element bearings. Therefore, each of the 8 frequency-band signals is further demodulated using the Hilbert transform. 7×8 frequency-domain statistical features are further extracted from each envelope spectrum of each of 8 frequency-band signals. Here, 7 frequency–domain features not only contain 4 common frequency–domain statistical features (X_{mf}, X_{fc}, X_{rmsf}, X_{stdf}) but also include 3 special frequency–domain statistical features for rolling element bearings (SPPI, SPPO, SPPR) as shown in Table 3.3 in Chapter 3. Thus, we obtain altogether 150 features finally, which contain 90 time–domain statistical features and 60 frequency–domain statistical features.

Although the aforementioned 150 detected features may recognize faults occurring in a machine from different aspects, they have different importance to recognize different faults. Some features may be sensitive and closely associated with the faults, but others are not. Thus, before a feature set is fed into a classifier, sensitive features providing bearing fault-related information should be chosen as a new feature set from the original feature set and irrelevant or redundant features in original feature set should be discarded. Thereby, the new feature set can improve the classifier performance and avoid the curse of dimensionality. The motivation of feature selection is to screen out sensitive features from each original feature set. Feature selection has been studied extensively in pattern recognition fields. Therefore, there are a lot of technologies available to choose effective or sensitive feature set. In this diagnosis method, a CDET, seen in Section 4.2.3 in Chapter 4, is employed to sift sensitive features from 150 features. This technique mainly calculates the average distance criteria between samples in different conditions and further selects several first features as sensitive features from large to small in terms of distance evaluation criteria. Finally, some sensitive features are obtained from the whole feature set. They are labeled as a new input feature set, which will be fed into multiple W*K*NN classifiers with different neighbor parameter *K*.

Therefore, a new hybrid intelligent fault diagnosis method based on multiple W*K*NN combination is shown in Fig. 5.1. The main procedures are explained as follows. (1) The vibration signals

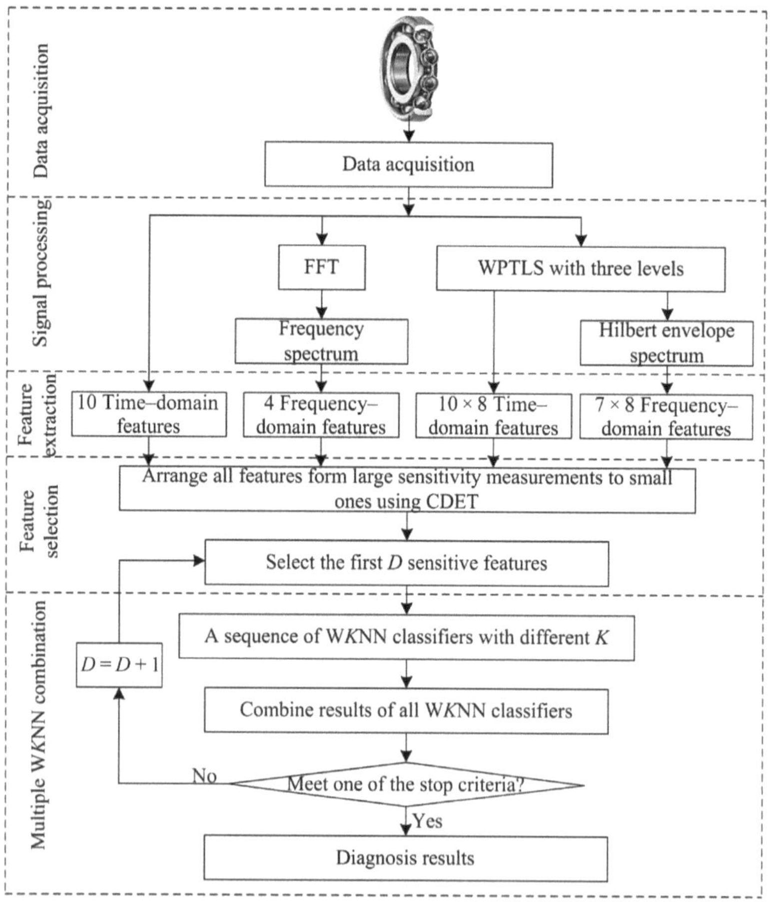

Figure 5.1. Flowchart of hybrid intelligent fault diagnosis method based on multiple W*K*NN combination.

collected from the rotating machinery are decomposed by WPTLS with level 3 and therefore 8 frequency-band signals are obtained. Eight frequency-band signals are further demodulated to calculate their envelope spectra. (2) Both time– and frequency–domain features, totally 150, are extracted not only from both the original vibration signals and frequency spectra but also from the 8 frequency-band signals and the corresponding envelope spectra. (3) One hundred and fifty features are evaluated and several sensitive features are selected from them utilizing CDET. (4) The sensitive features are fed into a finite sequence of multiple W*K*NN classifiers with the different neighborhood parameter *K*. (5) The weighted averaging technique based on GAs is employed to aggregate the results of different W*K*NN classifiers to achieve the combined or final results. Two stop criteria are given to terminate the training process: one is that the training accuracy increases to 100% and the other is that the training accuracy hardly increases when the number of the continuously added sensitive features is up to *F*, which is chosen as 5 in this method. When one of the two stop criteria is satisfied in the training process, the training process ends and the final combined results are obtained. In order to verify the effectiveness of the proposed method by experiments, next subsection will give a case study of rolling element bearings.

5.2.4 Intelligent Diagnosis of Fault Categories: A Rolling Element Bearing Case Study

The experimental system for rolling element bearings has been explained completely in Section 3.2.2.4 in Chapter 3. In order to describe the data sets conveniently, we make a compact review. The faults were introduced into the drive-end deep groove ball bearing of the motor by utilizing the electrical discharge machining (EDM) technique. Each of the three fault categories not only includes outer race, inner race, and ball faults but also contains three different fault levels: 0.007, 0.014, and 0.021 in. Each bearing was tested under four different loads: 0, 1, 2, and 3 hp, respectively. Data set of the bearing was acquired from the experimental system under four different operating conditions: (1) normal condition, (2) with outer race fault, (3) with inner race fault, and (4) with ball fault. Therefore, the vibration signals were acquired under different operating loads and bearing conditions.

Firstly, two experiments were implemented to acquire two different data sets A and B. The two data sets will be used to estimate the effectiveness of hybrid intelligent diagnosis method based on the multiple W*K*NN combination for different fault categories with the same fault levels under various loads. Both data sets A and B are depicted in Table 5.1.

Table 5.1 Data Sets A, B, C, D, E and F

Data Set	The Number of Training Samples	The Number of Testing Samples	Fault Size of Training/Testing Samples (In.)	Operating Condition	Label of Classification
A	30	30	0/0	Normal condition	1
	30	30	0.007/0.007	Outer race fault	2
	30	30	0.007/0.007	Inner race fault	3
	30	30	0.007/0.007	Ball fault	4
B	30	30	0/0	Normal condition	1
	30	30	0.021/0.021	Outer race fault	2
	30	30	0.021/0.021	Inner race fault	3
	30	30	0.021/0.021	Ball fault	4
C	30	30	0/0	Normal condition	1
	30	30	0.007/0.007	Outer race fault	2
	30	30	0.014/0.014	Outer race fault	3
	30	30	0.021/0.021	Outer race fault	4
D			0/0	Normal condition	1
	30	30	0.007/0.007	Inner race fault	2
	30	30	0.014/0.014	Inner race fault	3
	30	30	0.021/0.021	Inner race fault	4
E	30	30	0/0	Normal condition	1
	30	30	0.007/0.007	Ball fault	2
	30	30	0.014/0.014	Ball fault	3
	30	30	0.021/0.021	Ball fault	4
F	30	30	0/0	Normal condition	1
	30	30	0.007/0.007	Outer race fault	2
	30	30	0.007/0.007	Inner race fault	3
	30	30	0.007/0.007	Ball fault	4
	30	30	0.014/0.014	Outer race fault	5
	30	30	0.014/0.014	Inner race fault	6
	30	30	0.014/0.014	Ball fault	7
	30	30	0.021/0.021	Outer race fault	8
	30	30	0.021/0.021	Inner race fault	9
	30	30	0.021/0.021	Ball fault	10

Data set A contains 240 data samples from four different working conditions with the same fault size (0.007 in.) under four different loads. Each of the four working conditions contains 60 data samples. Data set A is separated into two data subsets: 120 samples for training and 120 for testing. It is a simple four-class classification problem corresponding to the four different fault categories. Data set B is similar to data set A and the only distinction is the fault size which is 0.021 in. for B. The intention of the two experiments over data sets A and B is to test the classification capability of the multiple W*K*NN combination to different fault categories with the same fault sizes or levels under different loads.

For data sets A and B, there are 120 training samples, respectively. According to Section 5.2.2, we can calculate the number K of nearest neighbor for the sequence of *K*NN classifiers, which is from 1 to 10 $\left(\left\lfloor\sqrt{N_{tr}}\right\rfloor=\left\lfloor\sqrt{120}\right\rfloor=10\right)$. When data set A is fed into the hybrid intelligent fault diagnosis method based on multiple W*K*NN combination, the first criterion, that is, the training accuracy reaches 100%, is satisfied during the training. The corresponding number of the selected features is two. The two selected features are the clearance factor ($X_{clearance}$) and shape factor (X_{shape}) of the fourth frequency-band signal obtained by utilizing WPTLS technique, respectively. The high accuracy (96.76%) of the testing result suggests that instead of using entire feature sets only using these two features in the hybrid intelligent fault diagnosis method may diagnose the early faults in bearings. The testing results are exhibited in Table 5.2.

Table 5.2 Comparison of Testing Accuracies (%) Between Multiple W*K*NN Combination and Individual *K*NN Classifier

	Single *K*NN Classifier			
Data Set	**Min**	**Mean**	**Max**	**W*K*NN Combination**
A	89.12	91.50	92.42	97.67
B	84.62	86.20	87.75	95.83
C	87.33	89.20	90.33	100
D	87.83	89.18	90.50	100
E	82.75	83.74	84.67	93.33
F	79.76	81.77	84.00	98.67

In order to demonstrate how well the hybrid intelligent fault diagnosis method based on multiple W*K*NN combination works, in each experiment, the performance of the single classifier based on *K*NN with the same number of input features was also tested. The features fed into the individual *K*NN classifier are chosen randomly from 150 features rather than utilizing the feature selection method which is, CDET. To make a sufficient comparison with hybrid intelligent diagnosis method based on multiple W*K*NN combination, the features are chosen randomly 10 times and the average testing results of single *K*NN classifier are shown in Table 5.2. In addition, the testing accuracies of the 10 *K*NN classifiers are also shown in Table 5.3.

When data set B is fed into multiple W*K*NN combination classifiers, the 100% training accuracy is first satisfied and the training process of the hybrid intelligent fault diagnosis method is terminated. The corresponding number of the selected features is one. The selected feature is crest factor of the original signals. Thus, the crest factor of testing sample is fed into the hybrid intelligent fault diagnosis method to test and the corresponding testing accuracy highly reaches 95.83% as shown in Table 5.2. It indicates that the method based on the multiple W*K*NN combination by using only one feature may discriminate the different fault categories under various loads with high classification accuracy. The diagnostic

Table 5.3 Testing Accuracies (%) of 10 *K*NN Classifiers for Different Fault Categories or Levels

Data set/Classifier	A	B	C	D	E
Classifier 1	92.2	84.7	88.2	87.9	82.8
Classifier 2	92.2	84.6	88.2	87.9	82.8
Classifier 3	92.4	85.3	90.0	87.8	83.5
Classifier 4	92.4	86.0	90.3	88.7	84.4
Classifier 5	91.7	85.9	90.0	89.1	83.9
Classifier 6	92.0	86.9	90.2	90.1	84.2
Classifier 7	90.9	86.5	89.4	90.3	84.7
Classifier 8	91.4	86.8	89.8	89.7	84.5
Classifier 9	89.1	87.5	88.6	90.5	83.3
Classifier 10	90.8	87.8	87.3	89.9	83.4

results of the hybrid intelligent fault diagnosis method based on multiple W*K*NN combination and single *K*NN classifier for data set B are shown in Table 5.2 and Table 5.3, respectively.

To verify the performance of hybrid intelligent fault diagnosis method based on multiple W*K*NN combination for different fault levels with the same fault categories under various loads, three experiments were implemented on three different data sets C, D, and E which correspond to the three fault categories: outer race fault, inner race fault, and ball fault, respectively. Detail introductions of these three data sets are depicted in Table 5.1, respectively. Data set C consists of 60 samples for the normal bearings and 60 for every fault level with the outer race fault under different loads. A half data of every condition are applied to train and the rest to test. This is also a simple four-class classification problem to diagnose the fault emergence and level of the rolling element bearing. The distinctions for data sets D and E from C are the fault categories, where data set D is with the inner race fault and data set E is with the ball fault. Both data sets D and E consist of 60 normal samples and 60 fault samples in every fault level under four different loads. These three data sets are employed to verify the performance of this multiple W*K*NN combination method on identifying different fault levels with the same fault categories.

When one of these two criteria in multiple W*K*NN combination method is satisfied, the training process stops. Consequently, the crest factor (X_{crest}) of the third frequency-band signal and the shape factor (X_{shape}) of the fourth frequency-band signal of WPTLS are chosen as sensitive features. The two sensitive features are used to diagnose the different fault levels in the bearing outer race on data set C. As for data set D, only one feature, that is, root mean square frequency X_{rmsf} of the original signals, is chosen to diagnose the inner race fault of rolling element bearings. However, it is relatively difficult to discriminate different fault levels of the roller on data set E. As a result, more features, including X_{crest} and X_{rmsf} of the original signals and SPRO, SPRI, SPRR of the fourth frequency-band signal of WPTLS, are applied to achieve 100% training accuracy. The final testing results of both the multiple W*K*NN combination and corresponding single *K*NN classifier for these three data sets are depicted in Table 5.2 and Table 5.3, respectively.

To further demonstrate the performance of the hybrid intelligent fault diagnosis method based on multiple W*K*NN combination for different fault categories and fault levels, last verification is implemented on data set F as exhibited in Table 5.1. The data set F consists of 600 data samples, which are extracted from four different working conditions including normal condition, outer race fault, inner race fault, and ball fault under four different loads. Moreover, every fault condition contains three different fault levels: 0.007, 0.014, and 0.021 in., respectively. The 600 data samples are also separated into two subsets: 300 samples for training and 300 for testing. For data set F, to recognize the categories and levels of faults simultaneously, 10-class classification problem needs to be tackled.

Apparently, it is a difficult and typical classification problem in fault diagnosis of rolling element bearings. We not only need to diagnose different fault categories but also to diagnose all kinds of damage degrees under different loads. The multiple W*K*NN combination method, therefore, requires more features to achieve a satisfactory diagnosis result. In this experiment, when the number of the selected sensitive features reaches to 9, we can obtain 100% training accuracy and consequently the corresponding training process is also terminated. The result of the sensitivity measurements by utilizing CDET for data set F is exhibited in Fig. 5.2. The corresponding testing result of the hybrid intelligent fault diagnosis method is shown in Table 5.2. In order to make a comparison, the testing results with single *K*NN classifier are also depicted in Table 5.4.

In the experiments performed on data sets A and B, when each single *K*NN classifier is applied, their testing accuracies range from 89.12% to 92.42% with average 91.50% and from 84.62% to 87.75% with average 86.20%, respectively. Through the hybrid intelligent fault diagnosis method, however, the testing accuracies for data sets A and B reach up to 97.67 and 95.83%, respectively.

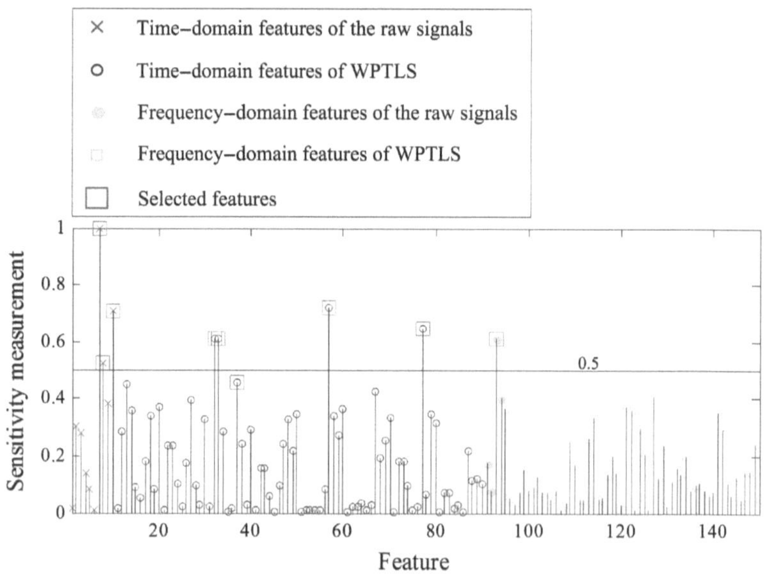

Figure 5.2. Sensitivity measurements of 150 features of data set F.

Table 5.4 Testing Accuracies (%) of Single *K*NN Classifier for Different Fault Categories and Levels (Data Set F)

Classifier	K1	K2	K3	K4	K5	K6	K7	K8	K9
Accuracy	83.00	84.00	83.53	83.77	82.63	82.40	81.97	82.23	81.83
Classifier	**K10**	**K11**	**K12**	**K13**	**K14**	**K15**	**K16**	**K17**	
Accuracy	81.43	80.97	81.07	80.67	80.63	80.00	80.23	79.76	

The diagnosis results suggest that the multiple W*K*NN combination method can diagnose the different fault categories under different loads better than individual *K*NN classifier.

For data sets C, D, and E, the testing accuracies of the hybrid intelligent diagnosis method are apparently higher than those of single *K*NN classifier. The accuracies of the multiple W*K*NN combination are 100% for data set C, 100% for D, and 93.33% for E, respectively. For individual *K*NN classifier, the accuracies of these three data sets are in the range of 87.33–90.33% with average 89.20%, 87.83–90.50% with average 89.18%, and 82.75–84.67% with average 83.74%, respectively. The experimental results indicate that the multiple W*K*NN combination method can identify various fault sizes with a better accuracy than single classifier.

For the last data set F from different fault categories and levels, however, it is a 10-class classification task and therefore more difficult than others. Consequently, the correct rate of each single classifier based on *K*NN obviously declines and the highest correct rate is only 84.00% and the lowest is 79.76%. The fault diagnosis method based on multiple W*K*NN combination, however, can obtain the highest testing accuracy, 98.67%. The classification results imply that the multiple W*K*NN combination method can not only recognize the different fault categories but also identify the different fault sizes better than single classifier.

Based on the aforementioned analytical results, it is reasonable to deduce that the multiple W*K*NN combination method should possess higher correct rate in identification accuracy and characterize better generalization ability in comparison with the single classifier based on *K*NN. The success created by the multiple W*K*NN combination can be attributed to the following three points. (1) Several sensitive features are chosen and other fault-unrelated features are discarded. These unrelated features would confuse the classifiers and thereby result in the decrease of classification accuracy. (2) The W*K*NN classifier tradeoffs different importance of each sensitive feature for diagnosing different faults of rolling element bearings. Consequently, feature weighting is conducted in the classification process to evaluate the different sensitive features. (3) It would generally be more reasonable to integrate the classification results of multiple W*K*NN classifiers with different neighbor parameter K than applying individual value of K.

It is also found that several selected sensitive features are detected from the frequency-band signals of WPTLS. The results suggest that processing the original vibration signals by employing advanced signal processing techniques can uncover more characteristic information reflecting health conditions of rolling element bearings. For example, in the experiment based on data set F, five of nine sensitive features derive from the frequency-band signals of WPTLS. This result can be observed from Fig. 5.2. Another phenomenon is also observed in the process of feature selection that the third and fourth frequency-band signals are relatively sensitive ones among all the eight frequency-band signals obtained by utilizing the WPTLS.

5.2.5 Epilog

In this section, a hybrid intelligent fault diagnosis method based on multiple W*K*NN combination is investigated by using different data sets including different fault categories with the same fault levels, different fault levels with the same fault categories and different fault categories and levels under four different loads. The experimental results suggest that hybrid intelligent diagnosis method based on multiple W*K*NN combination can achieve higher classification accuracy than single classifier based on *K*NN. Through the aforementioned experiments, it is known that the success may be attributed to three aspects: feature mining, feature selection, and multiple classifier combination.

Some valuable features are difficult to be extracted directly from original vibration signals. However, we can obtain some unexpected features by employing suitable signal processing tools. For example, several sensitive features are detected from frequency-band signals of WPTLS. Therefore, it is necessary that using advanced signal processing techniques uncovers the physical implications of original vibration signals, thereby mining better features.

From this experimental result with selected features randomly, it is noticed that some features in all extracted features are useless for fault diagnosis and they even can play a negative role, for example, in the curse of dimensionality. Therefore, before these features are fed into a classifier, they should be screened to improve classification accuracy and reduce dimensions of features. In addition, multiple W*K*NN classifiers with different neighborhood parameter *K* are combined and moreover the same selected feature sets are fed into them. This results show that under the same input feature sets different classifiers are able to compensate each other as we expected. From the inspiration of aforementioned summarizations, subsequently we investigate the effect of different input feature sets on final classification accuracy in next section.

5.3 Multiple ANFIS Hybrid Intelligent Fault Diagnosis

5.3.1 Motivation

Different from W*K*NN, the ANN has the capability of self-leaning and nonlinear approximation, but it lacks the inference which is common in human beings. Therefore, it requires massive amounts of training samples, which is an intensive time-consuming process. On the contrary, the fuzzy logic inference is a fast method for fuzzy and uncertain problems, but the traditional fuzzy systems largely depend on the knowledge and experience of experts and operators. Therefore, it is difficult to obtain a satisfactory prediction result, especially when this information is incomplete or insufficient. To solve aforementioned problems, adaptive network based on fuzzy inference system (ANFIS) is first developed by Jang (1993). ANFIS is a fuzzy inference system based on the Sugeno model. It combines the self-learning capability of ANN and the linguistic expression function of fuzzy inference, whose membership functions and fuzzy rules are obtained from a large number of existing data instead of experience or intuition. Obviously, ANFIS fuses the advantage of both ANN and fuzzy inference and overcomes their own shortcomings simultaneously. Therefore, it is supposed to be a promising classification method.

The combination of multiple classifiers has been extensively studied to break the limitations of individual classifier and to obtain higher accuracy. Therefore, based on the advantages of ANFIS, a hybrid intelligent fault diagnosis method based on multiple ANFIS combination is presented in this section. Different from the diagnosis method introduced in Section 5.2, the presented fault diagnosis method employs the same classification engine and different input feature sets. This method mainly utilizes the complementary classification performance between different input feature sets to further increase classification accuracy instead of different classification engines as shown in Section 5.2. The classification result of each ANFIS is combined to yield the final classification result by integration techniques and the final classification performance should be superior to the best performance of each single classifier on one feature set. In the process of multiple classification combination, multiple classifiers, that is ANFIS, will be combined by the weighted averaging technique, which assigns a nonnegative weight to each single classifier. Then weight of each classifier will be optimized by utilizing GAs. Therefore, this hybrid intelligent fault diagnosis method based on multiple ANFIS combination is expected to perform better than single ANFIS. Moreover, it also investigates hybrid intelligent fault diagnosis from other aspect: the same classifier combination with different input feature sets.

5.3.2 Multiple ANFIS Combination with GAs

The ANFIS maps inputs dependent on input membership functions and corresponding related parameters to hidden layer and then to outputs dependent on output membership functions. The original membership functions and rules of the fuzzy inference system are constructed to be modeled through utilizing human expertise about the objective system. Then the fuzzy if–then rules and membership functions of ANFIS are refined to represent the input/output relationship of a complicated system. Even if human expertise is unavailable it is probable to subjectively construct reasonable membership functions and then utilize the ANNs training process to produce a series of fuzzy if–then rules that approach an expected data set. Therefore, the performance of ANFIS is superior to each of ANN and fuzzy inference system. It is supposed to achieve good classification accuracy. However, single classifier is not good enough for complicated fault diagnosis. As shown in Section 5.2, multiple classifier combination will exhibit higher classification accuracy.

Combining multiple classifiers has recently obtained significant achievements in the mechanical fault diagnosis because it can largely ameliorate the behavior of single classifier. It is an aggregation of lots of classifiers to classify the testing samples by aggregating the single decision. The different classifiers may provide complementary information. Once they are combined, performance improvement may be achieved as shown in Section 5.2. In addition, utilizing different feature sets can also achieve the same improvement. Therefore, if the classification results of multiple classifiers utilizing different feature sets are integrated by integration techniques to obtain final classification result, the final classification accuracy may be better than that of single classifier. Among the all kinds of integration techniques presented in the literature, the weighted averaging technique is applied to the multiple classifier combination, which is the simplest and most frequently applied technique.

Therefore, multiple ANFIS combination method with different feature sets is proposed in this section. In this method, six different feature sets have been detected from both time–and frequency–domains and the six fault-related sensitive feature sets have been chosen as new feature sets from them by utilizing CDET. The number of ANFIS classifiers is determined by the number of selected sensitive feature sets. Then the combination of the classification results of six ANFIS classifiers is performed by utilizing the weighted averaging technique, and the final classification result is written as below

$$\hat{y}_n = \sum_{k=1}^{6} w_k \hat{y}_{n,k}, \ n=1,2,\ldots,N', \ k=1,2,\ldots,6 \tag{5.4}$$

subject to

$$\sum_{k=1}^{6} w_k, \ w_k \geq 0 \tag{5.5}$$

where $\hat{y}_n$ and $\hat{y}_{n,k}$ denote the classification results of the nth sample utilizing the classifier combination and the kth individual classifier respectively, w_k denotes the weight related with the kth individual classifier and N' denotes the number of all samples.

In the process of this multiple ANFIS combination, the weights are evaluated by using GAs to optimize the objective function. Real coded genomes are performed and a population size of 10 individuals is initialized with produced genomes at random. The maximum number of generations is selected as the termination condition for the solution process. In this method, the maximum number is 100. Nonuniform-mutation function and arithmetic crossover operator are applied, where the mutation probability and the crossover probability are 0.01 and 0.8, respectively. The objective function is written as:

$$f = \frac{1}{1+E} \tag{5.6}$$

where E denotes root mean square training errors represented as

$$E = \left[\frac{1}{N''} \sum_{n=1}^{N''} \left(y_n - \hat{y}_n \right)^2 \right]^{\frac{1}{2}}, n=1,2,\ldots,N'' \tag{5.7}$$

where y_n represents the real result of the nth training sample and N'' denotes the number of the training samples.

In this subsection, a scheme of multiple ANFIS combination based on GAs has been presented in detail, but a complete intelligent diagnosis fault method not only contains the classification method, that is, multiple ANFIS combination, but also includes effective feature extraction and selection. Therefore, next subsection will completely describe this fault diagnosis method based on multiple ANFIS combination from feature extraction to diagnosis results.

5.3.3 Fault Diagnosis Method Based on Multiple ANFIS Combination

At the beginning, 24 statistical features are extracted from original vibration signals, which consist of 11 common time–domain statistical features in Table 2.1 and 13 common frequency–domain statistical features in Table 2.2 in Chapter 2. The original signals are separated into some subsignals with 4096 data points. Each of these subsignals is further processed to detect 11 common time–domain statistical features from it and 13 common frequency–domain statistical features from its FFT spectrum, respectively. The common time–domain and frequency–domain statistical features detected from original vibration signals are labeled as feature set 1 and feature set 2, respectively. Thus, feature sets 1 and 2 include 11 and 13 features, respectively.

It is noticed that the low-frequency interference is contained in the acquired vibration signals. Thus, these signals are filtered by utilizing either high-pass or band-pass filters to remove the low-frequency interference components. In this method, three band-pass (BP1–BP3) and one high-pass (HP) filters are adopted. The band-pass frequencies (in kHz) of the BP1–BP3 filters are selected as: BP1 (2.2–3.8), BP2 (3.0–3.8), and BP3 (3.0–4.5), respectively. The cut-off frequency of the HP filter is selected as 2.2 kHz. These frequencies are chosen to cover the signal components consisting of the most of the bearing frequency-band energy. The 11 common time–domain statistical features as shown in Table 2.1 are detected from each of these 4 filtered signals. Total 11×4 time–domain statistical features are acquired and referred as feature set 3. Thus, feature set 3 includes 44 features. The influences of interference components in the chosen frequency band are able to be eliminated by using Hilbert demodulation technique. Since demodulation focuses on the low-amplitude and high-frequency broadband signals characterizing bearing conditions, it can make the fault diagnosis process a little more independent for a special machine. The Hilbert envelope spectra of these four filtered signals are further used to detect another feature set which consists of 13×4 common frequency–domain statistical features. This feature set is defined as feature set 4 and it totally includes 52 features.

To detect more characteristic information, each of these original signals is therefore decomposed by using empirical mode decomposition (EMD) technique illustrated in Chapter 2 to uncover physical nature buried in original vibration signals. Generally, the first eight intrinsic mode functions (IMFs) include almost all effective information, so they are selected to further extract some features. Similar to the feature extraction method of the original vibration signals, the 11 common time–domain statistical features are detected from each of first eight IMFs. Then, an additional feature set containing 11×8 common time–domain statistical features is acquired and defined as feature set 5. Thus, feature set 5 totally includes 88 features.

In addition, each IMF is demodulated by utilizing the Hilbert transform and corresponding Hilbert envelope spectrum is also calculated. Therefore, we can obtain the envelope spectra of first eight IMFs and further 13 common frequency–domain statistical features are detected from each envelope spectrum. Finally, another feature set including 13×8 frequency–domain features is obtained and referred as feature set 6. Therefore, feature set 6 includes 104 features totally.

Although the aforementioned extracted features may recognize the mechanical running conditions from different aspects, in the recognition process they have different importance for different fault types and sizes. It is possible that some of them are sensitive and closely associated to the corresponding fault conditions, but others are not. So aforementioned features extracted by all kinds of signal processing techniques should not be directly fed into a classifier without selection, which will result in the curse of dimensionality. Therefore, sensitive features related with bearing faults must be selected from these feature sets and simultaneously fault-unrelated features must be discarded to increase the classifier performance and avoid the curse of dimensionality. In this method, CDET is applied to choose the sensitive or effective features from all the feature sets.

Based on aforementioned principles, we utilize lots of technologies, such as statistics, EMD, CDET, ANFIS, and GAs, to propose a novel hybrid intelligent fault diagnosis method based on multiple ANFIS combination for rotating machinery as shown in Fig. 5.3. It contains the following several main procedures. (1) The original vibration signals are filtered by utilizing three band-pass and one

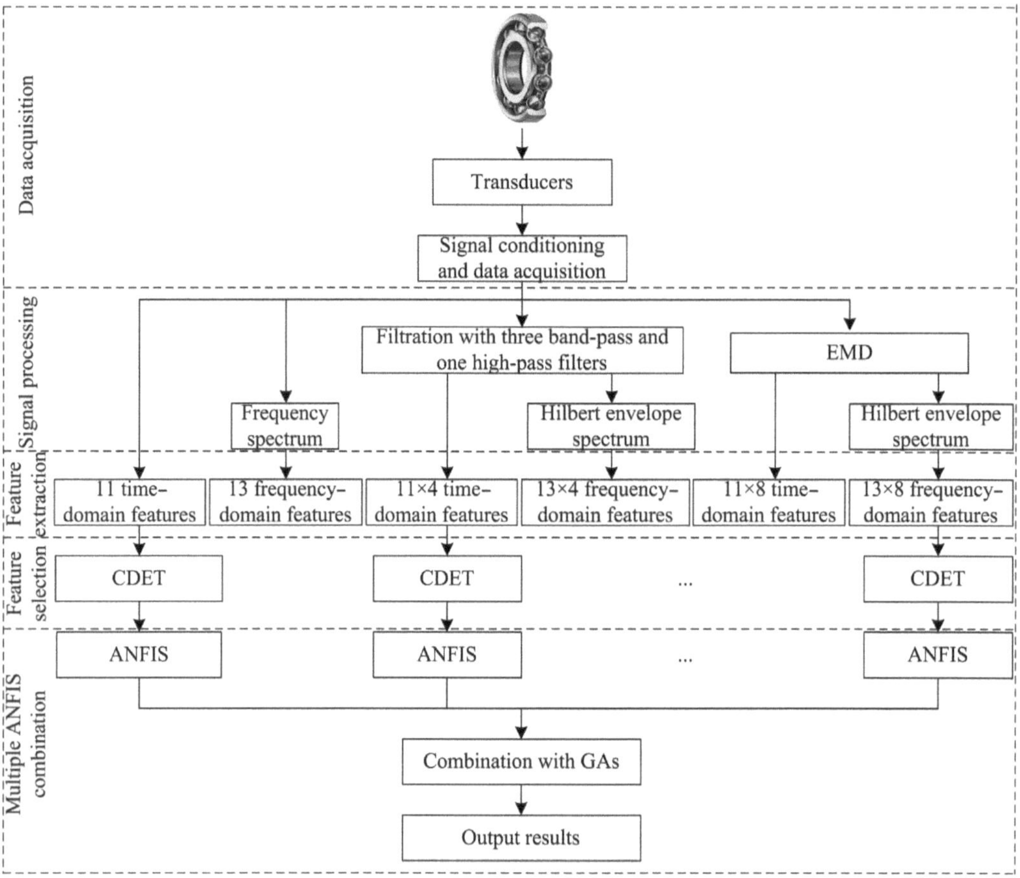

Figure 5.3. Flowchart of hybrid intelligent fault diagnosis method based on multiple ANFIS combination.

high-pass filters and four filtered signals are obtained. At the same time, original vibration signals are also decomposed by utilizing the EMD technique and first eight IMFs are obtained. The four filtered signals and eight IMFs are further demodulated by using Hilbert transform to compute their Hilbert envelope spectra. (2) We will acquire six feature sets including time–domain and frequency–domain from both the original and preprocessed signals. (3) Every feature set is estimated and several sensitive features are chosen from it by using the CDET. Correspondingly, six sensitive feature sets are acquired. (4) Every sensitive feature set is fed into the classifier based on ANFIS for training and testing. There are totally six classifiers based on ANFIS corresponding to the six sensitive feature sets. (5) The weighted averaging technique with GAs is utilized to incorporate the output results of the six classifiers based on ANFIS and generate the final classification results.

Therefore, a complete hybrid intelligent fault diagnosis method based on multiple ANFIS combination is exhibited in this subsection. The effectiveness and performance of this method will be further demonstrated by a rolling element bearing case in next subsection.

5.3.4 Intelligent Diagnosis of Fault Severities: A Rolling Element Bearing Case Study

The experimental system for rolling element bearings was explained completely in Section 3.2.2.4 in Chapter 3. The bearing data set was acquired from the experimental system under the four different operating conditions: (1) normal condition, (2) with an outer race fault, (3) with an inner race fault, and (4) with a ball fault. The details are referred to Section 5.2.

To estimate the effectiveness of the hybrid intelligent fault diagnosis method based on multiple ANFIS combination for different fault categories and levels, four experiments are implemented via four different data sets (A–D) from all data set of the rolling element bearings. The details of the four data sets are described in Table 5.5.

Data set A contains 240 data samples from four different operating conditions: normal condition, outer race fault, inner race fault, and ball fault, respectively. The fault severity is the same in data set A, where the fault defect sizes are all 0.007 in. under four different loads. Each of the four working conditions contains 60 data samples. Data set A with 240 data samples is divided into two subsets: 120 samples for training and 120 samples for testing. It is a simple four-class classification problem corresponding to the four different working conditions. The experiment over data set A is conducted to demonstrate the classification performance of multiple classifier combination method based on ANFIS for different fault types.

Data set B also includes 240 data samples. Here, 120 samples with the fault defect size 0.007 in. are the same with the 120 training samples of data set A and finally constitute the training samples of data set B. Another 120 samples with the fault defect size of 0.021 in. are the testing samples of data set B. The experiment via this data set is implemented to further study the diagnosis capability for the developing faults when the classifier is trained by the early faulty samples.

It is similar between data set C and data set B. They have the identical samples. However, the training set of C is the testing set of B and the testing set of C is the training set of B. The intention of utilizing this data set is to investigate the classification behavior

Table 5.5 Description of Four Data Sets (A–D)

Data Set	The Number of Training Samples	The Number of Testing Samples	Defect Size of Training/Testing Samples (In.)	Operating Condition	Label of Classification
A	30	30	0/0	Normal	1
	30	30	0.007/0.007	Outer race	2
	30	30	0.007/0.007	Inner race	3
	30	30	0.007/0.007	Ball	4
B	The training samples are the same to A above.	30	0/0	Normal	1
		30	0.007/0.021	Outer race	2
		30	0.007/0.021	Inner race	3
		30	0.007/0.021	Ball	4
C	The training samples are the testing samples of B.	The testing samples are the training samples of B.	0/0	Normal	1
			0.021/0.007	Outer race	2
			0.021/0.007	Inner race	3
			0.021/0.007	Ball	4
D	30	30	0/0	Normal	1
	30	30	0.007/0.007	Outer race	2
	30	30	0.007/0.007	Inner race	3
	30	30	0.007/0.007	Ball	4
	30	30	0.014/0.014	Outer race	5
	30	30	0.014/0.014	Inner race	6
	30	30	0.014/0.014	Ball	7
	30	30	0.021/0.021	Outer race	8
	30	30	0.021/0.021	Inner race	9
	30	30	0.021/0.021	Ball	10

of hybrid intelligent diagnosis method based on multiple ANFIS combination for the early faults when the classifier is trained by the samples with large defect sizes.

Data set D consists of 600 data samples from four kinds of working conditions and loads. Each fault condition also contains three kinds of fault levels of 0.007, 0.014, and 0.021 in., respectively. The 600 data samples are split into 300 samples for training and 300 for testing. For data set D, the purpose is to recognize the severe degree of the faults and 10-class classification task should be tackled in this experiment.

Usually, a single output is applied to stand for multiple classes. Therefore, it is easier to generate false diagnosis results and may result in lower classification capability. However, the single output may simplify the neural network architecture in a multiple class task. Thus, the number of the updated parameters is reduced and thereby the speed in the training and testing process is increased.

To test the performance of the hybrid intelligent fault diagnosis method based on multiple ANFIS combination, in other words how significant the generalization capability is increased by applying the multiple classifier combination based on ANFIS, in these experiments the classification capability of the single classifier based on ANFIS is also investigated to make a comparison.

Considering the computational time, in every case we should not select too many sensitive features to train and test the classifier. Therefore, only first four sensitive features are picked out from each of the six feature sets by employing the CDET and then are fed into the corresponding classifier based on ANFIS. For data set A, the distance evaluation criteria $\bar{\alpha}_j$ of the six feature sets are plotted in Fig. 5.4. In Fig. 5.4 (A), feature set 1 consists of

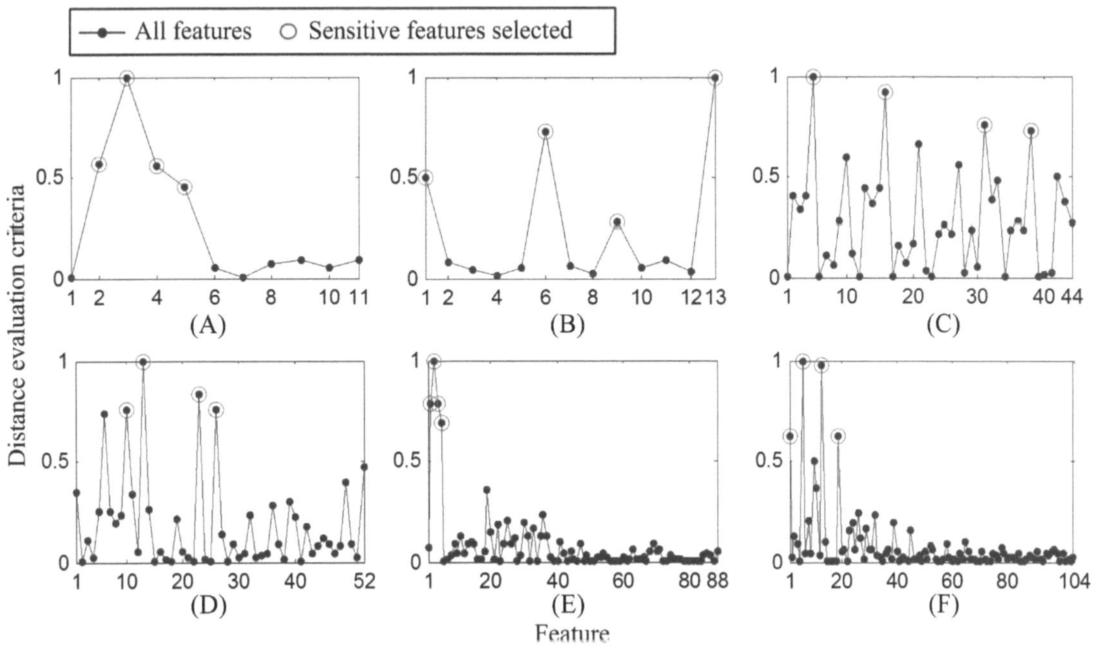

Figure 5.4. Distance evaluation criteria of six feature sets of data set A. (A) Feature set 1, (B) feature set 2, (C) feature set 3, (D) feature set 4, (E) feature set 5, (F) feature set 6.

11 common time–domain statistical features extracted from original vibration signals. However, it is observed that four selected sensitive features are standard deviation (X_{sd}), root amplitude (X_{root}), root mean square (X_{rms}), and peak (X_{peak}), respectively. Feature set 2 contains 13 common frequency–domain statistical features from FFT spectra of original vibration signals. It is found from Fig. 5.4 (B) that four selected sensitive features are p_1, p_6, p_9, and p_{13}, respectively, as defined in Table 2.2 in Chapter 2. In addition, feature set 3 consists of four sensitive features: three peaks (X_{peak}) form three filtered signals by utilizing BP1, BP2, and HP filters and clearance ($X_{clearance}$) from the filtered signal by utilizing BP3 filter. Feature set 4 includes two p_{10} and two p_{13} from two filtered signals of BP1 and BP2, respectively. Feature set 5 contains X_{sd}, X_{root}, X_{rms}, and X_{peak} from first intrinsic mode function (IMF1) of EMD. Feature set 6 includes p_1, p_6, and p_{13} from envelope spectrum of IMF1 and p_6 from envelope spectrum of IMF2. Obviously, each feature has different importance for different fault types and sizes. Feature selection can discard redundant or useless features and further achieve feature dimension reduction. Moreover, we can also find that many sensitive features can be extracted from the processed signals by employing advanced signal processing techniques. In other words, some valuable features can be produced by utilizing proper signal processing techniques. Therefore, it is necessary for improving classification accuracy to preprocess the original vibration signals by certain signal processing techniques. Finally, the classification results of the four data sets are exhibited in Table 5.6 and Fig. 5.5.

Analyzing the classification results from Table 5.6 and Fig. 5.5 we can draw several following conclusions. First, for data set A, since this classification problem is relatively easy, both the six single classifiers and their combination obtain the nice training and testing correct rates which reach up to 100%, as shown in Fig. 5.5 (A). However, it is noticed from Fig. 5.6 that the classification error of the multiple classifier combination is less than that of each single classifier.

Second, although 100% training accuracy is achieved by both the single classifier and multiple classifier combination on data set B, the testing accuracy of the single classifier is apparently lower than that of the multiple classifier combination as shown in Fig. 5.5 (B). The testing correct ratio of the single classifier is in the range of 60.83–83.33% with average 70%, while it is 92.5% for the multiple classifier combination. This experimental result suggests that the incorporation of the multiple classifiers trained by the early fault samples can identify the developing faults more effectively than individual classifier.

Table 5.6 Classification Accuracy Comparison (%) Between Single Classifier and Multiple Classifier Combination for Four Different Data Sets (A–D)

Data Sets/Classifier		A	B	C	D
Classifier1	Training	100	100	100	65.67
	Testing	100	66.67	62.5	61
Classifier2	Training	100	100	100	90
	Testing	100	75	79.17	87.67
Classifier3	Training	100	100	100	72.67
	Testing	100	61.67	74.17	68
Classifier4	Training	100	100	100	80.33
	Testing	100	83.33	80	77
Classifier5	Training	100	100	100	67.67
	Testing	100	60.83	55.83	67
Classifier6	Training	100	100	100	87.33
	Testing	100	72.5	81.67	81
Average of six classifiers	Training	100	100	100	77.28
	Testing	100	70	72.22	73.61
Classifier combination	Training	100	100	100	93.67
	Testing	100	92.5	90.83	91.33

Third, for data set C, it is found from Fig. 5.5 (C) that the training correct ratios of all classifiers are 100%. However, the multiple classifier incorporation generates higher testing correct ratio with 90.83% than every single classifier ranging from 55.83% to 81.67% with average 72.22%. This result indicates that the incorporation of the multiple classifiers trained by the fault samples with large defect sizes can recognize the early faults with higher classification accuracy than the single classifier.

Finally, for data set D it is a 10-class classification task and thereby relatively difficult. Therefore, the training accuracies of all classifiers start to decline gradually ranging from 65.67% to 93.67% as shown in Fig. 5.5 (D). However, the multiple classifier incorporation still obtains the highest training accuracy with 93.67%. For testing samples, the classification accuracy of the six

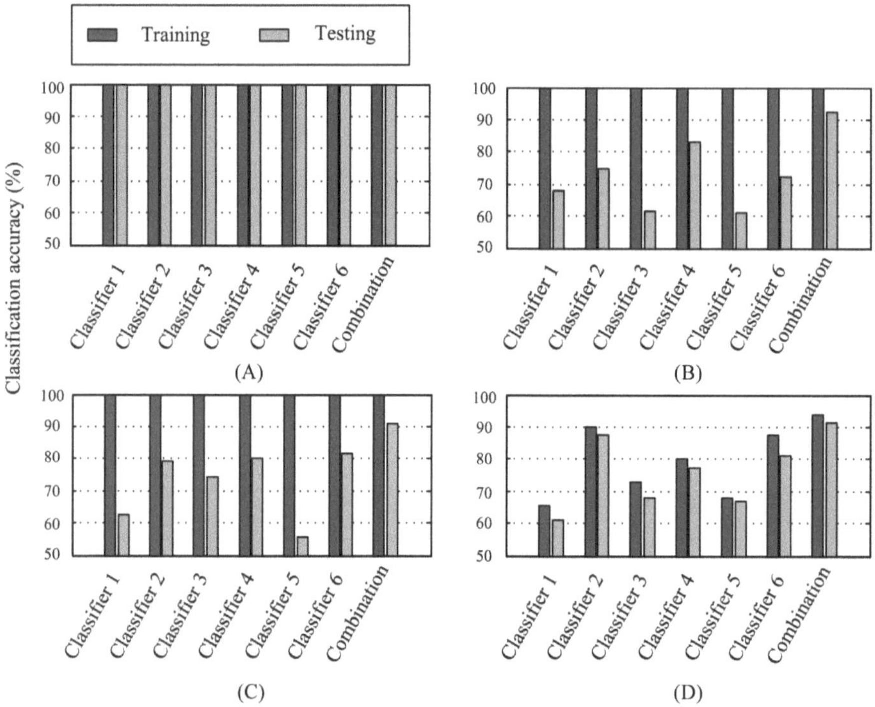

Figure 5.5. Classification accuracy comparisons between single classifier and multiple classifier combination. (A) Data set A, (B) data set B, (C) data set C, (D) data set D.

single classifiers is in the range of 61–87.67% with average 73.61%, whereas the classification accuracy of the multiple classifier incorporation is higher, which reaches up to 91.33%. These results indicate that the multiple classifier combination can recognize not only the different fault types but also the different fault levels better than each single classifier.

According to aforementioned analytical results, it is clear that the results combining multiple classifiers for various selections of feature sets usually are better than those using an individual classifier trained with one feature set. The hybrid intelligent fault diagnosis method based on multiple ANFIS combination acquires significant achievements in classification success rates and reveals better generalization ability than the single classifier.

To further verify the effect of the feature selection method on final classification accuracy in this experiment, aforementioned data set D as shown in Table 5.5 is taken for an example and another experiment is also performed on it. Here, four features, the identical feature number with the above experiments, are chosen from each of the six feature sets randomly rather than utilizing the feature selection strategy. The performance of feature selection strategy cannot be estimated through making a comparison between the classification accuracy from chosen features and that from all unselected features. Since the whole unselected features are fed into the ANFIS classifier, very long time will be required to calculate this classification result, which cannot be accepted by real applications. Therefore, the presented method is implemented to investigate the behavior of this feature selection method. Though the presented comparison is not very normative, it is possible for us to attain the intention of estimating the behavior of a feature selection method. Moreover, it is acceptable for us to take the time to achieve the presented comparison scheme.

We can repeat the experiment ten times and the average results are exhibited in Table 5.6 and Fig. 5.7, respectively. In order to make a comparison, Table 5.7 and Fig. 5.7 also depict the classification results of the whole classifiers, in which the sensitive features are selected utilizing the CDET method.

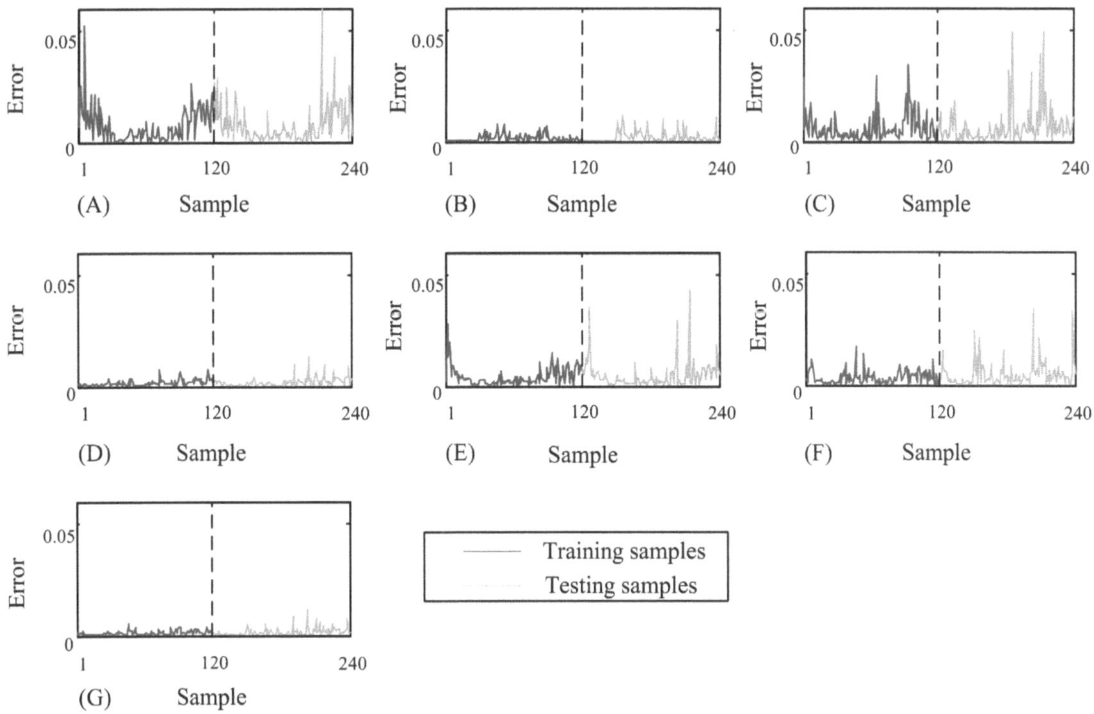

Figure 5.6. Classification errors of single classifier and multiple classifier combination. (A) Classifier 1, (B) classifier 2, (C) classifier 3, (D) classifier 4, (E) classifier 5, (F) classifier 6, (G) classifier combination.

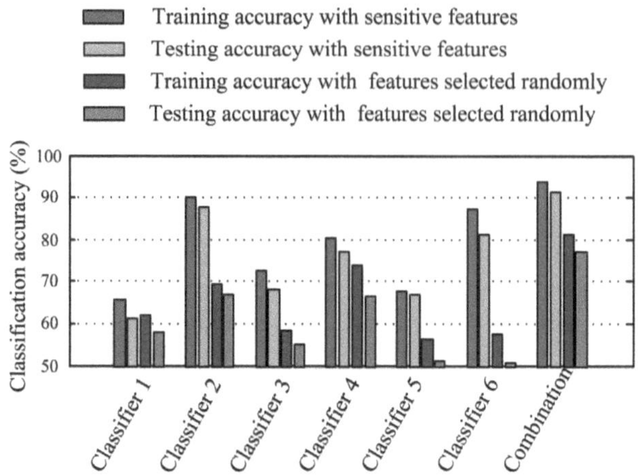

Figure 5.7. Classification accuracy comparisons for different features.

From Table 5.7 and Fig. 5.7, when four features chosen at random are applied to train and test, it can be observed that the training and testing correct rates of the six single classifiers are in the range of 56.5–73.73% with average 62.89% and 50.8–66.93% with average 58.02%, respectively. In comparison with those utilizing the sensitive features selected by CDET, it is obvious that their average training correct rate and average testing correct rate reduce by 14.39 and 15.59% severally. When these features chosen at random are applied to train and test, the training and testing correct rates of the multiple classifier combination fall from 93.67% to 81.19% and 91.33% to 77.15%, respectively. This result indicates that some of these features chosen at random may include too much useless or unrelated information. Moreover, there is a high degree of overlap between these features of 10-class classification task. These features may result in the confusion of the classifiers and further result in the decrease of classification correct rates. These results demonstrate that the feature selection method based on the CDET can discard the useless and insensitive features in the original feature set and consequently it is a powerful feature selection method.

In addition, it can also be found from Table 5.7 and Fig. 5.7 that although the six single classifiers with these features chosen at random have low classification accuracies: 56.5–73.73% for training and 50.8–66.93% for testing, they are complementary between each other. Therefore, their incorporation can achieve higher classification accuracy than each of them. The training and testing success rates based on multiple classifier combination reach up to 81.19 and 77.15%, respectively. These results demonstrate that

Table 5.7 Classification Accuracy (%) Comparison for Different Features

Feature/Classifier		Sensitive Features	Features Selected Randomly
Classifier1	Training	65.67	62
	Testing	61	58.03
Classifier2	Training	90	69.27
	Testing	87.67	66.93
Classifier3	Training	72.67	58.5
	Testing	68	55.1
Classifier4	Training	80.33	73.73
	Testing	77	66.3
Classifier5	Training	67.67	56.5
	Testing	67	50.93
Classifier6	Training	87.33	57.37
	Testing	81	50.8
Average of six classifiers	Training	77.28	62.89
	Testing	73.61	58.02
Classifier combination	Training	93.67	81.19
	Testing	91.33	77.15

the multiple classifier combination method is effective in the incipient fault diagnosis of the rolling element bearings once again.

5.3.5 Epilog

Inspired by Section 5.2, therefore, in Section 5.3 we investigate the effect of different input feature sets on the final classification result based on multiple classifier combination method. This proposed hybrid intelligent fault diagnosis method based on multiple ANFIS combination is verified by employing bearing experiments with different fault types and sizes. The experimental results demonstrate that the classification results under different input feature sets are complementary between each other. This conclusion broadens the definition of multiple classifier combination from the view of different feature sets rather than classifiers. Therefore, Section 5.3 is a extension for Section 5.2.

In addition, this experimental result also demonstrates that feature mining is necessary by virtue of some advanced signal processing techniques, especially when the acquired data is not long enough. Moreover, the majority of the extracted features are fault-unrelated and therefore feature selection should be performed by reasonable techniques or methods. Consequently, a powerful hybrid intelligent fault diagnosis method is supposed to be constructed.

From Sections 5.2 and 5.3, we can know that if an intelligent diagnosis method is constructed by combining the complementary effect of both different classifiers and different input features, it would achieve higher classification accuracy than each of aforementioned two combinations. Based on this motivation, a multidimensional hybrid intelligent diagnosis method is proposed in next subsection.

5.4 A Multidimensional Hybrid Intelligent Method

5.4.1 Motivation

Section 5.2 investigates the classification accuracy for rolling element bearings based on multiple W*K*NN combination with the same input feature sets. Additionally, Section 5.3 examines the classification accuracy for rolling element bearings based on multiple ANFIS combination with different input feature sets. However, in this section we will analyze the classification accuracy for gearboxes based on multiple different classifier combination with different input feature sets. Therefore, it is called "multidimensional" hybrid intelligent fault diagnosis method. Here, the word "multidimensional" means that we not only consider different input feature sets but also consider different classifiers. Thus, there are two tasks in this section: one examines the effect of different classifier combination with different input feature sets on final classification accuracy and the other is that the method is verified by the experiments of gearboxes instead of rolling element bearings because the faults of gears are more complicated than those of bearings.

The *K*NN is the simplest classifier and suitable to engineering applications. Therefore, it is considered as one member of multiple classifiers of multidimensional hybrid intelligent fault diagnosis method. The purpose of this consideration is to verify whether multiple different classifier combination can overcome their disadvantages.

Multilayer perception (MLP) neural networks are the most commonly applied feed-forward neural networks. It is characterized by fast operation, ease of implementation and smaller training set requirements (Übeyli, 2009). The MLP contains three sequential layers: input, hidden, and output layers. The common problem is that it is difficult to determine the number of neurons in the hidden layer. The insufficient or excessive number of neurons in the hidden layer most possibly results in the bad generalization and overfitting (Orhan et al., 2011). Usually, it is only obtained by trial and error (Oğulata et al., 2008). Moreover, it also has different structural characteristics from *K*NN classifier. Therefore, it is selected as one member of multiple classifiers in multidimensional hybrid intelligent fault diagnosis method. The intention of this selection is identical to that of *K*NN.

The structure of a RBF neural network is similar to that of an MLP. However, the activation function of the hidden layer is different from that of the MLP. Compared with other types of artificial neural network (ANN), the RBF requires less computational time. Moreover, it possesses a powerful framework for representing nonlinear mappings from several inputs to one or more outputs. Therefore, it is chosen as final member of multiple classifier combination.

The similarity between three classifiers is that they all belong to intelligent algorithm. Moreover, MLP and RBF are both feed-forward neural networks. But these three classifiers have different classification principles or algorithms. Therefore, they are supposed to exhibit complementary classification performance and overcome their own shortcomings to achieve higher classification accuracy than each of them. Finally, the classification results of these three different classifiers with different input features are integrated by GAs to obtain the final diagnosis result.

5.4.2 Multiple Classifier Combination

Commonly, the MLP neural network just includes three layers: input, hidden, and output layers. The number of nodes in the input layer is decided by the number of input features, whereas the number of output layers is determined by output classes. The number of hidden nodes usually is decided by utilizing the geometric pyramid rule written in the following equation where the active function of the hidden nodes is sigmoid function (Yang et al., 2002).

$$h = \sqrt{mn} \tag{5.8}$$

where h, m, and n denote the numbers of the hidden nodes, input, and output layers, respectively. The training of an MLP neural network is to search values of the connection weights. This process is achieved by minimizing an error function between the actual neural network outputs and the corresponding target output values in the training set. Commonly, the error function is the mean square error and the training algorithm is based on back-propagation.

The usually used activation function in RBF neural network is the Gaussian function defined as below

$$f(x) = \exp\left(-\frac{(x-c)^2}{2b^2}\right) \tag{5.9}$$

where b and c denote the width and center of the Gaussian function, respectively. The distance criteria between the input features and the center c of the data cluster are decided by the outputs of the hidden nodes. In addition, the parameter b really denotes the radius of the hypersphere. The training process of the RBF neural network is as: (1) according to the information included in the input samples, we can determine the nodes of the hidden layer and the center and the width of the Gaussian function. (2) Based on these parameters in the hidden layer, input samples, and these target values, we can decide the weights between the hidden and the output layers. Moreover, they can be adjusted by the principle of least squares minimization.

The *K*NNs is usually decided by calculating the Euclidean distance between the testing sample and every training sample. The Euclidean distance between the testing sample TE_d and the mth training sample $TR_{m,d}$ is written as:

$$D_m = \left[\sum_{d=1}^{D}(TE_d - TR_{m,d})^2\right]^{1/2}, \quad d = 1,2,\ldots,D, \; m = 1,2,\ldots,M \tag{5.10}$$

where D and M denote the numbers of input features and training samples, respectively. Determination of the neighborhood parameter K is a difficult problem as described in Section 5.2. Usually, the maximum value of the neighborhood parameter K is less than $\sqrt{M}$ according to rule of thumb (Ghosh et al., 2005). In this method, the K value is chosen as $\lfloor \sqrt{M} \rfloor$, where $\lfloor \sqrt{M} \rfloor$ denotes the largest integer less than or equal to $\sqrt{M}$.

From earlier description, it is noted that MLP, RBF, and *K*NN have distinct classification algorithms and reveal the complementary effect between each other. In addition, we also detect three different feature sets from time, frequency, and time–frequency domains, respectively. Therefore, a multidimensional hybrid intelligent diagnosis method is presented by incorporating MLP, RBF, *K*NN and different input feature sets to increase the recognition correct rates of gear incipient faults. The final diagnosis result is obtained by the weighted averaging technique with GAs

$$\begin{cases} y_m = \sum_{c=1}^{9} w_c y_{m,c}, \quad m = 1, 2, \ldots, M', \; c = 1, 2, \ldots, 9 \\ \sum_{c=1}^{9} w_c = 1, \quad w_c \geq 0 \end{cases} \tag{5.11}$$

where y_m and $y_{m,c}$ denote the classification results of the mth testing sample by utilizing the multidimensional hybrid intelligent method and the cth individual classifier respectively, w_c denotes the weight related with the cth individual classifier and M' is the number of the testing samples.

In this method, the weights in the weighted averaging technology are evaluated by employing optimization algorithm, that is GAs. The maximum number of generations is set up as 100 and seen as the termination criterion for the solution process. The mutation and crossover probabilities are selected as 0.01 and 0.8, respectively. The fitness or objective function is written as

$$\begin{cases} f = \dfrac{1}{1 + E_{tr}} \\ E_{tr} = \left[\dfrac{1}{M} \sum_{m=1}^{M} (y_m - \overline{y_m})^2 \right]^{\frac{1}{2}}, \quad m = 1, 2, \ldots, M \end{cases} \tag{5.12}$$

where E_{tr} denotes the root mean square training errors, $\overline{y_m}$ represents the target value of the mth training sample and M denotes the number of the training samples. According to aforementioned multiple classifier combination schemes, therefore a complete multidimensional hybrid intelligent fault diagnosis method for gearboxes is proposed in next subsection.

5.4.3 Diagnosis Method Based on Multiple Classifier Combination

Feature extraction is the first procedure of intelligent fault diagnosis. In this subsection, some features from three different domains are extracted to reflect the damage conditions of gears. Generally, the time–domain signal acquired from a gear often changes when a fault occurs. Both its amplitude and distribution may be different from those of the time–domain signal of a normal gear. Therefore, five common time–domain statistical features, including standard deviation (X_{sd}), root mean square (X_{rms}), kurtosis (X_{kurtosis}), crest factor (X_{crest}), and shape factor (X_{shape}), are extracted from original vibration signals and detail expressions are referred to Chapter 2. In addition, six special time–domain statistical features are also detected, which are specifically developed for gear fault extraction by the NASA. These features are FM0, FM4, NA4, NB4, ER, EOP, and detail expressions can be found in Chapter 4. Therefore, we obtain altogether 11 time–domain statistical features from original vibration signals. In addition, four common frequency–domain statistical features are also detected from the FFT spectra of original vibration signals. These features are mean frequency (p_1), frequency center (p_5), root mean square frequency (p_7), and standard deviation frequency (p_{14}) as shown in Table 2.2 in Chapter 2.

To obtain more characteristic information reflecting gear conditions, some advanced signal processing techniques are employed to process the original vibration signals. In this method, two advanced signal processing techniques are applied to preprocess the vibration signals before feature extraction. WT possesses good local characteristic in both time and frequency spaces. However, the high frequency parts in a signal cannot be decomposed in WT algorithm. Since the modulation information of mechanical faults indeed exists in the high frequency bands, therefore WT is unsuitable. However, wavelet packet transform (WPT) can further decompose the high frequency parts and thus may generate a more accurate frequency band partition. Consequently, a vibration signal x is decomposed by utilizing WPT and the decomposed frequency-band signal x_{2^k+m} is produced, where k represents the number of decomposition levels ($k = 0, 1, \ldots, J$, where J is the final number of decomposition levels) and m denotes the mth frequency-band signal in the kth level, where $m = 0, 1, \ldots, 2^k - 1$. The WPT decomposition process of x is described in Fig. 2.16 (B) in Chapter 2.

EMD can decompose a signal into a series of complete and almost orthogonal components, namely IMFs. IMFs denote simple oscillatory modes buried in a signal. The EMD detail decomposition process of a signal x is described in Table 2.3 in Chapter 2.

In order to compare WPT with EMD, it is noticed that the WPT is able to decompose a signal into independent frequency bands with equal frequency width and without overlap or gap. Therefore, they do not have any redundant or omitted information. However, the decomposition results of WPT deeply depend on the selection of wavelet basis function. Different wavelet basis function can result in different analysis results. However, EMD is completely self-adaptive and does not require any basis functions. Unfortunately, the number of IMFs cannot be controlled subjectively. Therefore, WPT and EMD process the signal from different views and possess their own merits. Consequently, the fault conditions of gear can be revealed from different aspects by employing WPT and EMD simultaneously to process the vibration signals. Moreover, the complementary effect between them may be achieved.

When a fault occurs, the energy of the faulty vibration signal is variable, which is determined by whether the fault tooth is in meshing or not. This variation can be represented by computing the energy entropy of the frequency-band signals of WPT and IMFs of EMD. Moreover, the variation can also be found by directly watching the energy distribution of the component signals of WPT and EMD.

In order to intuitively display this distinction and variation, Fig. 5.8 illustrates these waveforms of vibration signals of the normal and damaged gears, the decomposed frequency-band signals of WPT with levels 3 and the energy distribution, where the wavelet basis function is selected as db10. Fig. 5.9 exhibits the decomposed results of EMD. In practice, the energy of first six IMFs almost covers the energy of the whole signal. Therefore, other IMFs and residual are neglected. From the energy distribution of WPT and EMD, it is noticed that they are obviously distinct between the normal gear and the damaged one. Moreover, the energy entropy is depicted in

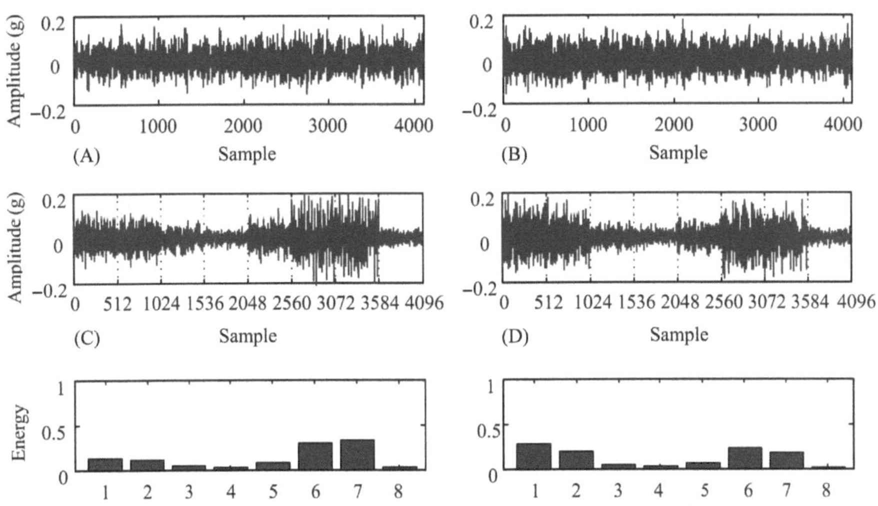

Figure 5.8. Wavelet packet transform (WPT) decomposition results and energy distribution: normal gear and damaged gear. (A) Original signals of the normal gear, (B) original signals of the damaged gear, (C) frequency band signals of the normal gear, (D) frequency-band signals of the damaged gear, (E) energy distribution of frequency-band signals, (F) energy distribution of frequency-band signals.

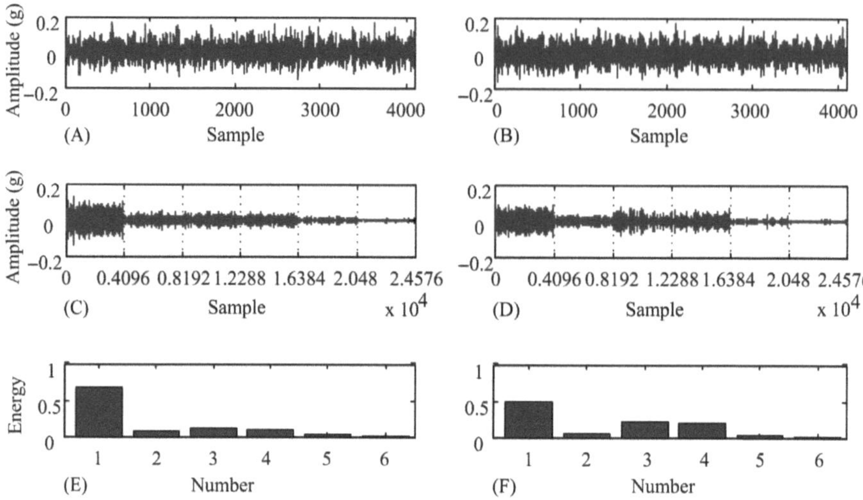

Figure 5.9. Empirical mode decomposition (EMD) decomposition results and energy distribution: normal gear and damaged gear. (A) Original signals of the normal gear, (B) original signals of the damaged gear, (C) IMFs of the normal gear, (D) IMFs of the damaged gear, (E) energy distribution of IMFs, (F) energy distribution of IMFs.

Table 5.8, which is calculated by using Eq. (5.13). It is clear that the energy entropy of the damaged gear is distinct from that of the normal one for both WPT and EMD. Here, the energy entropy is defined as

$$H_e = -\sum_{j=1}^{J} p_j \log p_j \tag{5.13}$$

where $p_j = E_j/E$ is the energy percentage of the jth frequency-band signal of WPT or the jth IMF of EMD in the whole signal energy E, where $E = \sum_{j=1}^{J} E_j$. Here, the energy E_j of jth frequency-band signal or

Table 5.8 Energy Entropy of the Vibration Signals of the Normal and Damaged Gears by Utilizing WPT and EMD

Method	Normal Gear	Damaged Gear
WPT	1.70	1.74
EMD	1.06	1.30

IMF can be obtained by using the computation method of time–frequency–domain features in Chapter 2.

According to the aforementioned analytical results, it is found that the energy distribution of the either frequency-band signals of WPT or the IMFs of EMD can vary with the variation of gear health conditions. Therefore, the energy entropy of the frequency-band signals and the IMFs is able to be selected as additional time–frequency–domain features to extract gear damages. Dependent on the characteristics of the vibration signals in gear experiments in this subsection, we choose db10 wavelet basis with three decomposition levels. Similarly, in all IMFs of EMD the first six IMFs are selected to detect energy entropy features. Finally, we can obtain eight energy entropy features from the frequency-band signals of WPT and six energy entropy features from the first six IMFs of EMD. Thereby, 14 energy entropy features are detected totally by virtue of EMD and WPT techniques.

Feature extraction has been completed in aforementioned parts. 29 features have been detected from three different domains: time, frequency, and time–frequency domains. According to common procedures of the intelligent fault diagnosis, next step should be feature selection. However, the features in each classifier are too little when they are separated into different classifiers. Therefore, it is unnecessary for us to implement the feature selection.

Different classifiers including MLP, RBF, and *K*NN can exhibit complementary classification performance since they have different classification principles and algorithms. Additionally, different feature sets also exhibit complementary effect between them. Based on these two motivations, therefore, a multidimensional hybrid intelligent diagnosis method for gearboxes is presented to improve the recognition accuracy and increase diagnosis reliability. The method is illustrated in Fig. 5.10 and consists of the following procedures. (1) Original vibration signals collected from gears are processed by employing the Hilbert transform. At the same time, the difference and residual signals are also acquired. (2) Moreover, 11 time–domain statistical features are detected from the original, the difference and the residual signals respectively, which contain five common and six special time–domain statistical features. Furthermore, four common frequency–domain statistical features are also detected from the FFT spectra of vibration signals. (3) In addition, 14 energy entropy features are extracted from the eight frequency-band signals of WPT and six IMFs of EMD. (4) Finally, the weighted averaging technique with GAs is utilized to incorporate the outputs of the nine classifiers and further achieve the final diagnosis result of gear conditions.

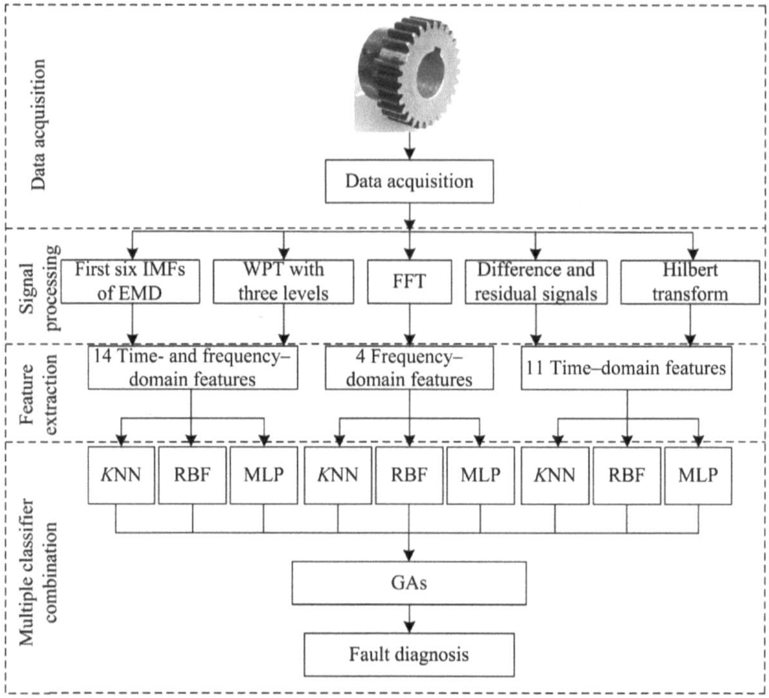

Figure 5.10. Flowchart of multidimensional hybrid intelligent method based on multiple classifiers combination.

5.4.4 Intelligent Diagnosis of Damage Categories: A Gearbox Case Study

In order to verify the performance of the multidimensional hybrid intelligent diagnosis method based on multiple classifier combination for different fault categories, two experiments on a gearbox test rig were performed. One is that gears with a chipped tooth and a missing tooth were machined and then corresponding gear vibration signals were collected under different conditions consisting of normal condition, the two damage conditions and the combination of these two damage modes. The other is that the cracks with different damage degrees were machined at the tooth root of a gear. The data were applied to estimate the classification behavior of this method for the incipient crack.

The test rig has been illustrated in Chapter 4. Therefore, a simple review is given. The system consists of a gearbox, a 3-hp ac motor for driving the gearbox and a magnetic brake for loading. Additionally, the system has three shafts inside the gearbox. Gear1 on shaft 1 is labeled as #1, which has 16 teeth and meshes with gear 2 (#2) with 48 teeth. Gear 3 on shaft 2 is labeled as #3, which has 40 teeth and meshes with gear 4 (#4) on the output shaft (shaft 3). Gears 1′ (#1′) and 4′ (#4′) are the tested gears. #1′ is the identical as #1 but it has a chipped tooth. #4′ is the identical as #4 but it has a missing tooth.

The vibration signals can be captured under the following gear conditions: (1) all gears are normal, (2) #1 is replaced with #1′, (3) #4 is replaced with #4′, and (4) both #1 and #4 are replaced by #1′ and #4′, respectively. In addition, the vibration signals were collected under 3 different loads and 5 different motor speeds from 1200 to 2400 rpm with an increment of 300 rpm. Among these 3 different loads, one had no torque and the other two had different torques. The sampling frequency was 5120 Hz and the data point of every sample was 4096. Under the same working condition, 4 samples were acquired. Finally, 60 samples were acquired for every gear condition and thereby altogether 240 samples for the four gear conditions were acquired. Vibration signals acquired from four gear conditions are plotted in Fig. 5.11.

Among aforementioned acquired 240 samples, 120 data samples are chosen as training samples and the remaining 120 samples as testing samples. Obviously, it is a relatively simple

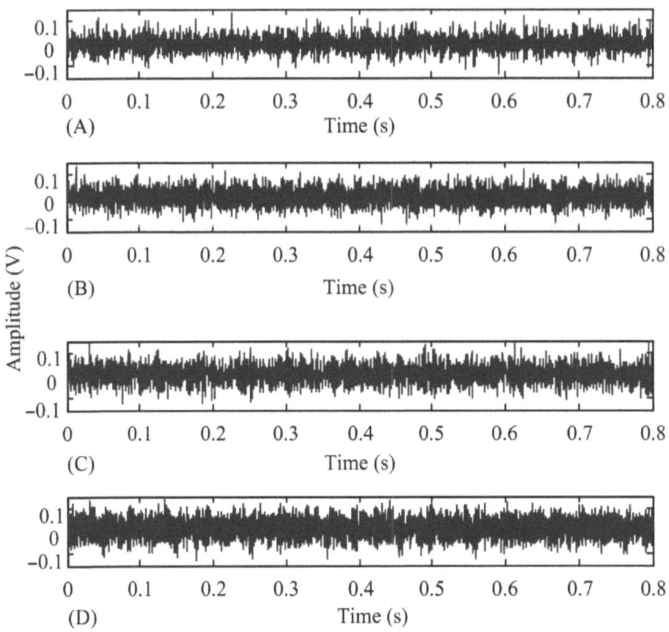

Figure 5.11. Vibration signals under different gear conditions. (A) Normal tooth, (B) chipped tooth, (C) missing tooth, (D) chipped tooth and missing tooth.

four-class classification task corresponding to the four different fault categories. The multidimensional hybrid intelligent fault diagnosis method is applied to recognize the fault categories of the gear. In order to make a comparison, the performance of the nine single classifiers is also tested by employing the same training and testing samples. The classification correct rates of both the multidimensional hybrid intelligent fault diagnosis method and the single classifier are exhibited in Table 5.9. For convenience, this experiment is labeled as Experiment #1 in Table 5.9. Form this diagnosis results, it is found that the testing accuracies of the nine single classifiers change from 71.67% to 88.33% with the average 80.37%. However, for multidimensional hybrid intelligent fault diagnosis method, the classification correct rate reaches up to 98.33%. Compared with each of nine single classifiers, the recognition accuracy is largely improved. The experimental results demonstrate that multidimensional hybrid intelligent fault diagnosis method is superior to individual classifier for identifying different fault categories of gears.

Table 5.9 Classification Accuracy (%) Comparison Between Single Classifier and the Multidimensional Hybrid Intelligent Method

Experiment	#1	#2		
Data set	\	A	B	C
Classifier K1	73.33	95.56	83.33	61.67
Classifier K2	71.67	100.00	84.72	60.00
Classifier K3	87.50	100.00	90.28	75.00
Classifier K4	72.50	92.22	61.11	85.00
Classifier K5	85.00	100.00	81.94	68.33
Classifier K6	88.33	100.00	87.50	55.00
Classifier K7	80.83	96.67	86.11	61.67
Classifier K8	75.83	100.00	76.39	80.00
Classifier K9	88.33	100.00	83.33	50.00
Average of the 9 classifiers	80.37	98.27	81.63	66.30
The proposed method	98.33	100.00	100.00	98.33

Table 5.10 Geometry of the Crack Faults

	Geometry of Fault			
Crack Fault Mode	Depth (mm)	Width (mm)	Thickness (mm)	Crack Angle
F0	0	0	0	\
F1	(1/4)*a*	(1/4)*b*	0.4	45°
F2	(1/2)*a*	(1/2)*b*	0.4	45°

To evaluate the classification accuracy of the multidimensional hybrid fault diagnosis method for different fault levels, another experiment was conducted, where a timing belt was added between the motor and the gearbox. Additionally, the configuration of the gearbox system was modified and illustrated Fig. 4.4 in Chapter 4. From the figure, it is observed that the gearbox is driven by the motor through a timing belt. Gears 1 ~ 4 (#1 ~ #4) have 48, 16, 24, and 40 teeth, respectively. The marked gear 3 is the tested gear. In order to simulate crack faults, the crack sizes of small proportions were machined in the gearbox experiment. The corresponding geometric properties of crack are illustrated in Chapter 4. The gears with different crack sizes are listed in Table 5.10. Three gears with different operating conditions (F0 ~ F2) consisting of one normal gear and two gears with damages were tested in the experiment. In gear condition F1, there is a very small crack at a tooth root and therefore it is considered as an incipient crack. In gear condition F2, there is a bigger crack at a tooth root but not serious and therefore it is considered as a developing crack.

The vibration signals were captured for each of the three gears by employing the identical data acquisition system. The vibration signals were also collected under both three different loads and five different motor speeds from 1200 to 2000 rpm with an increment of 200 rpm. Three different loads, including one no torque and two different torques, are marked as 0, 1, and 2, respectively. The sampling frequency is still 5120 Hz and the data point of every sample is 4096. Four data samples were obtained under the same operating condition and altogether there are 180 data samples for these three sizes of gear cracks. Fig. 5.12 depicts vibration waveforms of the normal and cracked gears. The data acquired here were marked as Experiment #2 as listed in Table 5.9.

Three examinations are performed on three different data sets (A–C) acquired from experiment #2. For the first examination on data set A, the compositions of the training and testing samples are

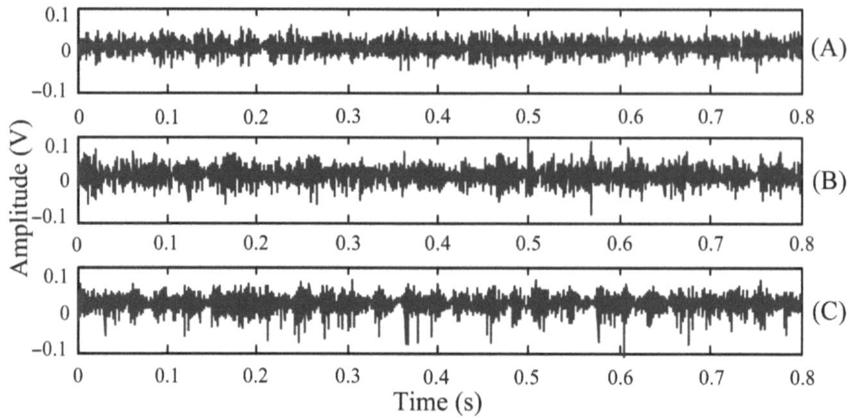

Figure 5.12. Original waveforms of gears. (A) F0, (B) F1, and (C) F2.

similar to those of experiment #1. The obtained 180 data samples were split into 90 samples for training and 90 for testing. The diagnosis results by using both nine single classifiers and multidimensional hybrid intelligent method for the three different sizes of gear cracks are exhibited in Table 5.9. Since this classification task is relatively simple, both six single classifiers and the multidimensional hybrid intelligent method all obtain the highest recognition accuracy with 100%. Among them, however, there are still three single classifiers failing to generate the ideal recognition accuracy.

Another examination is conducted on data set B. Here, the training and testing samples are recombined as described in Table 5.11. The 108 training samples were acquired under the lower motor speeds 1200, 1400, and 1600 rpm, whereas the 72 testing samples were acquired under the higher motor speeds 1800 and 2000 rpm. The examination for these training and testing samples is implemented to further verify the robustness of the multidimensional hybrid intelligent method for different motor speeds.

The diagnosis results for both the single classifier and the multidimensional hybrid intelligent method are exhibited in Table 5.9. It is observed that the multidimensional hybrid intelligent method still achieves the highest classification accuracy with 100%, whereas the highest accuracy of the nine single classifiers is 90.28%. The accuracies of the nine single classifies change from 61.11% to 90.28% with the average 81.63%. Compared with the average accuracy, the multidimensional hybrid intelligent method enhances the identification correct rate by 18.37%.

Further, another examination is conducted on data set C. The training and testing samples are exhibited in Table 5.11. The samples acquired under the gear with a developing crack (F2) are used

Table 5.11 Description of three data sets

Data Set	Number of Training/ Testing Samples	Fault Modes of Training/ Testing	Motor Speeds of Training/Testing Samples (r/min)	Loads of Training/Testing Samples	Label of Classification
A	30/30	F0/F0	1200 ~ 2000/1200 ~ 2000	0, 1, 2/0, 1, 2	1
	30/30	F1/F1	1200 ~ 2000/1200 ~ 2000	0, 1, 2/0, 1, 2	2
	30/30	F2/F2	1200 ~ 2000/1200 ~ 2000	0, 1, 2/0, 1, 2	3
B	36/24	F0/F0	1200 ~ 1600/1800, 2000	0, 1, 2/0, 1, 2	1
	36/24	F1/F1	1200 ~ 1600/1800, 2000	0, 1, 2/0, 1, 2	2
	36/24	F2/F2	1200~1600/1800, 2000	0, 1, 2/0, 1, 2	3
C	30/30	F0/F0	1200 ~ 2000/1200 ~ 2000	0, 1, 2/0, 1, 2	1
	30/30	F2/F1	1200 ~ 2000/1200 ~ 2000	0, 1, 2/0, 1, 2	2

to train and the samples of an early crack (F1) to test. The intention of this examination is to evaluate the generalization capability of the multidimensional hybrid intelligent method for an incipient fault when it is trained by fault-developing samples.

The multidimensional hybrid intelligent diagnosis method and single classifier are performed once again on these faults and the corresponding diagnosis results are exhibited in Table 5.9. The diagnosis results indicate that the multidimensional hybrid intelligent diagnosis method obtains much higher recognition correct rate with 98.33% than that of every single classifier. The accuracies of the nine single classifiers range from 50.00% to 85.00% with average 66.30%. Obviously, the multidimensional hybrid intelligent method possesses stronger robustness for different fault levels than single classifier.

In the two aforementioned experiments, the data were acquired from the normal gear and damaged gears with a chipped tooth, a missing tooth, and combination of both a chipped and missing teeth in experiment #1. They are used to train and test the multidimensional hybrid intelligent method. In Experiment #2, three investigations are conducted on three different data sets. The first investigation employs the data of the normal gear and the damaged gear with different crack levels to train and test the multidimensional hybrid intelligent method. The second investigation utilizes the collected data under the lower motor speeds of both the normal and the damaged gears to train, while the collected data under the higher motor speeds to test the multidimensional hybrid intelligent method. The third investigation adopts the data of a developing crack to train and the data of an early crack to test. Analyzing the two experiment results, it is noticed that only in the first investigation of experiment #2, six of the nine single classifiers obtain the same high classification accuracies as the multidimensional hybrid intelligent method since it is a relatively simple classification task. For all other results, the multidimensional hybrid intelligent method achieves the best result in the gear fault diagnosis. It increases the classification accuracy by 17.96, 18.37, and 32.03% respectively in comparison with the average accuracy of the single classifier. For the second investigation of experiment #2, the highest accuracy is produced by the multidimensional hybrid intelligent method, which suggests that it has better robustness under various motor speeds than the single classifier. In addition, the diagnosis results of the third investigation of experiment #2 confirm that the generalization ability of the multidimensional hybrid intelligent method is superior to that of the single classifier in gear early fault diagnosis.

All the aforementioned diagnosis results demonstrate that the multidimensional hybrid intelligent method achieves significant improvements in identification correct rate and possesses better generalization capability in comparison with the single classifier. It can recognize not only the different fault types but also the different fault sizes better than single classifier. The success of the multidimensional hybrid intelligent method may be attributed to the following three aspects. (1) Using Hilbert transform, WPT, and EMD can uncover more fault-related features from the original vibration signals of the gear. (2) Different features from time, frequency, and time–frequency domains are able to represent the health conditions of the gear better. (3) The weighted averaging technique based on GAs is employed to come up with the final classification result and thereby increasing classification correct rate.

It is observed from the recognition success rates of the nine single classifiers in the two experiments that the single classifier with the time–frequency domain features can mostly achieve better results than those with time or frequency domain features alone. It is the reason that the time–frequency domain features are able to provide more fault-related information than each of the time and frequency domains. These problems studied in the two experiments contain individual fault diagnosis (a chipped tooth or a missing tooth), multifault diagnosis (combination of chipped and missing teeth), and incipient fault diagnosis (incipient crack). In addition, the vibration signals were acquired under the all kinds of both operating loads and motor speeds. Therefore, they simulate vividly the fault diagnosis of gears. The satisfactory diagnosis results prove the effectiveness and generalization capability of the multidimensional hybrid intelligent diagnosis method.

5.4.5 Epilog

In this section we examine the effect of different both input feature sets and classifiers on the final classification results. This multidimensional hybrid intelligent fault diagnosis method based on MLP, *K*NN, and RBF combination is demonstrated by employing gear experiments with different fault categories and levels. Here, the input feature sets are extracted from time, frequency, and time–frequency domains, respectively. Therefore, this method not only considers the complementary behavior of different input feature sets but also tradeoffs that of different classifiers. Consequently, it is supposed to possess higher recognition accuracy than the combination of either different input feature sets or different classifiers alone. Finally, the experimental results of the gears demonstrate that the identification accuracy under different both input feature sets and classifiers are superior to those of nine individual classifiers. This conclusion further verifies the validity and reasonability of the conclusions in Section 5.2 and 5.3 since the advantage in Section 5.4 is the combination of those in Section 5.2 and 5.3.

In addition, the feature sets extracted from multidimensional space can more effectively reflect the health conditions of machinery, especially the features in time–frequency domain can contain more fault-related characteristic information than that in time or frequency domains alone.

5.5 Conclusions

In this chapter, the topic mainly focuses on hybrid intelligent fault diagnosis method. Based on three motivations of multiple classifier combination in Section 5.1, first in Section 5.2 the intelligent fault diagnosis method based on multiple W*K*NN combination with the same input feature sets is investigated and analyzed by the different bearing experiments. Then, Section 5.3 studies a multiple ANFIS combination method with different input feature sets for bearing fault diagnosis. Finally, Section 5.4 examines the multidimensional hybrid intelligent diagnosis method based on different classifier (MLP, RBF, and *K*NN) combination with different input feature sets for complex gear damages. The experimental results demonstrate that either different classifiers or different input feature sets combination can improve the output recognition accuracy. Here, different classifiers have two meanings: one is that the same classification engine but different parameter setting, for example different neighborhood parameters K in WKNN classifier and the other is that different classification engines, such as MLP, RBF, and KNN. The experimental results in Section 5.4 further suggest that double complementation from both different classifiers and different input feature sets can achieve higher success rate than each of them.

Among these hybrid intelligent fault diagnosis methods, we not only pay the attention to combining multiple classifiers but also concern on feature extraction and selection. Some fault-related characteristic information is often buried in the collected vibration signals and therefore many features extracted directly from their time–and frequency–domains are insensitive or useless. However, lots of effective and suitable signal processing techniques can uncover the fault characteristics embedded in vibration signals. Thus, hybrid intelligent fault diagnosis should employ lots of advanced signal processing techniques to reveal valuable features and in addition it can also add the number of feature for sample shortage. Since in whole extracted features some features are insensitive to the faults and moreover too many input features may cause the curse of dimensionality, feature selection should be performed on the whole features to screen out sensitive features. Then, the sensitive features are fed into multiple classifiers and the final classification results are further obtained. It is noticed that the feature selection not only increases final classification accuracy but also overcomes the dimension curse of classifiers. Therefore, a robust hybrid intelligent diagnosis method not only depends on suitable feature extraction and effective feature selection but also depends on the compensation between multiple classifiers.

References

Abu-Mahfouz, I.A., 2005. A comparative study of three artificial neural networks for the detection and classification of gear faults. Int. J. Gen. Syst. 34, 261–277.

Bi, Y., Guan, J., Bell, D., 2008. The combination of multiple classifiers using an evidential reasoning approach. Artif. Intel. 172, 1731–1751.

Boyacioglu, M.A., Avci, D., 2010. An Adaptive Network-Based Fuzzy Inference System (ANFIS) for the prediction of stock market return: the case of the Istanbul Stock Exchange. Expert Syst. Appl. 37, 7908–7912.

Chen, S., Cowan, C.F., Grant, P.M., 1991. Orthogonal least squares learning algorithm for radial basis function networks. IEEE Trans. Neural Netw. 2, 302–309.

Duan, C., He, Z., Jiang, H., 2007. A sliding window feature extraction method for rotating machinery based on the lifting scheme. J. Sound Vib. 299, 774–785.

Fumera, G., Roli, F., 2005. A theoretical and experimental analysis of linear combiners for multiple classifier systems. IEEE Trans. Pattern Anal. Mach. Intell. 27, 942–956.

Ghosh, A.K., Chaudhuri, P., Murthy, C., 2005. On visualization and aggregation of nearest neighbor classifiers. IEEE Trans. Pattern Anal. Mach. Intell. 27, 1592–1602.
Halim, E.B., Shoukat Choudhury, M.A.A., Shah, S.L., Zuo, M.J., 2008. Time domain averaging across all scales: a novel method for detection of gearbox faults. Mech. Syst. Signal Process. 22, 261–278.
Hansen, J.V., 1998. Comparative performance of back propagation networks designed by genetic algorithms and heuristics. Int. J. Intel. Syst. Acc. Finan. Manag. 7, 69–79.
Huenupán, N.F., Yoma, N.B., Molina, C., Garretón, N.C., 2008. Confidence based multiple classifier fusion in speaker verification. Pattern Recogn. Lett. 29, 957–966.
Jafarizadeh, M.A., Hassannejad, R., Ettefagh, M.M., Chitsaz, S., 2008. Asynchronous input gear damage diagnosis using time averaging and wavelet filtering. Mech. Syst. Signal Process. 22, 172–201.
Jang, J.-S.R., 1993. ANFIS: adaptive-network-based fuzzy inference system. IEEE Trans. Syst. Man Cybernetics 23, 665–685.
Kittler, J., Hatef, M., Duin, R.P., Matas, J., 1998. On combining classifiers. IEEE Trans. Patt. Anal. Mach. Intel. 20, 226–239.
Lei, Y., He, Z., Zi, Y., Hu, Q., 2007. Fault diagnosis of rotating machinery based on multiple ANFIS combination with GAs. Mech. Syst. Signal Process. 21, 2280–2294.
Lei, Y., He, Z., Zi, Y., 2008a. Application of a novel hybrid intelligent method to compound fault diagnosis of locomotive roller bearings. J. Vib. Acoust. 130, 034501.
Lei, Y., He, Z., Zi, Y., 2008b. A new approach to intelligent fault diagnosis of rotating machinery. Expert Syst. Appl. 35, 1593–1600.
Lei, Y., He, Z., Zi, Y., Chen, X., 2008c. New clustering algorithm-based fault diagnosis using compensation distance evaluation technique. Mech. Syst. Signal Process. 22, 419–435.
Lei, Y., He, Z., Zi, Y., 2009. A combination of WKNN to fault diagnosis of rolling element bearings. J. Vib. Acoust. 131, 064502.
Lei, Y., Zuo, M.J., He, Z., Zi, Y., 2010. A multidimensional hybrid intelligent method for gear fault diagnosis. Expert Syst. Appl. 37, 1419–1430.
Lei, Y., He, Z., Zi, Y., 2011a. EEMD method and WNN for fault diagnosis of locomotive roller bearings. Expert Syst. Appl. 38, 7334–7341.
Lei, Y., Lin, J., He, Z., Zi, Y., 2011b. Application of an improved kurtogram method for fault diagnosis of rolling element bearings. Mech. Syst. Signal Process. 25, 1738–1749.
0Loutridis, S.J., 2008a. Gear failure prediction using multiscale local statistics. Eng. Struct. 30, 1214–1223.
Loutridis, S.J., 2008b. Self-similarity in vibration time series: application to gear fault diagnostics. J. Vib. Acoust. 130, 031004.
Lu, Y., Sundararajan, N., Saratchandran, P., 1998. Performance evaluation of a sequential minimal radial basis function (RBF) neural network learning algorithm. IEEE Trans. Neural Netw. 9, 308–318.
Mallat, S., Hwang, W.L., 1992. Singularity detection and processing with wavelets. IEEE Trans. Inf. Theory 38, 617–643.
Oğulata, S.N., Şahin, C., Erol, R., 2008. Neural network-based computer-aided diagnosis in classification of primary generalized epilepsy by EEG signals. J. Med. Syst. 33, 107–112.
Orhan, U., Hekim, M., Ozer, M., 2011. EEG signals classification using the K-means clustering and a multilayer perceptron neural network model. Expert Syst. Appl. 38, 13475–13481.

Rafiee, J., Arvani, F., Harifi, A., Sadeghi, M.H., 2007. Intelligent condition monitoring of a gearbox using artificial neural network. Mech. Syst. Signal Process. 21, 1746–1754.

Rasheed, S., Stashuk, D., Kamel, M., 2008. Fusion of multiple classifiers for motor unit potential sorting. Biomed. Signal Process. Control 3, 229–243.

Riedmiller, M., 1994. Advanced supervised learning in multi-layer perceptrons—from back propagation to adaptive learning algorithms. Comp. Stand. Interfaces 16, 265–278.

Samanta, B., 2004. Gear fault detection using artificial neural networks and support vector machines with genetic algorithms. Mech. Syst. Signal Process. 18, 625–644.

Sun, Q., Tang, Y., 2002. Singularity analysis using continuous wavelet transform for bearing fault diagnosis. Mech. Syst. Signal Process. 16, 1025–1041.

Sweldens, W., 1998. The lifting scheme: a construction of second generation wavelets. SIAM J. Math. Anal. 29, 511–546.

Übeyli, E.D., 2009. Combined neural network model employing wavelet coefficients for EEG signals classification. Digital Signal Process. 19, 297–308.

Wang, J., Neskovic, P., Cooper, L.N., 2007. Improving nearest neighbor rule with a simple adaptive distance measure. Patt. Recog. Lett. 28, 207–213.

Yang, D.M., Stronach, A.F., Macconnell, P., Penman, J., 2002. Third-order spectral techniques for the diagnosis of motor bearing condition using artificial neural networks. Mech. Syst. Signal Process. 16, 391–411.

Yilmaz, A.S., Özer, Z., 2009. Pitch angle control in wind turbines above the rated wind speed by multi-layer perceptron and radial basis function neural networks. Expert Syst. Appl. 36, 9767–9775.

Zhou, C.Y., Chen, Y.Q., 2006. Improving nearest neighbor classification with cam weighted distance. Patt. Recog. 39, 635–645.

Zio, E., Baraldi, P., Gola, G., 2008. Feature-based classifier ensembles for diagnosing multiple faults in rotating machinery. Appl. Soft Comput. 8, 1365–1380.

Zuo, M.J., Lin, J., Fan, X., 2005. Feature separation using ICA for a one-dimensional time series and its application in fault detection. J. Sound Vib. 287, 614–624.

6

REMAINING USEFUL LIFE PREDICTION

6.1 Background

Various definitions of prognostics have been proposed in the literature. One of the most comprehensive definitions is presented by ISO13381-1 (2004). It defines prognostics as "an estimation of time to failure and risk for one or more existing and future failure modes." From this definition, it is recognized that prognostics involve predicting the time of a specific fault mode progressing from its incipience to the final failure. Therefore, the diagnostic process, that is, the fault detection, isolation and identification, must be conducted before the prognostic process. The relationship between diagnostics and prognostics is described as follows. Diagnostics involve identifying and quantifying the damage that has occurred, while prognostics attempt to predict the damage that has not occurred. Prognostics are conducted based on the diagnostic outputs (e.g., fault mode, degradation rates, etc.) and therefore is unable to be done in isolation.

Remaining useful life (RUL), also named as remaining service life, residual life or remnant life, refers to the time left before machinery loses its operation ability given the current age and condition, and the past operation profile. The RUL is generally defined as a conditional random variable which is specified in relative or absolute time units, for example, load cycles, flight hours, etc. The RUL l_k at current time t_k is denoted as

$$l_k = \inf\left\{l_k : x\left(l_k + t_k\right) \geq \lambda \mid x_{0:k}\right\} \tag{6.1}$$

where $x\left(l_k + t_k\right)$ represents the health state at the future time $l_k + t_k$, $x_{0:k}$ represents the health states observed at previous time index $t_{0:k}$, and λ is a prespecified failure threshold.

RUL is also defined as follows (Jardine et al., 2006):

$$l_k = t_{\mathrm{EoL}} - t_k \tag{6.2}$$

where t_{EoL} denotes the random variable of time at the end of life (EoL). Since RUL is a random variable, the probability density function (PDF) of the RUL $f_{l_k|x_{0:k}}(l_k \mid x_{0:k})$ would be of interest for full understanding of the RUL.

An illustration of the RUL prediction process is shown in Fig. 6.1. In this figure, a common degradation process in real cases is taken, for example, which includes a normal operation stage (stage I) and an accelerated degradation stage (stage II) during the whole lifetime. The degradation indicator in stage I is stable, which means that the machine is healthy. Once a fault occurs, the degradation process switches from stage I to stage II where the degradation indicator increases with an accelerating rate as the fault gets severer. The task in stage I is to monitor the heath state and detect incipient faults, thus diagnostics play major roles during this stage. When a fault is detected, the prognostics are triggered, and the RUL is predicted during stage II. The time to start predicting is defined as the first predicting time (FPT), which is denoted by t_{FPT} in Fig. 6.1. Obviously, the FPT is the boundary between stage I and stage II.

In Fig. 6.1, (A) presents the RUL prediction process at a single time point t_k, and (B) gives the real-time RUL prediction result during stage II. The former is just a particular case of the latter. In other words, the real-time RUL prediction result is obtained by conducting the RUL prediction process from t_{FPT} to t_{EoL} continuously.

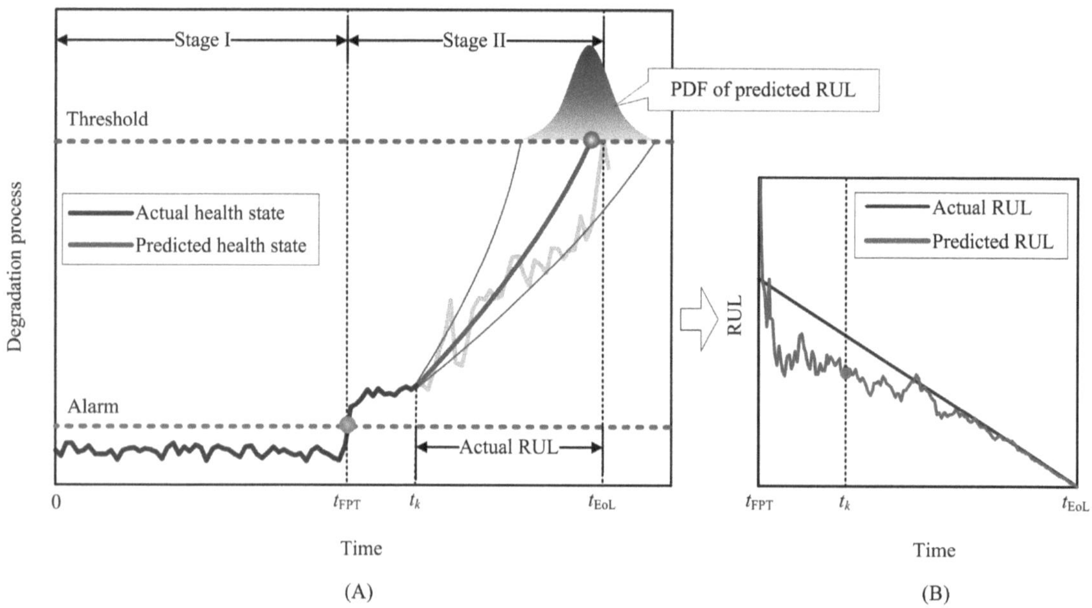

Figure 6.1. Illustration of the RUL prediction process. (A) RUL prediction at a single time index and (B) real-time RUL prediction during stage II.

As a significant part of prognostics, RUL prediction contributes to reducing risks and maintenance costs and increasing the maintainability, availability, reliability, and productivity of machines. Consequently, the machine RUL prediction has attracted more and more attention in recent years (Liu et al., 2009, Li et al., 2014, Park and Padgett, 2005). In general, RUL prediction methods in the literature are categorized into date-driven methods and model-based methods. Data-driven methods attempt to derive degradation processes of machinery from measurement signals using machine learning techniques instead of building models based on comprehensive system physics and human expertise. These methods rely on the assumption that the statistical characteristics of data are relatively consistent unless a fault occurs. They produce RUL prediction results based on historical measurement signals. Therefore, the prediction accuracy of data-driven methods depends on both the quantity and the quality of the historical measurements. However, it is generally difficult to collect enough qualified measurements in real cases.

Model-based methods attempt to set up mathematical or physical models to describe degradation processes of machinery, and update model parameters using measured data. Model-based methods could incorporate both expert knowledge and real-time information from machinery. Consequently, they may work well in RUL prediction of machinery. Model-based methods also require less data than data-driven methods. Model-based methods, however, may not be suitable to deal with the prognostics of rotating machinery in the following scenarios: (1) machinery fault patterns vary under different operation conditions; and (2) the mechanical system is so complex that the principles of the system operation are not straightforward and it is prohibitively expensive to develop an accurate model.

In the rest of this chapter, basic concepts and fundamental theories about data-driven methods and model-based methods will be introduced. Then, some typical techniques and their case studies in RUL prediction of rotating machinery will be described in detail.

6.2 Data-driven Prediction Methods

6.2.1 Introduction

Data-driven methods derive the trend of degradation processes from the measured data using machine learning techniques. The raw measured data always includes the interference of noises. To reveal the degradation trend of machinery, health indicators which are supposed to incorporate useful information of degradation, are generally extracted from the raw measured data. The health indicators at time series $t_1, t_2, \ldots, t_{k-1}, t_k$ are represented by a set of variables $\{x_1, x_2, \ldots, x_{k-1}, x_k\}$, where x_k is the health indicator at the current time t_k. The task of prognostics is to predict the future health condition of machinery, that is, the value of the health indicator in the future, according to the available health indicators.

In data-driven methods, the predictor can be represented by a predicting function $f(\cdot)$, and the predicting process is denoted as follows:

$$\hat{x}_{i+p} = f\left(x_{i-np}, \ldots, x_{i-2p}, x_{i-p}, x_i\right) \tag{6.3}$$

where $\{x_{i-np}, \ldots, x_{i-2p}, x_{i-p}, x_i\}$ are the input variables, that is, the health indicators at time points $t_{i-np}, \ldots, t_{i-2p}, t_{i-p}, t$, n is the order of the input variables, p is the time interval, and $\hat{x}_{i+p}$ is the predicted health indictor at time point t_{i+p}.

Take $n = 3$ and $p = 1$, for example, Fig. 6.2 shows the predicting process of the data-driven methods. Data-driven prediction methods generally include two steps: training and predicting. In

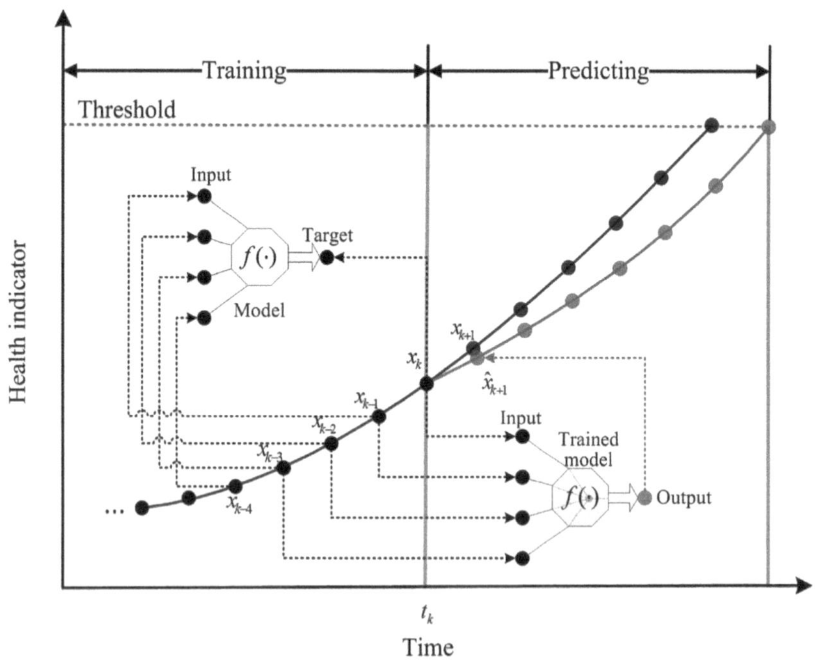

Figure 6.2. Predicting process of data-driven methods.

the training step, available health indicators are organized in a specified form $\left\{x_{i-np},\ldots,x_{i-2p},x_{i-p},x_i,x_{i+p}\right\}_{i=np}^{k-p}$, where i is a history time index which satisfies the following condition:

$$np \le i \le k-p. \tag{6.4}$$

The former $n + 1$ health indicators are the input variables and the last one is the target variable. The predictor function is trained according to the recurrence relationship between the input variables and the target value. The commonly used training strategy is updating the parameters or weights of the predictor with the feedback of prediction errors. After the training step, the prediction function is supposed to have learned the degradation trend of machinery from the available health indicators.

In the predicting step, the trained prediction function $f(\cdot)$ is utilized to forecast the future health indicators. In the special case of Fig. 6.2, the predicting step is described as follows. First, the health indicators $\left\{x_{k-3},x_{k-2},x_{k-1},x_k\right\}$ are input into the prediction function, and the health indicator at the next time t_{k+1} is predicted. This is the one-step prediction process which is denoted as

$$\hat{x}_{k+1} = f\left(x_{k-3},x_{k-2},x_{k-1},x_k\right). \tag{6.5}$$

Then, the predicted health indicator $\hat{x}_{k+1}$ is input into the prediction function, and the two-step prediction result is calculated by

$$\hat{x}_{k+2} = f\left(x_{k-2},x_{k-1},x_k,\hat{x}_{k+1}\right). \tag{6.6}$$

Long-term prediction is able to be recursively performed in this way until the predicted value reaches a predefined failure threshold.

From the predicting process of the data-driven methods it is seen that, the significant mission in data-driven methods is to select and train a suitable predictor. The machine learning techniques are benefit for leaning the potential nonlinear degradation trend of machinery. Therefore, they have been widely used as the predictors by many researchers, such as the artificial neural network (ANN) (Gebraeel et al., 2004), adaptive neural fuzzy interference system (ANFIS) (Chen et al., 2011), support vector machine (SVM) (Liu et al., 2013), and relevance vector machine (RVM) (Zio and Di Maio, 2012). In the following subsections, we will take the RVM, for example, and introduce the knowledge about how to apply the RVM for RUL prediction, the problems in the application, and how to deal with these problems. At last, an enhanced RVM-based RUL prediction method is proposed and demonstrated using a case study of rotating machinery.

6.2.2 RVM-Based Prediction Methods

6.2.2.1 RVM-Based Predictors

RVM is based on the Bayesian framework. It is developed from the SVM technique. Compared with SVM, RVM has the following advantages: (1) RVM is able to provide the membership probability of output members; (2) The kernel functions of the RVM are no longer constrained by the Mercer theorem and is able to be constructed freely; (3) The parameters, like the penalty factor, are assigned automatically rather than subjectively by users. Thanks to the above advantages, RVM has been studied widely and used in various fields, like biomedicine, image processing, and fault classification of machinery.

In Section 3.3, we have applied RVM into the fault diagnosis of rotating machinery, where RVM is used as a classifier. In this section, RVM will be used as a predictor and introduced into the field of RUL prediction. Actually, the major difference between the RVM-based classifier and the RVM-based predictor is the format of the input and target/output variables. In the RVM-based classifier, the input variables are $\{\boldsymbol{x}_i\}_{i=1}^{N}$, where $\boldsymbol{x}_i \in R^D$ are D dimensional features and N is the number of the input variables. The target variable is $\{\boldsymbol{t}_i\}_{i=1}^{N}$, where $\boldsymbol{t}_i \in \{1,2,3,\ldots,C\}$ are the class labels. In the RVM-based predictor, the input variables are $\{\boldsymbol{x}_i\}_{i=np}^{k-p}$, where $\boldsymbol{x}_i = x_{i-np},\ldots,x_{i-2p},x_{i-p},x_i$ are the health indicators at time $t_{i-np},\ldots,t_{i-2p},t_{i-p},t_i$, k is the current time index, n is the order of the input variables and p is the time interval. The target variable is $\boldsymbol{t} = \{x_{i+p}\}_{i=np}^{k-p}$, that is, the health indicator at time t_{i+p}.

In the RVM-based predictor, the relationship between the input variables and the target variable is given as the following function:

$$x_{i+p} = f(\boldsymbol{x}_i;\boldsymbol{w}) = \sum_{i=np}^{k-p} w_i K(\boldsymbol{x},\boldsymbol{x}_i) + w_0 \tag{6.7}$$

where $K(\cdot)$ is the kernel function, w_i is the weight of the ith training example. Considering the scalar-valued target functions only, it follows the standard probabilistic formulation and assumes that the new targets for RVM learning model are the original targets with additive noise.

$$x'_{i+p} = x_{i+p} + \omega_i = f(\boldsymbol{x}_i;\boldsymbol{w}) + \omega_i = \sum_{i=np}^{k-p} w_i K(\boldsymbol{x},\boldsymbol{x}_i) + w_0 + \omega_i \tag{6.8}$$

where ω_i is the random noise assumed to be Gaussian with mean 0 and variance σ^2.

The PDF of the target variables based on the input variables are denoted as

$$p\left(x'_{i+p} \mid x_i\right) = N\left(x'_{i+p} \middle| f\left(x_i; w\right), \sigma^2\right). \tag{6.9}$$

The target variables are assumed to be independent identically distributed (i.i.d.). The joint PDF of the target variables is

$$p\left(t \middle| w, \sigma^2\right) = \prod_{i=np}^{k-p} N\left(x_{i+p} \middle| f\left(x_i; w\right), \sigma^2\right) = \left(2\pi\sigma^2\right)^{-\frac{N}{2}} \exp\left(-\frac{\|t - \Phi w\|^2}{2\sigma^2}\right) \tag{6.10}$$

where $\Phi = \left[\Phi_{np}, \Phi_{np+1}, \ldots, \Phi_{k-p}\right]^T$, $\Phi_i = \left[1, K\left(x, x_{np}\right), K\left(x, x_{np+1}\right), \cdots, K\left(x, x_{k-p}\right)\right]^T$.

Let $t' = \left\{x'_{i+p}\right\}_{i=np}^{k-p}$, then the PDF of t' is expressed as

$$p\left(t' \middle| t\right) = \int p\left(t' \middle| w, \sigma^2\right) p\left(w, \sigma^2 \middle| t\right) dw d\sigma^2. \tag{6.11}$$

w_i is assumed to follow the normal distribution of $N(0, 1/\alpha_i)$, then the PDF of w based on α is

$$p\left(w \middle| \alpha\right) = \prod_{i=np}^{k-p} \sqrt{\frac{\alpha_i}{2\pi}} \exp\left(-\frac{\alpha_i^2 w_i^2}{2}\right). \tag{6.12}$$

Eq. (6.11) is further transformed into

$$p\left(t' \middle| t\right) = \int p\left(t' \middle| w, \alpha, \sigma^2\right) p\left(w, \alpha, \sigma^2 \middle| t\right) dw d\alpha d\sigma^2 \tag{6.13}$$

where

$$p\left(w, \alpha, \sigma^2 \middle| t\right) = p\left(w \middle| t, \alpha, \sigma^2\right) p\left(\alpha, \sigma^2 \middle| t\right) \tag{6.14}$$

$$p\left(w \middle| t, \sigma, \sigma^2\right) = \left(2\pi\right)^{-\frac{k-p-np+2}{2}} \left|\Sigma\right|^{-\frac{1}{2}} \exp\left(-\frac{1}{2}\left(w - \mu\right)^T \Sigma^{-1} \left(w - \mu\right)\right) \tag{6.15}$$

$$p\left(t \middle| \alpha, \sigma^2\right) = \left(2\pi\right)^{-\frac{k-p-np+1}{2}} \left|\sigma^2 \mathrm{I} + \Phi \mathrm{A}^{-1} \Phi\right|^{-\frac{1}{2}} \exp\left(-\frac{1}{2} t^T \left(\sigma^2 \mathrm{I} + \Phi \mathrm{A}^{-1} \Phi\right)^{-1} t\right) \tag{6.16}$$

with $\Sigma = \left(\sigma^{-2} \Phi^T \Phi + A\right)^{-1}$, $A = \mathrm{diag}\left(\alpha_0, \alpha_1, \cdots, \alpha_N\right)$, $\mu = \sigma^{-2} \Sigma \Phi^T t$.

Eq. (6.13) is transformed into the following one by substituting Eq. (6.14).

$$p\left(t' \middle| t\right) = \int p\left(t' \middle| w, \alpha, \sigma^2\right) p\left(w \middle| t, \alpha, \sigma^2\right) p\left(\alpha, \sigma^2 \middle| t\right) dw d\alpha d\sigma^2 \tag{6.17}$$

The closed-form expression of the previous equation is unable to be solved analytically. Therefore, the numerical solution is calculated using the maximum likelihood estimation algorithm and the delta-function algorithm. First, the partial derivatives of Eq. (6.16) with respect to $\boldsymbol{\alpha}$ and σ^2, and the updating process of these two parameters is as follows:

$$\begin{cases} \alpha_i^{\text{new}} = \dfrac{\gamma_i}{w_i^2} \\ (\sigma^2)^{\text{new}} = \dfrac{\|t - \Phi\mu\|^2}{N - \sum_{i=0}^{N} \gamma_i} \end{cases} \tag{6.18}$$

where $\gamma_i = 1 - \alpha_i \Sigma_{i,i}$, and $\gamma_i \in [0, 1]$, $\Sigma_{i,i}$ are the diagonal elements of Σ. The maximum likelihood values $\boldsymbol{\alpha}_{\mathbf{MP}}$ and σ_{MP}^2 are calculated by updating the two parameters $\boldsymbol{\alpha}$ and σ^2 using Eq. (6.18). With the calculated parameters $\boldsymbol{\alpha}_{\mathbf{MP}}$ and σ_{MP}^2, Eq. (6.17) is calculated using the delta-function.

$$p(t'|t) \approx \int p(t'|w,\alpha_{\text{MP}},\sigma_{\text{MP}}^2)\, p(w|t,\alpha_{\text{MP}},\sigma_{\text{MP}}^2)\, dw \tag{6.19}$$

As $p(t'|w,\alpha_{\text{MP}},\sigma_{\text{MP}}^2)$ and $p(w|t,\alpha_{\text{MP}},\sigma_{\text{MP}}^2)$ are normally distributed, the previous equation is further expressed using

$$p(t'|t) \approx N(t'|y',\sigma'^2) \tag{6.20}$$

where

$$\begin{cases} y' = \mu^{\mathrm{T}} \Phi(x') \\ \sigma'^2 = \sigma_{\text{MP}}^2 + \Phi(x')^{\mathrm{T}} \Sigma \Phi(x') \\ \Phi(x') = \left[1, K(x, x_{np}), K(x, x_{np+1}), \cdots, K(x, x_{k-p})\right]. \end{cases} \tag{6.21}$$

It is well known that RVM is a technique based on the kernel function, which attempts to reveal the mapping relationship in the high-dimensional space. There are numerous types of kernel functions and it is difficult to explain their individual characteristics. The commonly used kernel functions are classified into two categories, namely local kernels and global kernels (Smits and Jordaan, 2002). In local kernels, only the data points which are close or in the proximity of each other have an influence on the kernel values. The widely used local kernels include Gaussian function, Cauchy function, inverse multiquadric function, Laplace function, etc. In contrast, a global kernel allows data points

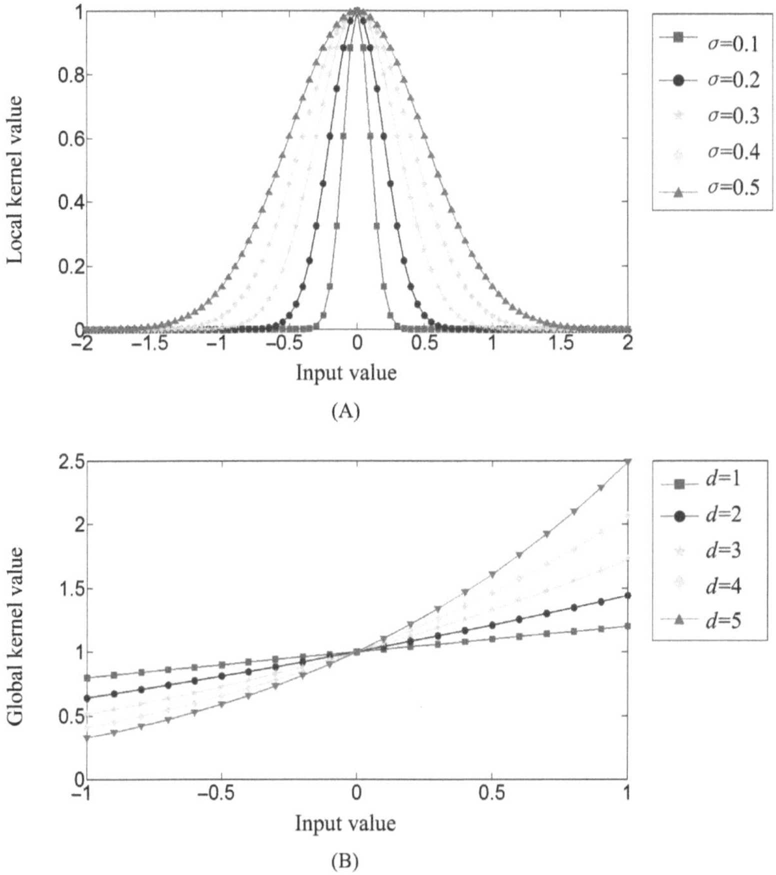

Figure 6.3. (A) Examples of local kernels (Gaussian function), and (B) examples of global kernels (polynomial function).

that are far away from each other to have an influence on the kernel values as well. And the commonly used global kernels are polynomial function, spline function, etc.

Examples of local kernels, that is, Gaussian function, and global kernels, that is, polynomial function, are shown in Fig. 6.3. The Gaussian function is a typical local kernel, which is expressed as

$$K(x \cdot x_i) = \exp\left(-\frac{(x - x_i)^2}{2\sigma^2}\right) \tag{6.22}$$

where the kernel parameter is the width σ of the radial basis function. In Fig. 6.3A, the local effect of the Gaussian function is shown for a chosen test input for different values of the width σ. It is seen that, a local kernel only has an effect on the data points in the neighborhood of the test point.

The polynomial function, a representative kernel in global kernels, is denoted as

$$K(x \cdot x_i) = \left(ax^{\mathrm{T}} x_i + c\right)^d \quad (6.23)$$

where the kernel parameter d is the degree of polynomial to be used. In Fig. 6.3B, the global effect of the polynomial function of various degrees is displayed. It is seen that, for each degree of polynomial, all data points in the input domain have nonzero kernel values. The test data point has a global effect on the other data points.

The quality of a kernel is determined not only by its ability to learn from the data but also its ability to predict unseen data. These two abilities are named as interpolation capacity and extrapolation capacity (Smits and Jordaan, 2002), respectively. From the two examples shown in Fig. 6.3 it is concluded that, a global kernel shows better interpolation ability, but fails to provide longer-range extrapolation. On the other hand, the global kernel has better extrapolation capacity but cannot provide good interpolation ability. Therefore, we will investigate whether the advantages of local kernels and global kernels can be combined through mixtures. With this intention, we choose a commonly used filtering algorithm, that is, particle filtering (PF), to mixture the advantages of different kinds of kernels. And a multikernel RVM-based prediction method is proposed. In the following subsection, the basic theory of PF is first introduced. Then the flowchart of the multikernel RVM-based prediction method is described.

6.2.2.2 Particle Filtering Algorithms

PF is developed from traditional filtering algorithms, such as Kalman filtering and Extended Kalman filtering (Arulampalam et al., 2002). PF is a filtering algorithm based on the Bayesian theory, which attempts to evaluate the state of a system with respect to measurements. A sequential importance sampling (SIS) algorithm is developed and applied in PF (Doucet et al., 2001a). Based on the Bayesian theory and SIS, PF is particularly effective in state estimation of nonlinear and/or non-Gaussian systems, and has been widely used in the fields of diagnostics and prognostics (Li et al., 2015).

In the PF algorithm, the state of a system x_k at t_k is described in the following state space model:

$$x_k = f_k\left(x_{k-1}, \omega_k\right) \quad (6.24)$$

where $f_k: R^{n_\alpha} \times R^{n_\omega} \to R^{n_\alpha}$ is the state transition function, and ω_k is the i.i.d. state noise.

Generally, it is difficult or impossible to directly obtain the system states. Therefore, the system states should be evaluated based on the measurements. The relationship between them is denoted by

$$z_k = h_k(x_k, v_k) \tag{6.25}$$

where $h_k: R^{n_\alpha} \times R^{n_v} \to R^{n_z}$ is the measurement function and v_k is the i.i.d. measurement noise.

Based on the knowledge of the transition function and the previous state estimation, the prior PDF of the present state is calculated as follows:

$$p(x_k \mid z_{1:k-1}) = \int p(x_k \mid x_{k-1}) p(x_{k-1} \mid z_{1:k-1})\, dx_{k-1}. \tag{6.26}$$

Once a new measurement is available, the prior PDF is updated to generate the posterior PDF using the following equation.

$$p(x_k \mid z_{1:k}) = \frac{p(z_k \mid x_k) p(x_k \mid z_{1:k-1})}{p(z_k \mid z_{1:k-1})}. \tag{6.27}$$

The recursive propagation of the posterior PDF is only a conceptual solution. It is impossible to be calculated analytically because it requires the evaluation of complicated high-dimensional integrals. Therefore, the posterior PDF is approximated numerically using the SIS algorithm. SIS approximates the state PDF using a set of particles $\{x_k^n\}_{n=1:N_s}$ sampled from an importance PDF $q(x_k \mid z_{1:k})$, where N_s is the number of particles. Using SIS, the state of the system is approximated using

$$p(x_k \mid z_{1:k}) \approx \sum_{n=1}^{N_s} w_k^n \delta(x_{0:k} - x_{0:k}^n) \tag{6.28}$$

where w_k^n is the weight of the particle $x_{0:k}^n$.

When a new measurement is captured, the particle weights are updated and normalized using

$$w_k^n = w_{k-1}^n \frac{p(z_k \mid x_k^n) p(x_k^n \mid x_{k-1}^n)}{q(x_k^n \mid x_{k-1}^n, z_{1:k})}, \quad \tilde{w}_k^n = \frac{w_k^n}{\sum_{n=1}^{N_s} w_k^n}. \tag{6.29}$$

In general, the importance PDF $q(x_k \mid z_{1:k})$ is chosen as $p(x_k \mid x_{k-1})$, and then Eq. (6.29) is simplified as:

$$w_k^n = w_{k-1}^n p\left(z_k \mid x_k^n\right), \ \tilde{w}_k^n = \frac{w_k^n}{\sum_{n=1}^{N_s} w_k^n}. \tag{6.30}$$

A common problem in the SIS algorithm is the particle degeneracy, that is, after a few iterations, all but one particle have negligible weights. To deal with this problem, a resampling step is utilized to eliminate particles with small weights and to concentrate upon particles with large weights. A suitable measure of the degeneracy of particles is the effective sample size introduced in (Liu and Chen, 1998) and defined as:

$$N_{eff} = \frac{N_s}{1 + \text{var}\left(w_k^{n*}\right)} \tag{6.31}$$

where $w_k^{n*} = p\left(x_k^n \mid z_{1:k}\right) / q\left(x_k^n \mid x_{k-1}^n, z_{1:k}\right)$ is referred to the true weight. The effective sample size is unable to be calculated exactly, but it can be estimated by

$$\hat{N}_{eff} = \frac{1}{\sum_{n=1}^{N_s} \left(w_k^n\right)^2}. \tag{6.32}$$

When $\hat{N}_{eff}$ falls below a threshold N_T, a new particle set $\left\{x_k^{n*}\right\}_{n=1:N_s}$ will be generated by resampling N_s times following the principle $P\left(x_k^{n*} = x_k^n\right) = w_k^n$, and the weights are reset to $w_k^n = 1/N_s$.

6.2.2.3 Multikernel RVM-Based Prediction Method

In Section 6.2.2.1, it has been recognized that the selected kernels and their parameters determine the prediction performance of the RVM-based predictor. What's more, different types of kernels perform distinctly in the interpolation ability and extrapolation capacity. Therefore, we try to make full use of the superiorities of different kernels by combining them using the PF algorithm. The combined kernel is named as the multikernel, and a new multikernel RVM-based prediction method is proposed. The flowchart of this method is shown in Fig. 6.4.

Measurements are captured using the accelerometers installed on the rotating machinery, and health indicators are extracted. The health indicators are input into the multikernel RVM models for training. The multikernel RVM models are constructed as follows.

1. Select M single kernels, which are represented by $\left\{K^m\right\}_{m=1:M}$.

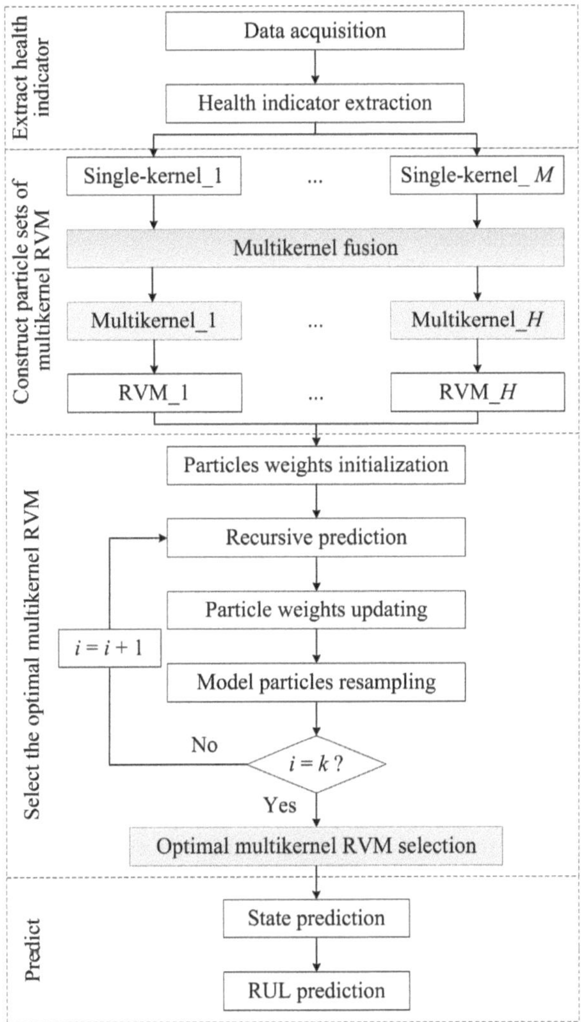

Figure 6.4. Flowchart of the multikernel RVM-base prediction method.

2. Construct H multikernels by fusing the M single kernels using the following equation:

$$K_h = \sum_{m=1}^{M} w_h^m K^m, \quad h = 1,\ldots H \tag{6.33}$$

where w_h^m is the weight of the mth single kernel in the hth multikernel. It is randomly assigned following the principles

$$\sum_{m=1}^{M} w_h^m = 1 \quad \text{and} \quad 0 \le w_h^m \le 1. \tag{6.34}$$

3. Set up particle sets of the multikernel RVM model using the H multikernels.

After the multikernel RVM model construction, these models are trained with the available health indicators. Then, these models are used to recursively predict the states of the rotating machinery. According to their prediction accuracy, the optimal one is selected from them using the PF algorithm. The detailed processes are as follows.

1. Initialize the particle weight of each multikernel RVM model as $v_1^h = \frac{1}{H}$.
2. At t_{i-1}, predict the value of the health indicator at the next time t_i using each model and acquire the predicted vector

$$z_i' = \left[\left(z_i^1\right)', \left(z_i^2\right)', \cdots, \left(z_i^H\right)' \right]^{\mathrm{T}} \tag{6.35}$$

where $\left(z_i^h\right)', h = 1, 2, \cdots, H$ is the predicted value of the hth model.
3. Update the particle weight of each multikernel RVM model according to its predicted value, and normalize the updated weights.

$$v_i^h = \frac{1}{\sqrt{2\pi}\sigma_v} \exp\left[-\frac{1}{2} \left(\frac{z_i^h - z_i}{\sigma_v} \right)^2 \right], \quad \tilde{v}_i^h = \frac{v_i^h}{\sum_{h=1}^{H} v_i^h} \tag{6.36}$$

where v_i^h is the particle weight of the hth model at t_i.
4. Resample the particles according to the particle weights, that is, replicate particles which have large weights and eliminate particles which have small weights.
5. Repeat step (2–4) until $i = k$, where k is the current time index. And select the particle which has the largest number as the optimal one.

6.2.2.4 RUL Prediction of Gears: a Case Study

Accelerated degradation tests of gears are conducted on an experimental system in Fig. 6.5. This system is designed in a back-to-back arrangement, which is composed of two gearboxes. One gearbox contains a pair of spur gears and the other holds a pair of helical gears. The spur gears are more likely to degrade during the tests. Therefore, the former gearbox is named as the tested gearbox and receives more attention during the testes. The other is labeled as the slave gearbox. They are connected through shafts and a clutch to form a closed loop. One of the shafts is divided into two sections for preloading a large torque inside the loop by rotating

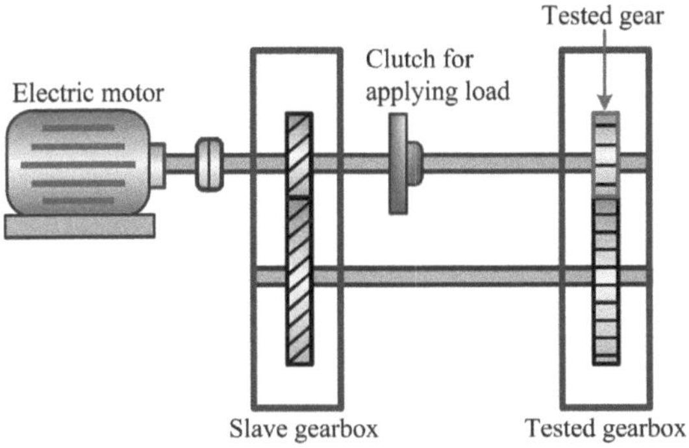

Figure 6.5. Experimental system for accelerated degradation tests of gears.

the two parts of the shaft against each other and then blocking them with a clutch. An electric motor provides power to overcome the friction of the system and to accommodate the start-up torques for the system acceleration. The maximum torque loaded on the tested gears is about 1300 Nm and the rotating speed keeps 1460 rpm. Accelerometers are placed on the top of the tested gearbox. The sampling frequency is 20 kHz and each sample contains 120,000 data points. Sampling is repeated every 6 s. The tested gears degrade naturally without seeding a fault in advance.

After the tests, a serious crack is found on a tooth of the tested gear. The vibration signals of tested gear during the whole lifetime are shown in Fig. 6.6. At the beginning of the lifetime, the amplitudes of the vibration signals fluctuate randomly. After a period of running-in, the amplitudes of vibration signals become stable. At the end of the lifetime, the vibration amplitudes increase gradually with the degeneracy of the gears. The vibration signals during the degradation stage are used for further analysis. In order to extract

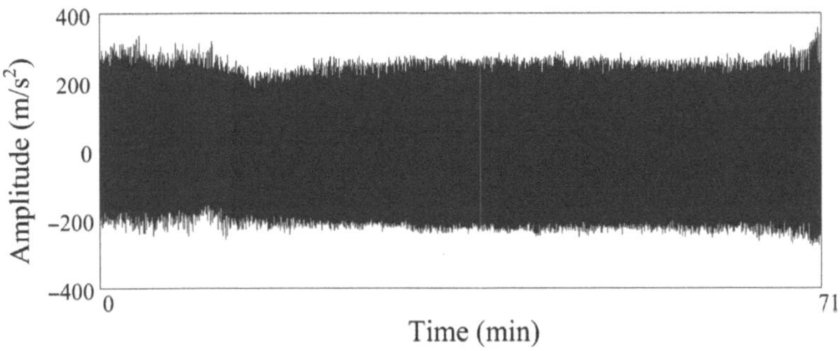

Figure 6.6. Vibration signals of the tested gear.

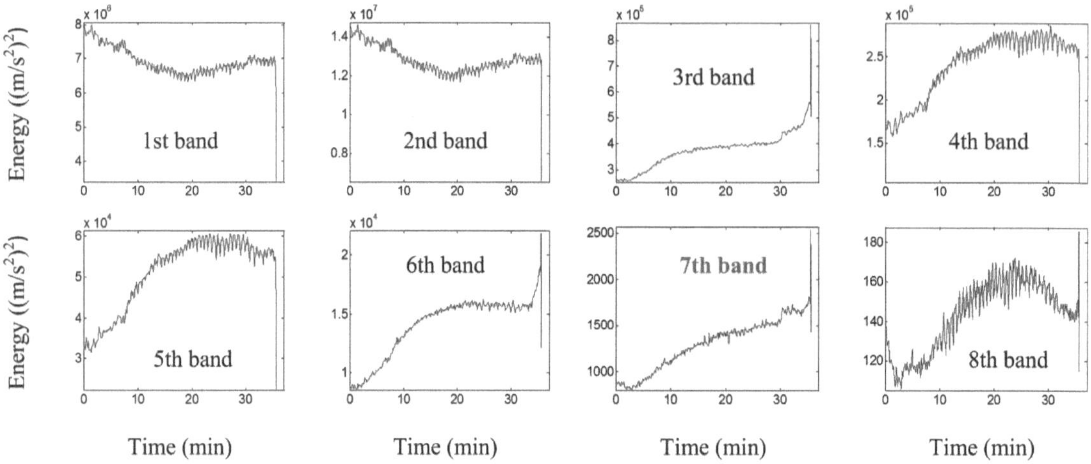

Figure 6.7. Energy curves of decomposed signals in eight frequency bands.

a health indicator which is supposed to reveal the degradation trend of the tested gear, the vibration signals are processed using the three-layer wavelet pocket decomposition. And the energies of the decomposed signals in eight frequency bands are extracted and displayed in Fig. 6.7. Among these energy curves, the curve in the 7th band has the most obvious degradation trend. Therefore, it is chosen as the health indicator of the tested gear. The health indicator is further smoothed to reduce random noises of the energy curve. And the smoothed health indicator is given in Fig. 6.8.

After the health indictor extraction, multikernel RVM models are constructed following the processes described in Section 6.2.2.3, where the single kernels are Gaussian, inverse multiquadric (inverseM), Cauchy, polynomial, and spline. Then the health indicators are input into the multikernel RVM models.

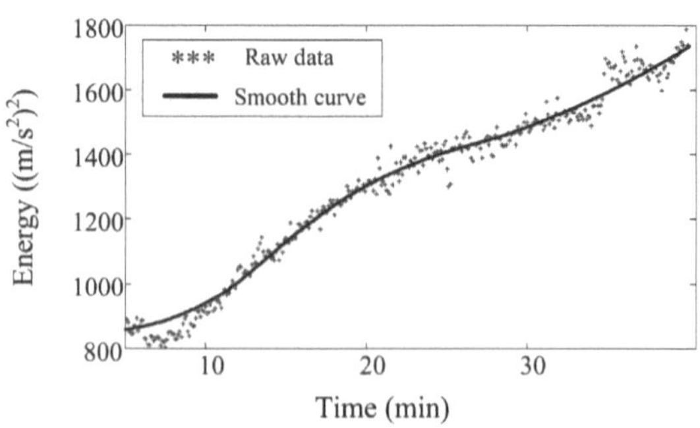

Figure 6.8. Smooth results of the energy curve in the 7th frequency band.

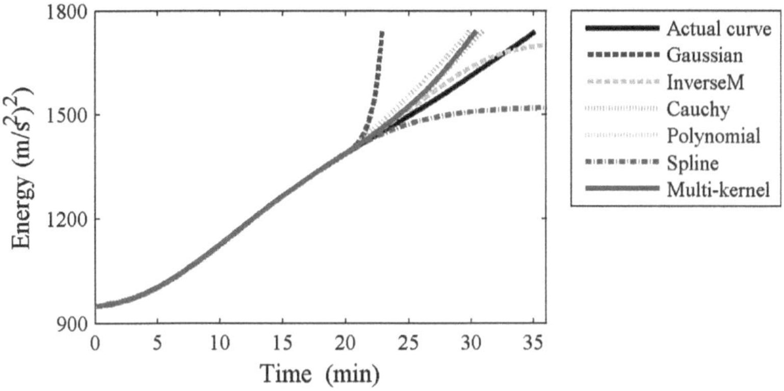

Figure 6.9. Predicted health indicator curves using the single and multikernel RVM models at 20 min.

At each time point, the health indicator curves in the future are recursively predicted. Taking t_i = 20 min and 30 min for example, the predicted health indicator curves using the single and multikernel RVM models are shown in Figs. 6.9 and 6.10, respectively. The maximum health indicator value is determined as the failure threshold, that is, 1735 $(m/s^2)^2$, and the RUL of the tested gear is predicted from 20 min to the end of the lifetime. The prediction results using the single and multikernel RVM models are given in Fig. 6.11.

From the health indicator curves and RUL prediction results it is seen that, the multikernel RVM model performs better than each single kernel RVM model. The multikernel RVM model is

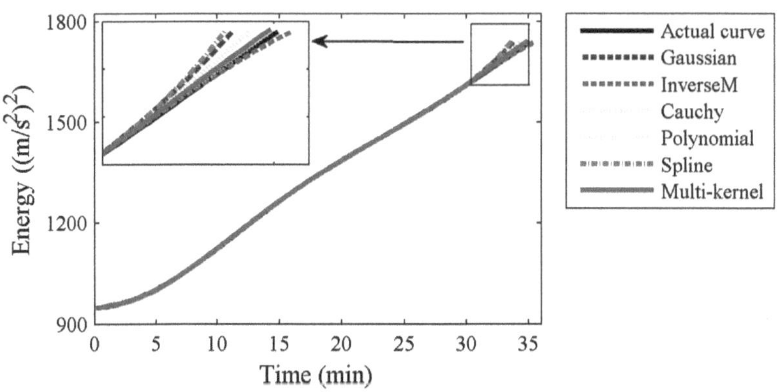

Figure 6.10. Predicted health indicator curves using the single and multikernel RVM models at 30 min.

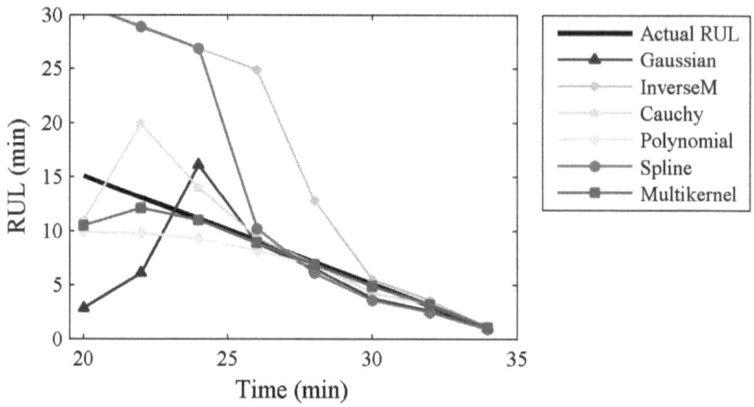

Figure 6.11. RUL prediction results of the single and multikernel RVM models.

not only able to converge to the actual RUL in high speed but also present stable and accurate prediction results. Other single kernel RVM models, however, may acquire a good prediction result at certain time points, but fail to predict the RUL accurately during the whole lifetime. In order to quantitatively compare the prediction accuracy of these RVM models, three criterions of prediction accuracy are calculated including absolute percentage error (APE), mean of absolute percentage error (MAPE) and root mean square of error (RMSE). APE evaluates the accuracy of prediction results at each time point. It is expressed as

$$\mathrm{APE}_k = \frac{l_k - \mathrm{ActRUL}_k}{\mathrm{ActRUL}_k} \times 100\% \tag{6.37}$$

where l_k is the predicted RUL and ActRUL_k is the actual RUL at t_k.

MAPE and RMSE demonstrate the prediction accuracy of a method at certain time indexes, which are denoted by

$$\mathrm{MAPE} = \frac{1}{K}\sum_{k=1}^{K}\left|\frac{l_k - \mathrm{ActRUL}_k}{\mathrm{ActRUL}_k} \times 100\%\right| \tag{6.38}$$

$$\mathrm{RMSE} = \sqrt{\frac{1}{K}\sum_{i=1}^{K}\left(l_k - \mathrm{ActRUL}_k\right)^2} \tag{6.39}$$

where K is the number of time points. Obviously, smaller APE, MAPE, and RMSE values mean lower prediction errors resulting in higher prediction accuracies.

APE values of the prediction results are calculated and illustrated in Fig. 6.12. MAPE and RMSE are displayed in Table 6.1. It

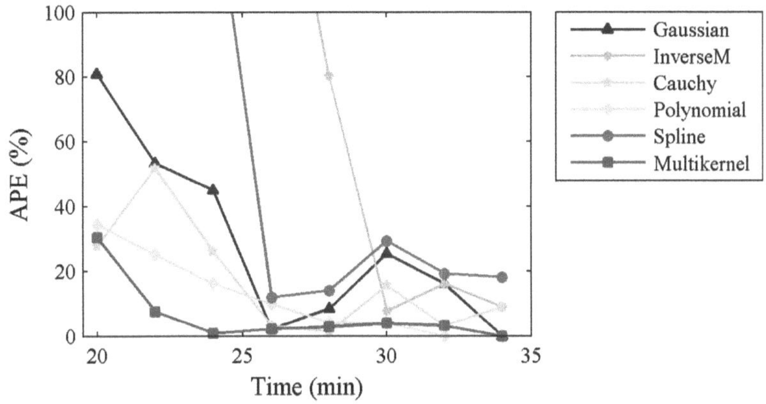

Figure 6.12. APE of the prediction results for the single and multikernel RVM models.

is observed that, the multikernel RVM model has smaller APE values during the whole lifetime than most of the single kernel RVM models. The MAPE and RMSE of the multikernel RVM model are the smallest among these models. It is demonstrated that the multikernel RVM model is more effective in the RUL prediction of the tested gears than the single kernel RVM models.

6.2.3 Epilog

In this section, the basic theory of data-driven prediction methods is first introduced. Then a typical machine learning technique, that is, the RVM, is employed as an example to predict the

Table 6.1 MAPE and RMSE of the Prediction Results for the Single and Multikernel RVM Models

RVM Model	MAPE (%)	RMSE (min)
Gaussian	28.9428	35.1960
InverseM	+∞	+∞
Cauchy	17.3197	45.8574
Polynomial	11.7351	33.3586
Spline	57.5887	145.3960
Multikernel	6.3950	26.4290

RUL of rotating machinery. The RVM was applied into the fault classification of rotating machinery in Section 3.3. There are few differences between the RVM-based classifier and the RVM-based predictor. Therefore the RVM-based predictor is introduced in Section 6.2.2.1. Through the discussion of the characteristics of different kernels, it is proven that different kernels generally present distinct advantages in the RUL prediction. So we use the PF algorithm to mixture the advantages of different kinds of kernels, and a multikernel RVM-based prediction method is proposed. This method first constructs a series of multikernel RVM models through random combination of different signal kernels. Then an optimal multikernel RVM model is selected according to their performances in the state prediction. Accelerated degradation tests of gears are used to demonstrate the effectiveness of the multikernel RVM prediction method. Through the comparison with the single kernel RVM methods, the multikernel RVM method shows better RUL prediction results.

6.3 Model-Based Prediction Methods

6.3.1 Introduction

Model-based methods are to set up mathematical or physical models to describe the degradation process of machinery, and to estimate model parameters using measurements. To construct proper models is critical in model-based methods. The models are supposed to accurately describe the degradation trend of the machinery and facilitate the model parameter estimation and RUL prediction. There are numerous prediction models in the literature and they are distinctly different in both the model sources and the application targets. Some of them are developed from the physical mechanism of degradation processes. Others are built according to the expert knowledge or the mathematical formulas. Many researchers have attempted to classify the prediction models in different terms (Sikorska et al., 2011, Baraldi et al., 2013, Si et al., 2011, Heng et al., 2009). In terms of the model sources, we categorize the commonly used models into the following three classes: statistical models, physical models, and stochastic process models (Fig. 6.13).

Statistical models are constructed based on the statistics of abundant failure events, such as the proportional hazards model (Cox, 1972), the proportional intensity model, the Markov-based model (Rausand and Hoyland, 2004) and the logistic regression model (Hosmer and Lemeshow, 1989). They estimate the damage emergence and progression based on previous inspection results

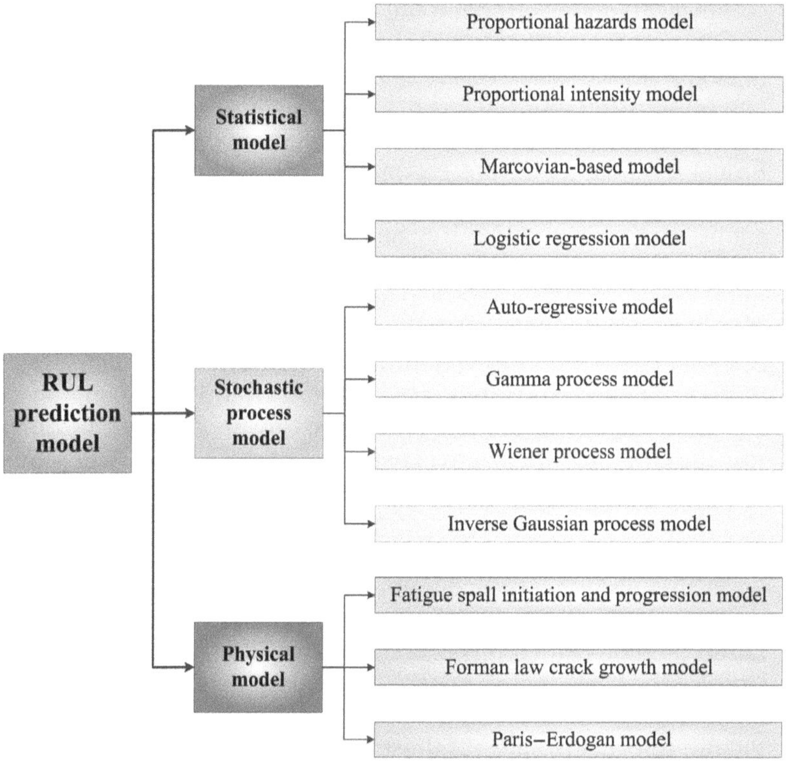

Figure 6.13. Categories of the RUL prediction models.

on similar machines. They are often used when no suitable dynamic model of the physical process is available. However, the model parameter estimation requires large numbers of event data from similar machines, which is hard to capture in real application.

Stochastic process models describe the degradation processes of machinery as stochastic processes. Since mechanical systems generally degrade stochastically, their degradation processes are naturally modeled as stochastic processes. So far, different kinds of stochastic process models have been developed and applied into the machine RUL prediction, such as the Gamma process model (Park and Padgett, 2005), the inverse Gaussian process model (Ye, 2013), the autoregressive model (Pang et al., 2014), the random coefficient model (Park and Bae, 2010), and the Wiener process model (Si et al., 2013b). They are based on the assumption that the failure time of identical components can be considered as statistically identical and independent random variables and thus can be described by a probability density function.

Physical models describe the degradation process of machinery using the degradation physical laws based on the mechanism of degradation processes, for example, the fatigue spall initiation and progression model (Orsagh et al., 2004), the Forman law crack growth model (Oppenheimer and Loparo, 2002) and the Paris–Erdogan model (Paris and Erdogan, 1963). These models imply a thorough understanding of the mechanical system behavior in response to the stress at both macroscopic and microscopic levels. Furthermore, it assumes that this behavior can be described accurately and analytically. Physical models estimate an output for the RUL of a component/machine by solving a deterministic equation or a set of equations derived from extensive empirical data. Some of data has been converted into common scientific and engineering knowledge, while other data needs to be acquired through specific laboratory or in situ experimentations. Normally, deriving physical models for a particular mechanical system will involve identifying one or more parameters specific to that system (e.g., exact physical properties, stress levels, and equation constants).

In this section, three different model-based RUL prediction methods are presented, including two stochastic process model-based methods and a physical model-based method. We first introduce the basic theories of the original models. Then the strategies of RUL prediction using these models are described in details. Finally, the effectiveness of these methods is demonstrated using accelerated degradation tests of rotating machinery.

6.3.2 Exponential Model-Based Methods

6.3.2.1 Motivation

The exponential model is one of the most widely used stochastic process models. It was first established by Gebraeel et al. (2005). In the first version, a Bayesian approach is employed to update the model parameters to incorporate the measured information. By now, many variants have been developed from the first version and been applied into the RUL prediction of machinery (Gebraeel, 2006, Elwany and Gebraeel, 2008, You et al., 2010, Si et al., 2013a). For example, Gebraeel (2006) improved the model by updating the model parameters using multiple historical degradation signals acquired from condition monitoring. Si et al. (2013a) combined the Bayesian approach with the expectation maximization (EM) algorithm for parameter estimation and offered a closed-form RUL distribution.

The exponential model works well for the RUL prediction of systems with exponential-like degradation processes. However, there are still two shortcomings in this model. One shortcoming is that the FPT is selected subjectively. The concept of FPT has been explained in Section 6.1. It is important to select an appropriate FPT value for the exponential model. If an inappropriate FPT value is selected, either interference noises may be included in the predicting process or critical information may be excluded from the predicting process. Both cases would decrease the RUL prediction accuracy. The appropriate FPT is supposed to be the time of fault occurrence. It is yet difficult to be detected because incipient faults of a bearing often occur randomly and the fault characteristics are quite weak. Several approaches for selecting FPT have been reported in the literature, such as the engineering norm ISO 10816 and the approach based on the longest time constant of a machine and statistical properties of a candidate baseline (Ginart et al., 2006, Tran and Yang, 2012). In these approaches, the FPT is selected based on an alarm, which is set according to the statistical properties of large numbers of systems. Therefore, the approaches fail to adjust the alarm adaptively according to the variation of an individual system.

The other shortcoming of the exponential model is that random errors of the stochastic process decrease the RUL prediction accuracy. This can be explained using a simulation shown in Fig. 6.14, where a stochastic degradation process of a machine is simulated

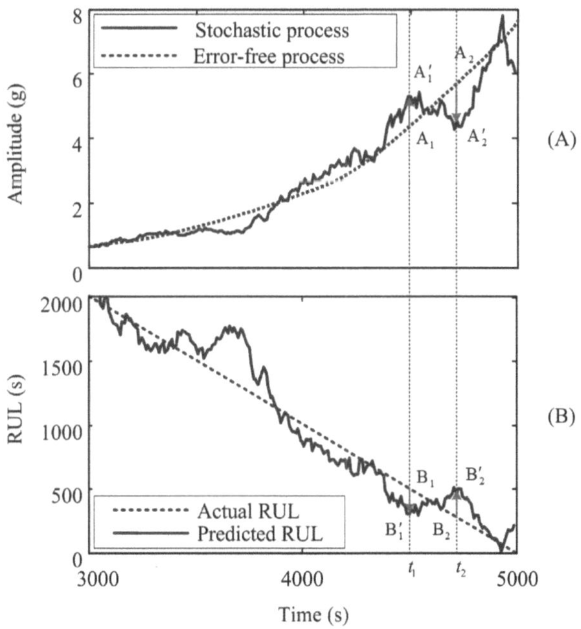

Figure 6.14. (A) Simulated degradation process and (B) RUL prediction result.

and the RUL of the machine is predicted using the exponential model. The error-free process describes the actual degradation process of the machine without the interference of random errors. It is noticed that there is an obvious negative correlation between random errors of the stochastic process and deviations of the prediction result. For example, the positive random error at time t_1 leads to the negative deviation between the predicted RUL B_1 and the actual RUL B_1'. Vice versa, the negative random error at time t_2 causes the positive deviation between B_2 and B_2'. This negative correlation can be explained as follows. Taking time t_1, for example, the value of the degradation process increases from A_1' to A_1 because of the positive random error. As a result, the fault severity is overestimated compared with the actual severity. A severer fault generally indicates a shorter RUL. Therefore, the overestimated severity leads to a shorter predicted RUL than the actual one.

To overcome these two shortcomings of the exponential model, an improved exponential model-based method is proposed in Li et al. (2015) for predicting the RUL of rotating machinery. The major improvement of this work includes the following two aspects: (1) an adaptive FPT selection approach is established based on the 3σ interval; and (2) random errors of the stochastic process are reduced utilizing PF and therefore the RUL prediction accuracy is enhanced. The detailed process of the improved exponential model-based method will be introduced in the following subsection.

6.3.2.2 Exponential Models

In this subsection, the basic theory of the exponential model in Si et al. (2013a) is introduced. In this variant of the exponential model, the degradation process is described using a stochastic process $\{x_k = x(t_k), t_k \geq 0\}$, where x_k is the system state at time t_k and

$$x_k = \varphi + \theta' \exp\left(\beta' t_k + \sigma B(t_k) - \frac{\sigma^2}{2} t_k\right) \tag{6.40}$$

where φ is a known constant, θ' and β' are random variables characterizing the stochastic part, which captures the variation of the degradation process of an individual system, σ is a constant representing the deterministic part, which implies a constant physical phenomenon of all systems of a given population, and $\sigma B(t_k)$ is a Brownian motion (BM) following a normal distribution of $N(0, \sigma^2 t_k)$, which represents random errors in the stochastic process.

For convenience, the exponential model is transformed into the logarithmic format and is defined as:

$$\begin{aligned} s_k &= \ln(x_k - \varphi) = \ln(\theta') + \left(\beta' - \frac{\sigma^2}{2}\right)t_k + \sigma B(t_k) \\ &= \theta + \beta t_k + \sigma B(t_k), \end{aligned} \tag{6.41}$$

where $\theta = \ln(\theta')$ follows $N(\mu_0, \sigma_0^2)$ and $\beta = \beta' - \sigma^2/2$ follows $N(\mu_1, \sigma_1^2)$. It is assumed that $\varphi = 0$ to simplify the analysis.

Suppose that we have observed a sequence of measurements $S_{1:k} = \{s_1, s_2, \ldots, s_k\}$. Since random errors $\sigma B(t_k)$ are i.i.d. normal random variables, if θ and β are known, the conditional joint PDF of $S_{1:k}$ is as:

$$\begin{aligned} p(S_{1:k} \mid \theta, \beta) = &\left(\frac{1}{\sqrt{2\pi\sigma^2\Delta t}}\right)^k \\ &\exp\left[-\frac{(s_1 - \theta - \beta t_1)^2}{2\sigma^2 t_1} - \sum_{j=2}^{k} \frac{(s_j - s_{j-1} - \beta\Delta t)^2}{2\sigma^2\Delta t}\right], \end{aligned} \tag{6.42}$$

where $\Delta t = t_j - t_{j-1}$ is the constant time interval. Then the joint posterior PDF of θ and β conditioned on $S_{1:k}$ is still normal resulted from the normal distribution of θ and β, that is, $\theta, \beta | S_{1:k} \sim N(\mu_{\theta,k}, \sigma_{\theta,k}^2, \mu_{\beta,k}, \sigma_{\beta,k}^2, \rho_k)$, with

$$\begin{aligned} \mu_{\theta,k} &= \frac{(s_1\sigma_0^2 + \mu_0\sigma^2 t_1)(\sigma^2 + \sigma_1^2 t_k) - \sigma_0^2 t_1 (s_k\sigma_1^2 + \mu_1\sigma^2 - 0.5\sigma^4)}{(\sigma_0^2 + \sigma^2 t_1)(\sigma_1^2 t_k + \sigma^2) - \sigma_0^2\sigma_1^2 t_1} \\ \sigma_{\theta,k}^2 &= \frac{\sigma_0^2\sigma^2 t_1(\sigma^2 + \sigma_1^2 t_k)}{(\sigma_0^2 + \sigma^2 t_1)(\sigma_1^2 t_k + \sigma^2) - \sigma_0^2\sigma_1^2 t_1} \\ \mu_{\beta,k} &= \frac{(s_k\sigma_1^2 + \mu_1\sigma^2 - 0.5\sigma^4)(\sigma_0^2 + \sigma^2 t_1) - \sigma_1^2(s_1\sigma_0^2 + \mu_0\sigma^2 t_1)}{(\sigma_0^2 + \sigma^2 t_1)(\sigma_1^2 t_k + \sigma^2) - \sigma_0^2\sigma_1^2 t_1} \\ \sigma_{\beta,k}^2 &= \frac{\sigma_1^2\sigma^2 t_1(\sigma_0^2 + \sigma^2 t_1)}{(\sigma_0^2 + \sigma^2 t_1)(\sigma_1^2 t_k + \sigma^2) - \sigma_0^2\sigma_1^2 t_1} \\ \rho_k &= \frac{-\sigma_0\sigma_1\sqrt{t_1}}{\sqrt{(\sigma_0^2 + \sigma^2 t_1)(\sigma_1^2 t_k + \sigma^2)}}. \end{aligned} \tag{6.43}$$

From Eq. (6.43) it is known that the posterior estimations of θ and β are easily updated once new measurements are available. However, there are unknown parameters needed to be estimated before updating, which are expressed as $\Theta = [\sigma^2, \mu_0, \sigma_0^2, \mu_1, \sigma_1^2]$. The unknown parameters will be estimated using the EM algorithm. Given $\hat{\Theta}_k^{(i)} = \left[\hat{\sigma}_k^{2(i)}, \hat{\mu}_{0,k}^{(i)}, \hat{\sigma}_{0,k}^{2(i)}, \hat{\mu}_{1,k}^{(i)}, \hat{\sigma}_{1,k}^{2(i)}\right]$ as the estimation in the ith step, let $\partial\left[\ell\left(\Theta_k \mid \hat{\Theta}_k^{(i)}\right)\right] / \partial\Theta_k = 0$, then the estimated parameters in the next step $\hat{\Theta}_k^{(i+1)}$ are obtained as

$$\begin{aligned}\hat{\sigma}_k^{2(i+1)} = \frac{1}{k}\Bigg(& \frac{s_1^2 - 2s_1\left(\mu_{\theta,k} + \mu_{\beta,k}t_1\right) + \sigma_{\theta,k}^2 + \sigma_{\beta,k}^2}{t_1} \\ & + 2\left(\rho_k\sigma_{\theta,k}\sigma_{\beta,k} + \mu_{\theta,k}\mu_{\beta,k}\right) + t_1\left(\mu_{\beta,k}^2 + \sigma_{\beta,k}^2\right) \\ & + \sum_{j=2}^{k}\frac{\left(s_j - s_{j-1}\right)^2 - \left(s_j - s_{j-1}\right)\Delta t\mu_{\beta,k} + (\Delta t)^2\left(\mu_{\beta,k}^2 + \sigma_{\beta,k}^2\right)}{\Delta t}\Bigg),\end{aligned} \tag{6.44}$$

$$\hat{\mu}_{0,k}^{(i+1)} = \mu_{\theta,k}, \hat{\sigma}_{0,k}^{2(i+1)} = \sigma_{\theta,k}^2,$$
$$\hat{\mu}_{1,k}^{(i+1)} = \mu_{\beta,k}, \hat{\sigma}_{1,k}^{2(i+1)} = \sigma_{\beta,k}^2.$$

The RUL L_k at t_k is defined as:

$$L_k = \inf\left\{l_k : s\left(l_k + t_k\right) \geq \gamma \mid S_{1:k}\right\} \tag{6.45}$$

where γ is the prespecified failure threshold, and

$$s\left(l_k + t_k\right) = s_k + \beta l_k + \sigma W\left(l_k\right), \quad \text{for} \quad l_k \geq 0 \tag{6.46}$$

with

$$W\left(l_k\right) = B\left(l_k + t_k\right) - B\left(t_k\right) \tag{6.47}$$

which is a standard BM. Given the estimated parameters from Eq. (6.44), the PDF of the RUL is be expressed by

$$f_{L_k|S_{1:k}}\left(l_k \mid S_{1:k}\right) = \frac{\gamma - s_k}{\sqrt{2\pi l_k^3\left(\sigma_{\beta,k}^2 l_k + \hat{\sigma}_k^2\right)}} \exp\left[-\frac{\left(\gamma - s_k - \mu_{\beta,k}l_k\right)^2}{2l_k\left(\sigma_{\beta,k}^2 l_k + \hat{\sigma}_k^2\right)}\right], \; l_k \geq 0. \tag{6.48}$$

6.3.2.3 An Improved Exponential Model-Based RUL Prediction Method

The flowchart of the improved exponential model for predicting RUL of rotating machinery is shown in Fig. 6.15. Vibration signals are acquired from rotating machinery. Then two features are

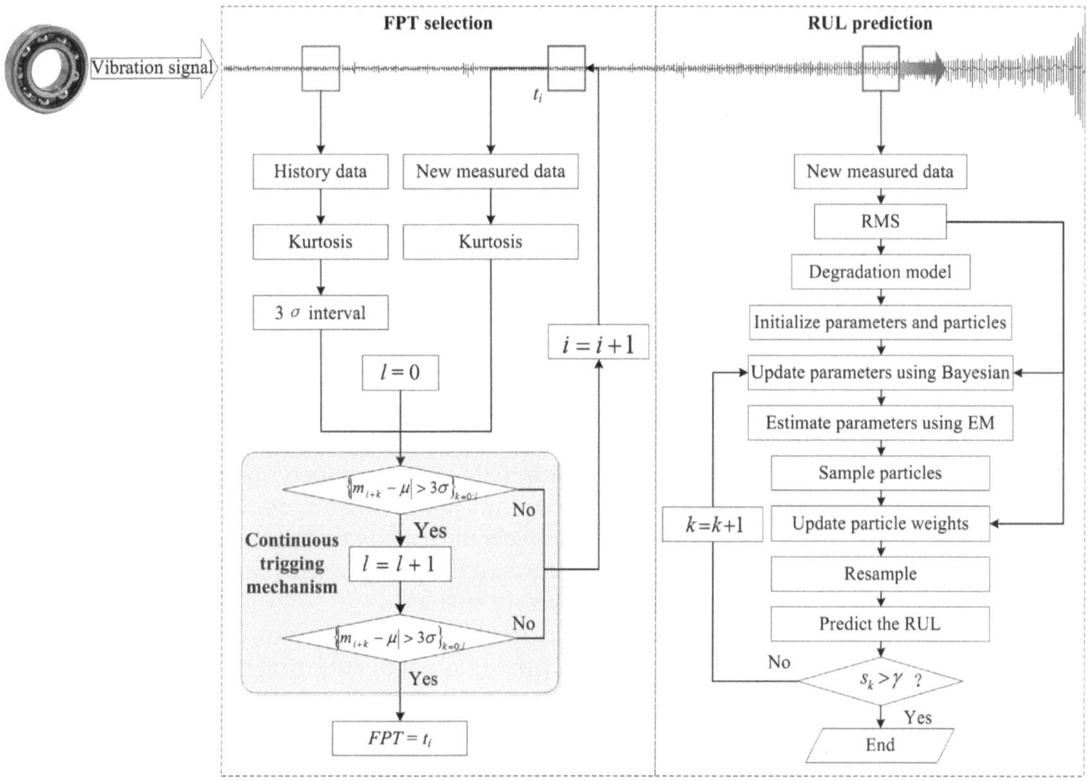

Figure 6.15. Flowchart of the improved exponential model-based method.

extracted from the vibration signals. One is used for health monitoring and the other is used for RUL prediction. It is supposed that the feature used for health monitoring is sensitive to incipient faults, and the feature used for RUL prediction is informative for the development of the faults. According to the requirements of the features for two different stages, Kurtosis and RMS are adopted as the monitoring indicator and predicting indicator, respectively (Lei et al., 2007). In the normal operation stage, Kurtosis is used for health monitoring and FPT selection. Once a FPT is decided, the predicting process is triggered, and RMS is utilized as the predicting indicator for the RUL prediction of rotating machinery.

In order to select an appropriate FPT, an adaptive approach is established based on the 3σ interval. At first, the history data of the bearing in the normal operation stage is utilized to decide the 3σ interval $[\mu-3\sigma,\mu+3\sigma]$ by calculating the mean μ and standard deviation σ of Kurtosis. Then, the 3σ interval is utilized to identify normal and abnormal states of the rotating machinery. When the new

Kurtosis m_i at t_i is available, it is compared with the 3σ interval. If m_i is beyond the 3σ interval, the health state is considered to be abnormal. The abnormal state may be caused by either a fault or random noises. In this approach a new trigger mechanism is applied to restrict the interference of random noises, that is, the predicting process is triggered when $l + 1$ consecutive Kurtosis values exceed the 3σ interval. The process of the trigger mechanism is as follows:

First, let $l = 0$, find the first time when the Kurtosis value exceeds the 3σ interval, and select the time as FPT_0.

Second, let $l = l + 1$, find the time t_i when the consecutive $l + 1$ Kurtosis values $\left\{m_{i+k}\right\}_{k=0:l}$ satisfy $\left\{\left|m_{i+k} - \mu\right| > 3\sigma\right\}_{k=0:l}$, and select the time t_i as FPT_l.

Third, let l increase from 1 until l satisfying $\text{FPT}_l = \text{FPT}_{l-1}$, and output FPT_l as the final FPT value.

The theory behind this trigger mechanism is that abnormal states caused by random noises are almost impossible to appear $l + 1$ times consecutively during the normal operation stage. When the selected FPT starts to keep stable with the increase of l, it may imply that a fault has occurred.

After the FPT is decided, RMS is input into the degradation model for the RUL prediction of the rotating machinery. In order to reduce random errors of the stochastic process, PF is employed in this study to estimate the health state. The RUL prediction process of the improved exponential model-based method is as follows.

First, the model parameters are initialized as $\sigma^2 = \hat{\sigma}_0^2$, $\mu_0 = \mu_{\theta,0}$, $\sigma_0^2 = \sigma_{\theta,0}^2$, $\mu_1 = \mu_{\beta,0}$ and $\sigma_1^2 = \sigma_{\beta,0}^2$ (Gebraeel et al., 2005). Then initial particles $\left\{s_0^n\right\}_{n=1:N_s}$ are sampled from $p\left(\tilde{s}_0 \mid s_0, \hat{\Theta}_0\right) \sim N\left(s_0, \hat{\sigma}_0^2 \Delta t\right)$. Once the measurement s_k at t_k is available, model parameters are updated using Eq. (6.43), and further estimated using Eq. (6.44). With the estimated parameters and the following variant of the degradation model

$$s_k = s_{k\text{-}1} + \beta \Delta t + \sigma W(\Delta t), \tag{6.49}$$

the PDF of the state is described as:

$$p\left(\tilde{s}_k \mid \hat{s}_{k-1}, \hat{\Theta}_k\right) = \frac{1}{\sqrt{2\pi\Delta t\left(\sigma_{\beta,k}^2 \Delta t + \hat{\sigma}_k^2\right)}} \exp\left[-\frac{\left(\tilde{s}_k - \hat{s}_{k-1} - \mu_{\beta,k}\Delta t\right)^2}{2\Delta t\left(\sigma_{\beta,k}^2 \Delta t + \hat{\sigma}_k^2\right)}\right] \tag{6.50}$$

where $\tilde{s}_k$ is the actual state at t_k, $\hat{\Theta}_k = \left[\hat{\sigma}_k^2, \mu_{\theta,k}, \sigma_{\theta,k}^2, \mu_{\beta,k}, \sigma_{\beta,k}^2\right]$ are the estimated parameters, $\hat{s}_{k-1}$ is the estimated state at t_{k-1} which is different from the measurement s_{k-1}. It should be noticed that if $\hat{s}_{k-1}$ is replaced by s_{k-1}, Eq. (6.50) will change to $p\left(\tilde{s}_k \mid s_{k-1}, \hat{\Theta}_k\right)$, which is the PDF of the state acquired using the original exponential model.

Now we need to focus on how to estimate the actual state $\tilde{s}_k$ at t_k based on the measurement s_k. Here we take $p\left(\tilde{s}_k \mid \hat{s}_{k-1}, \hat{\Theta}_k\right)$ as the importance PDF which contains the information of the state estimation at t_{k-1}. And then, particle sets $\left\{s_k^n\right\}_{n=1:N_s}$ are sampled from Eq. (6.50). These particles are only able to approximate the prior PDF of the state. In order to approximate the posterior PDF, particle weights should be updated according to the measurement s_k as follows:

$$w_k^n = w_{k-1}^n \frac{1}{\sqrt{2\pi\hat{\sigma}_k^2 t_k}} \exp\left[-\frac{\left(s_k - s_k^n\right)^2}{2\hat{\sigma}_k^2 t_k}\right], \quad \tilde{w}_k^n = \frac{w_k^n}{\sum_{n=1}^{N_s} w_k^n}. \tag{6.51}$$

With the particles and their corresponding weights, the posterior PDF of the system state is approximated using the following discrete density:

$$p\left(\tilde{s}_k \mid s_k, \hat{\Theta}_k\right) \approx \sum_{n=1}^{N_s} \tilde{w}_k^n \delta\left(\tilde{s}_k - s_k^n\right). \tag{6.52}$$

Through consecutive iteration and updating, the particle whose motion curve is closer to the actual degradation process acquires a higher weight. The health state of the rotating machinery is estimated by

$$\hat{s}_k = \sum_{n=1}^{N_s} \left(w_k^n s_k^n\right). \tag{6.53}$$

With the weighted average of updated particles, the estimated state $\hat{s}_k$ is closer to the actual state than the measurement s_k.

After the particle weights have been updated, the effective sample size $\hat{N}_{eff}$ is calculated using Eq. (6.33) in Section 6.2.3 to measure the degeneracy of particles. When $\hat{N}_{eff}$ falls below a threshold N_T, a new set of particles $\left\{s_k^{n*}\right\}_{n=1:N_s}$ are resampled, and the particle weights are reset to $w_k^n = 1/N_s$. Finally, the PDF of the RUL is calculated using

$$f_{L_k \mid S_{1:k}}\left(l_k \mid S_{1:k}\right) = \frac{\gamma - \hat{s}_k}{\sqrt{2\pi l_k^3 \left(\sigma_{\beta,k}^2 l_k + \hat{\sigma}_k^2\right)}} \exp\left[-\frac{\left(\gamma - \hat{s}_k - \mu_{\beta,k} l_k\right)^2}{2 l_k \left(\sigma_{\beta,k}^2 l_k + \hat{\sigma}_k^2\right)}\right], \quad l_k \geq 0. \tag{6.54}$$

Table 6.2 RUL Prediction Process Using the Improved Exponential Model-Based Method

Step 1: At initial time t_0, select initial parameters $\sigma^2 = \hat{\sigma}_0^2$, $\mu_0 = \mu_{\theta,0}$, $\sigma_0^2 = \sigma_{\theta,0}^2$, $\mu_1 = \mu_{\beta,0}$, and $\sigma_1^2 = \sigma_{\beta,0}^2$. Then sample random particles $\{s_0^n\}_{n=1:N_s}$ from $p(\tilde{s}_0 \mid s_0, \hat{\Theta}_0) \sim N(s_0, \hat{\sigma}_0^2 \Delta t)$ with particle weights $w_0^n = 1/N_s$.

Step 2: Once obtaining a measurement s_k at t_k for $k \geq 1$, let $\sigma^2 = \hat{\sigma}_{k-1}^2$, $\mu_0 = \mu_{\theta,k-1}$, $\sigma_0^2 = \sigma_{\theta,k-1}^2$, $\mu_1 = \mu_{\beta,k-1}$ and $\sigma_1^2 = \sigma_{\beta,k-1}^2$. Then $\mu_{\theta,k}$, $\sigma_{\theta,k}^2$, $\mu_{\beta,k}$, and $\sigma_{\beta,k}^2$ are calculated by Eq. (6.43).

Step 3: Based on $\mu_{\theta,k}$, $\sigma_{\theta,k}^2$, $\mu_{\beta,k}$, and $\sigma_{\beta,k}^2$, the parameters $\sigma^2 = \hat{\sigma}_k^2$, $\mu_0 = \mu_{\theta,k}$, $\sigma_0^2 = \sigma_{\theta,k}^2$, $\mu_1 = \mu_{\beta,k}$, and $\sigma_1^2 = \sigma_{\beta,k}^2$ are obtained by Eq. (6.44).

Step 4: Sample particles $\{s_k^n\}_{n=1:N_s}$ from the importance PDF $p(\tilde{s}_k \mid \hat{s}_{k-1}, \hat{\Theta}_k)$ using Eq. (6.50).

Step 5: Update the particle weight w_k^n according to Eq. (6.51).

Step 6: Calculate $\hat{N}_{eff}$ using Eq. (6.32). If $\hat{N}_{eff} < N_T$, resample new particles $\{s_k^{n^*}\}_{n=1:N_s}$ and reset the weights $w_k^n = 1/N_s$.

Step 7: Based on $\mu_{\theta,k}$, $\sigma_{\theta,k}^2$, $\mu_{\beta,k}$, and $\sigma_{\beta,k}^2$ from Step 2, $\hat{\sigma}_k^2$ from Step 3 and particles $\{s_k^n\}_{n=1:N_s}$ from Step 6, estimate the RUL by Eq. (6.54).

Step 8: Let $k = k + 1$, go back to Step 2 and repeat the above steps until $s_k > \gamma$.

The RUL prediction process using the improved exponential model is summarized in Table 6.2.

6.3.2.4 RUL Prediction of Rolling Element Bearings: a Case Study

In this subsection, the whole lifetime vibration signals are captured from accelerated degradation tests of rolling element bearings to verify the effectiveness of the improved model. The accelerated degradation tests are conducted on a test rig named PRONOSTIA (Nectoux et al., 2012). This test rig is specially designed to verify the methods for bearing diagnosis and RUL prediction. In order to conduct accelerated degradation tests of bearings in short time, a 4 kN radial force, which is equal to the bearing's maximum dynamic load, is applied on the tested bearings. The force is generated by a cylinder pressure, and the pressure is delivered through a pressure regulator. During the tests, the rotating speed of the bearing keeps 1,800 rpm. Accelerometers are fixed

Figure 6.16. Photographs of a tested bearing before and after test.

on the outer race of the bearing to capture vibration signals. The sampling frequency is 25.6 kHz. Each sample contains 2560 data points, that is, 0.1 s, and the sampling is repeated every 10 s. The vibration signals are transmitted into a PC for data visualization and storage through an NI DAQ card. The bearing useful-life ends at the time when the amplitude of the vibration signal exceeds 20 g.

Each bearing is naturally degraded during the tests without seeding a fault in advance. Due to the diversity of the damage mechanism for different bearings and the randomness of degradation processes, the fault modes may be slightly different for distinct bearings. As a result, the vibration signal pattern of each bearing is different from each other. Pictures of a tested bearing before and after a test are shown in Fig. 6.16. It is seen that different kinds of faults occur on the balls and inner race of the bearing. The vibration signals of four tested bearings during the whole lifetime are shown in Fig. 6.17. The signal amplitudes of bearings 1 and 2 perform gradually increasing trends, which indicates that the faults get severe gradually. While the signal amplitudes of bearings 3 and 4 show rapid increases at the end of lifetime, thus representing abrupt degradation processes. Taking bearing 4, for example, more detailed information about the vibration signals in the normal operation stage and the accelerated degradation stage is presented in Fig. 6.17. It is shown that there exist obvious impacts in the vibration signals of the accelerated degradation stage than those of the normal operation stage.

The Kurtosis and RMS are first extracted from the vibration signals. Based on the Kurtosis, the FPT selection approach proposed in Section 6.3.2.3 is used to decide the FPT of the four bearings.

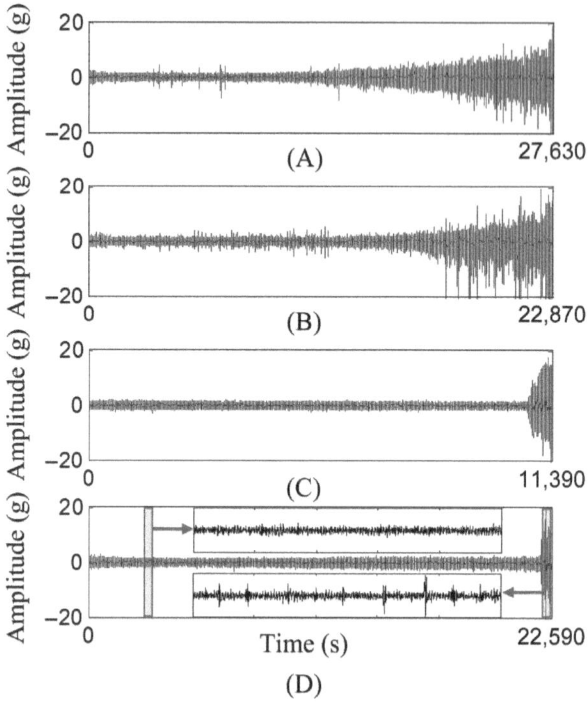

Figure 6.17. Vibration signals of four tested bearings. (A) Bearing 1, (B) bearing 2, (C) bearing 3, and (D) bearing 4.

For comparison, another approach proposed by Ginart et al. (2006) is employed to select the FPT, as well. The FPT selection results are given in Fig. 6.18, and all FPT values are displayed in Table 6.3, where "N/A" means that none FPT is selected during the whole lifetime. Fig. 6.18 shows that the alarm decided by Ginart's approach is much higher than the 3σ interval and fails to distinguish the normal and the abnormal states. On the contrary, the 3σ interval is able to detect the abnormal states. However, there are some abnormal states caused by random noises instead of faults in the normal operation stage, which are marked by "*" in Fig. 6.18. Due to the antiinterference ability of our approach, the time when faults occur is appropriately selected as the FPT.

Fig. 6.19 presents the RMS values of the four tested bearings and the corresponding FPT decided by the two approaches. It is seen that RMS has a more obvious degradation trend than Kurtosis in Fig. 6.18 with the development of faults. Since the faults of bearings 1 and 2 get severe gradually, it is a little difficult to exactly give the time of the fault occurrence. Consequently, the FPT values selected by our approach exhibit a little time delay as

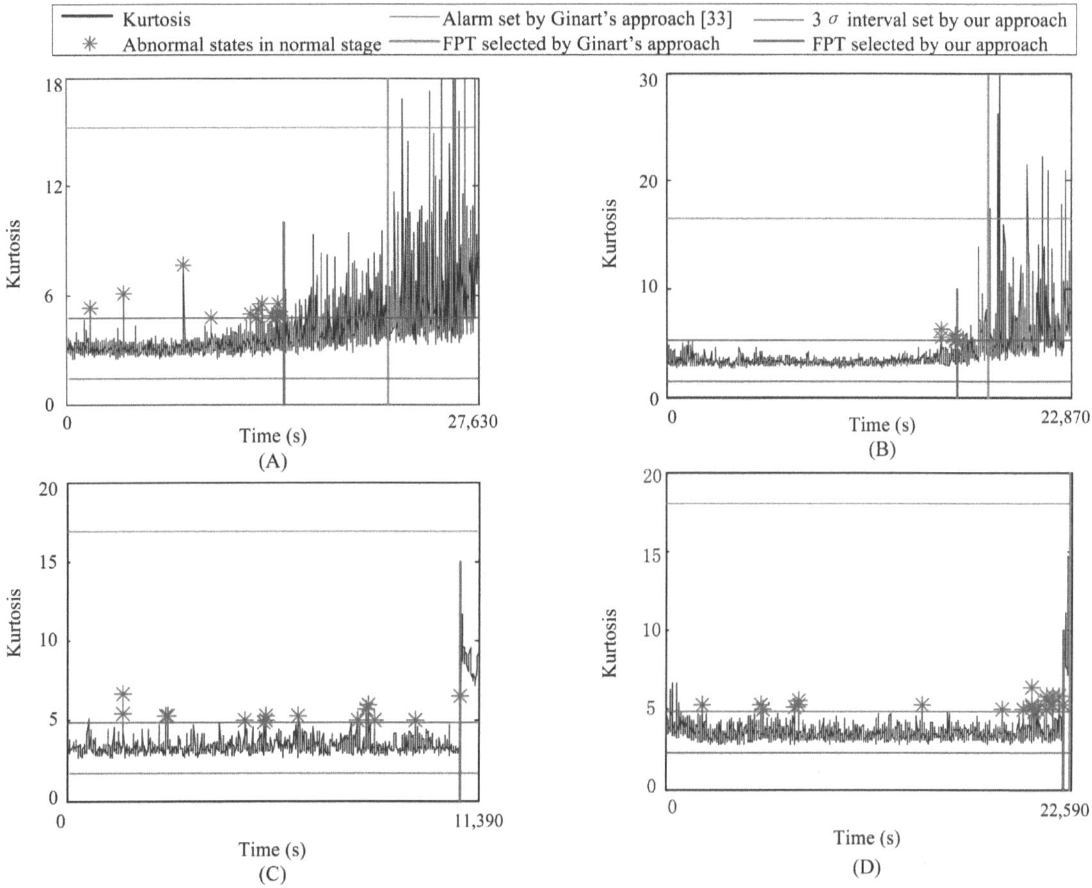

Figure 6.18. FPT selection results. (A) Bearing 1, (B) bearing 2, (C) bearing 3, and (D) bearing 4.

shown in Fig. 6.19. The delayed time, however, is much smaller than that of Ginart's approach, especially in the case of bearing 1. In addition, the small delay is acceptable for the RUL prediction of bearings. Since bearings 3 and 4 degrade abruptly, it is not as difficult as bearings 1 and 2 to select the FPT. Our approach

Table 6.3 Selected FPT of four Tested Bearings

Case	Bearing 1	Bearing 2	Bearing 3	Bearing 4
Ginart's approach	21,670 s	18,270 s	N/A	22,580 s
Our approach	14,630 s	16,440 s	10,910 s	22,130 s

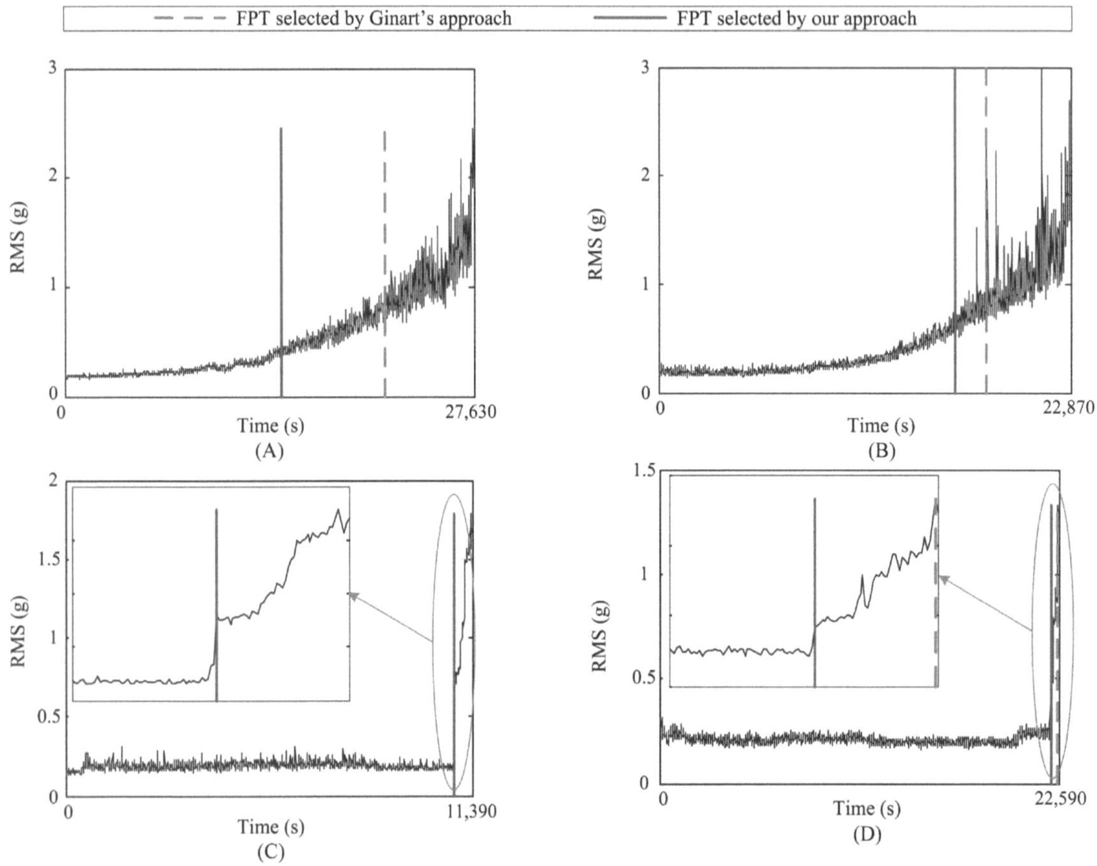

Figure 6.19. RMS division results using the selected FPT. (A) Bearing 1, (B) bearing 2, (C) bearing 3, and (D) bearing 4.

still performs better than Ginart's approach in selecting the FPT of bearings 3 and 4.

To further verify the suitability of the 3σ interval for the FPT selection, 1σ–4σ intervals are utilized in our approach to select the FPT of four tested bearings. Fig. 6.20 shows the selection results. We can see from the results that the 2σ and 4σ intervals select much too late FPT for bearing 1. For bearing 2, the 1σ and 2σ intervals offer much too early FPT. All of them have the same FPT for bearing 3. And for bearing 4, the 1σ interval presents much too early FPT. In conclusion, the 3σ interval has the best performance among the 1σ–4σ intervals.

After the FPT selection, the RUL of four tested bearings is predicted using the improved exponential model-based method with a particle number of 1000. The RMS estimation results are shown in

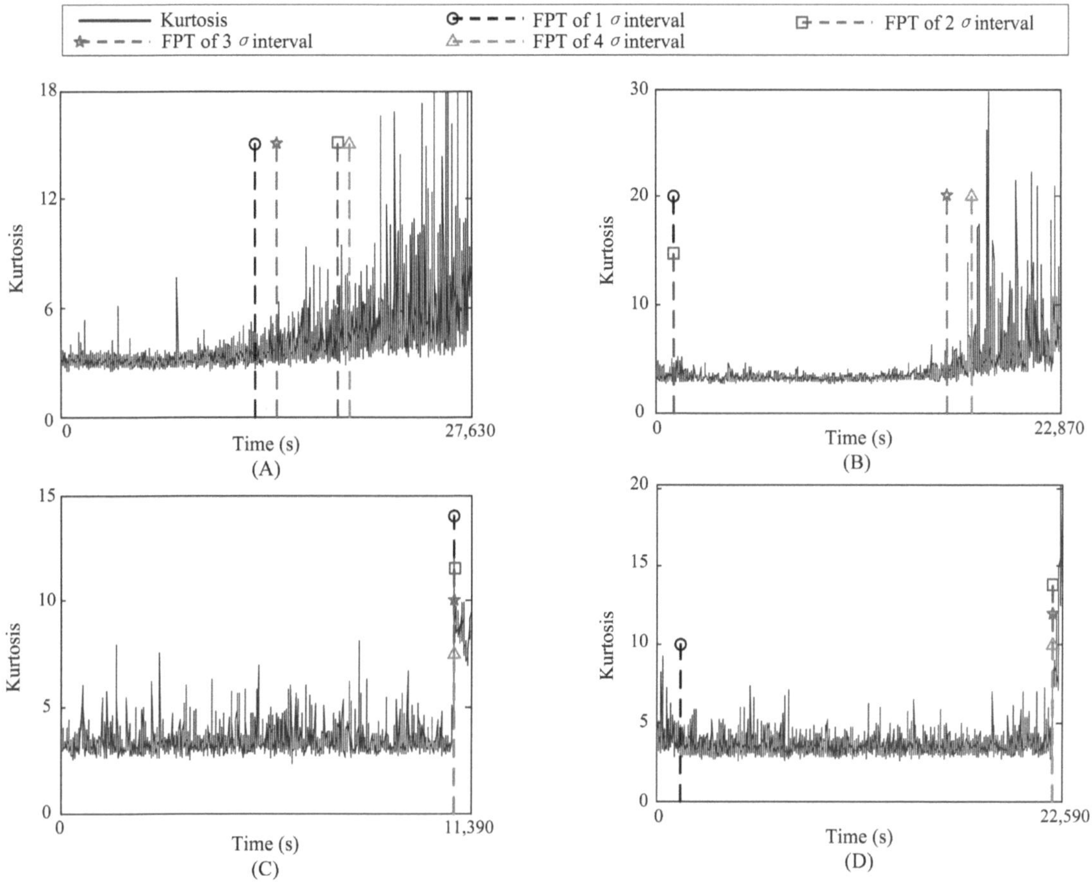

Figure 6.20. FPT selection results with different σ intervals. (A) Bearing 1, (B) bearing 2, (C) bearing 3, and (D) bearing 4.

Fig. 6.21, and the RUL prediction results are presented in Fig. 6.22. For comparison, the original exponential model (Si et al., 2013a) and the Paris–Erdogan model (An et al., 2013) are employed to predict the RUL of the four tested bearings as well. The same FPT and particle number are used for them and the estimation and prediction results are also presented in Figs. 6.21 and 6.22, respectively. Fig. 6.21 indicates that the estimation results of the original exponential model almost coincide with the actual RMS, while two PF-based methods, that is, the Pairs–Erdogan model and the improved exponential model, may reduce random errors of the stochastic processes and therefore reflect the global degradation processes of the bearings. In Fig. 6.22, all of them generate inaccurate results at the beginning because of the lack of RMS values. When enough

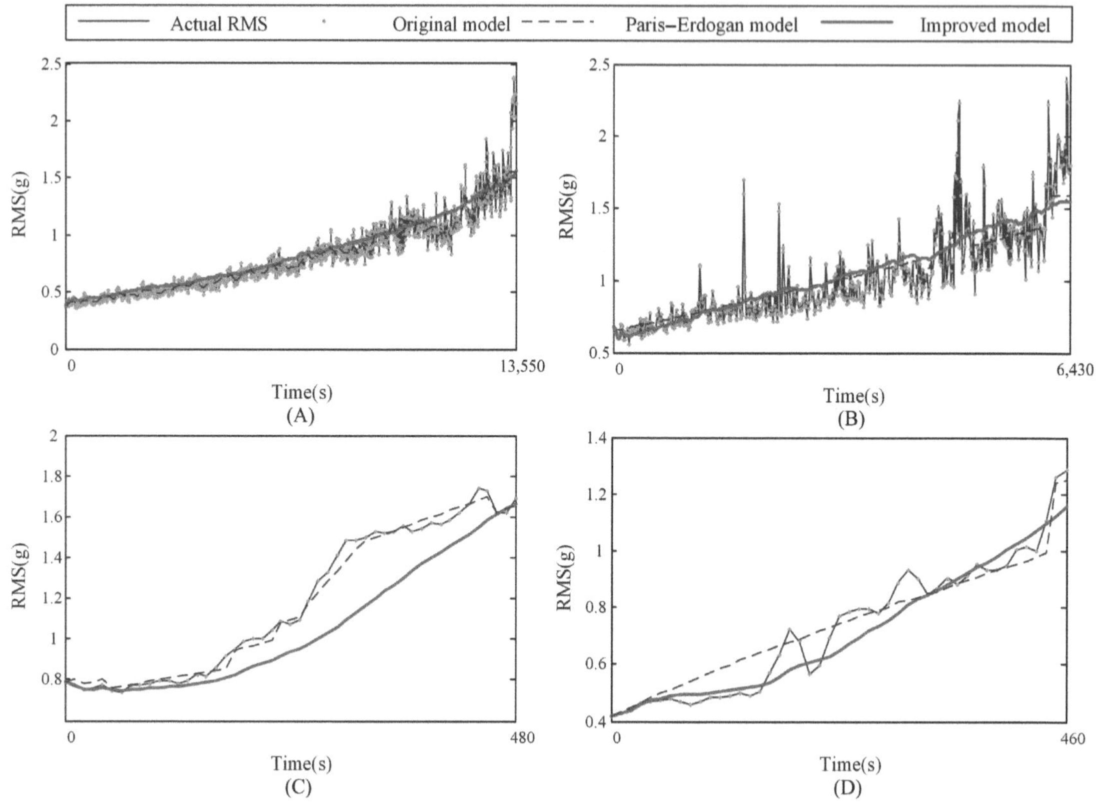

Figure 6.21. RMS estimation results of the tested bearings. (A) Bearing 1, (B) bearing 2, (C) bearing 3, and (D) bearing 4.

RMS values are available and used, the original exponential model and the Paris–Erdogan model still produce large prediction errors, especially in the case of bearing 2. The improved exponential model-based method, however, converges to the actual RUL fastest and presents the most accurate prediction results among them. This should be explained as follows. The original exponential model is unable to accurately estimate the health state of the bearings, thus the prediction results are influenced by random errors of the stochastic process. The Paris–Erdogan model performs better than the original exponential model in the state estimation with the help of PF, but it may not estimate the model parameters effectively. Therefore, the results of the Paris–Erdogan model still have large random errors. The improved exponential model-based method incorporates the superiority of PF in the state estimation and the advantage of the Bayesian updating and EM algorithm in the parameter estimation. Consequently, it presents the best prediction results among the three models.

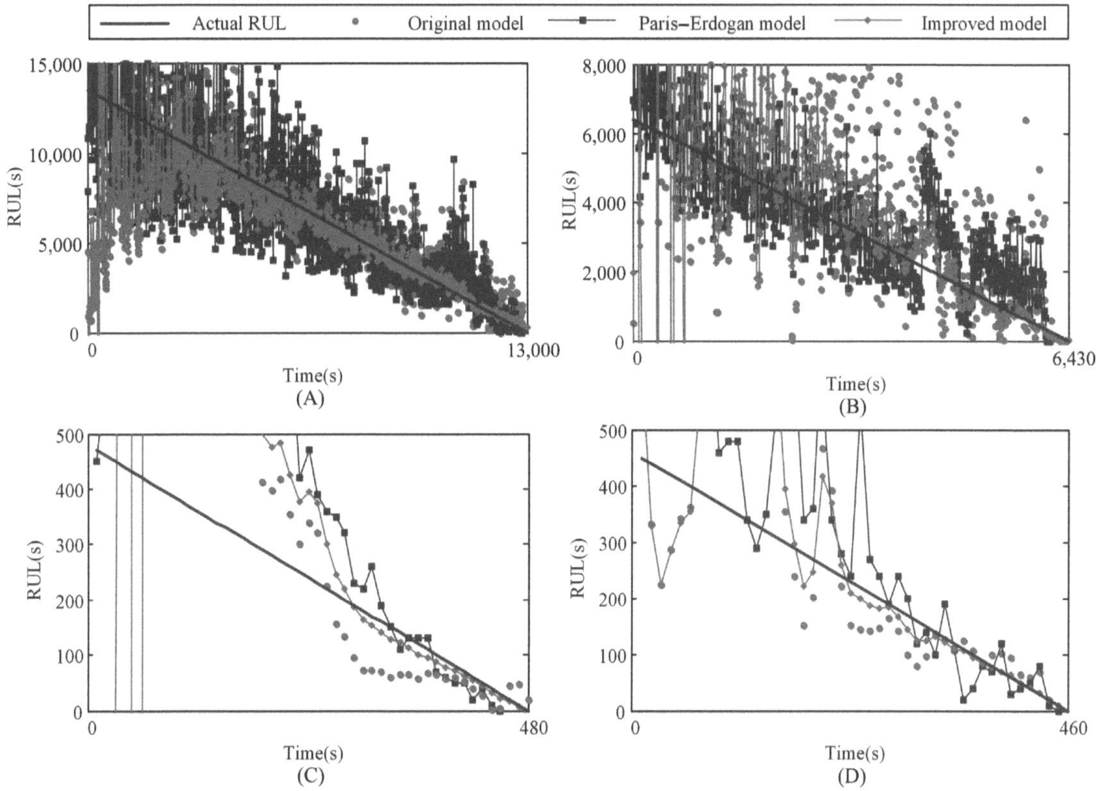

Figure 6.22. RUL prediction results of the tested bearings. (A) Bearing 1, (B) bearing 2, (C) bearing 3, and (D) bearing 4.

To compare the performance of the three models quantitatively, two commonly used performance metrics, cumulative relative accuracy (CRA) and Convergence proposed in Saxena et al. (2010), are calculated. CRA quantifies how accurately a prediction method performs at specific time indexes, and the perfect score is 1. Convergence quantifies how fast a method approaches the actual RUL. A lower score means a faster convergence. The CRA and Convergence are calculated at the time indexes from the half to the end of the lifetime. The scores of the two metrics are summarized in Table 6.4. It is noticed that the exponential model has the highest CRA scores and the lowest Convergence scores among the three models for all of the tested bearings. Therefore, the improved exponential model-based method has the most accurate prediction results and the fastest convergence speeds among them. In conclusion, the improved exponential model-based method performs best in the RUL prediction of the bearings.

Table 6.4 Scores of Two Performance Metrics for Texted Bearings

Metric	Case	Paris-Erdogan Model	Original Model	Improved Model
CRA	Bearing 1	0.6967	0.7111	0.8696
	Bearing 2	0.6074	0.5311	0.7623
	Bearing 3	0.6317	0.5420	0.8712
	Bearing 4	0.7443	0.7463	0.9324
Convergence	Bearing 1	9,234	9,382	8,993
	Bearing 2	4,406	4,320	4,284
	Bearing 3	307.3	329.9	275.4
	Bearing 4	315.3	296.7	290.2

6.3.3 Polynomial Model-Based Methods

6.3.3.1 Motivation

The uncertainty of stochastic degradation processes of rotating machinery is generally caused by the following four variability sources: (1) the temporal variability, (2) the unit-to-unit variability, (3) the nonlinear variability, and (4) the measurement variability. The temporal variability is the inherent uncertainty of the degradation processes over time, which is usually represented by random noises in the stochastic process models (Pandey et al., 2009). The unit-to-unit variability implies the diversity of the degradation processes among different machines or components due to their diversity in health states and operation conditions. The unit-to-unit variability is generally denoted by introducing random variances in specified model parameters in a stochastic process model (Peng and Tseng, 2009). The nonlinear variability denotes the variation of the degradation rate with the development of the fault severity (Si, 2015). To describe the nonlinear variability, the stochastic process model is supposed to incorporate some nonlinear structures. The measurement variability means the difference between the measurements and the actual health states of machines, which is usually described by introducing measurement noises in the stochastic process models (Ye et al., 2013). If the above four variability sources are not considered comprehensively, it would be difficult to handle the uncertainty of the machine degradation processes, resulting in unreliable RUL prediction results. Therefore, these four variability sources need to be considered simultaneously in modeling the machine degradation processes.

It is recognized from the previous analysis that an appropriate stochastic process model is supposed to incorporate the aforementioned four variability sources simultaneously. However, most studies reported in literature only consider some of the four variability sources. For example, the exponential model described in Section 6.3.2 just considers the temporal variability and the unit-to-unit variability, which could be transformed into a linear model using logarithm transformation. Peng and Tseng (2009) proposed a linear degradation model by incorporating the temporal variability, the unit-to-unit variability, and the measurement variability and used it to predict the RUL of a laser device. Ye et al. (2013) also constructed a linear degradation model considering the temporal variability, the unit-to-unit variability, and the measurement variability simultaneously. These earlier studies have successfully established stochastic process models considering some of the four variability sources. However, they pay little attention to the nonlinear variability in modeling the machine degradation processes. A research work relating to the nonlinear variability can be found in Si (2015). Unfortunately, this research work does not take the measurement variability into account. Through the earlier analysis, it is noticed that it is necessary to construct a stochastic process model considering the four variability sources simultaneously.

Constructing an appropriate stochastic process model is not enough for the issue of the RUL prediction of rotating machinery. Another vital task is to estimate model parameters and health states adaptively according to real-time information of measurements. The PF algorithm is superior to incorporate the real-time information of measurements into the empirical models (Arulampalam et al., 2002). Therefore, it is suitable to deal with this task. Until now, the PF algorithm has been widely used in the field of RUL prediction. Orchard et al. (2013) proposed a PF-based prognostic method and applied it into the RUL prediction of lithium-ion batteries. Qian and Yan (2015) improved the PF algorithm using the BP neutral network and proposed a RUL prediction method of rolling element bearings based on the enhanced PF algorithm. Chen et al. (2012) constructed a degradation model based on neuro-fuzzy systems and used a high-order PF algorithm to estimate the system states. Liu et al. (2012) combined the PF algorithm with data-driven algorithms and established a data-model-fusion prediction method. However, there exists a common problem in the PF algorithm, that is, the particle degeneracy (Arulampalam et al., 2002). The phenomenon of this problem is that after a few iterations most particle weights become negligible. As a result, the particles degrade into a poor distribution and plenty of computational resources are wasted to update particles which are ineffective to the state estimation. This problem is caused by an inappropriate importance PDF, that is, the source of the particles. If the importance PDF dramatically deviates far from the PDF of the system state, most particles sampled from it would locate far from the actual state, resulting in the particle degeneracy problem. An effective way for handling this problem is to sample a set of high quality particles from a superior importance PDF which is supposed to approximate the PDF of the actual state. How to construct such an importance PDF should be taken into account in the application of the PF algorithm.

Based on the previous discussion, a machine RUL prediction method based on a polynomial model is established. This polynomial model incorporates the aforementioned four variability sources simultaneously, that is, the temporal variability, the unit-to-unit variability, the nonlinear variability, and the measurement variability. The model parameters and the health states are estimated using the Kalman particle filtering (KPF) algorithm which improves the PF by reducing its particle degeneracy problem.

6.3.3.2 A Polynomial Model-Based RUL Prediction Method

1. *Polynomial model development*

   The state of a system at t_k in the polynomial model is described as follows (Si, 2015):

$$x_k = x_0 + \theta_k t_k^b + \sigma B(t_k) \tag{6.55}$$

   where x_0 is the initial state of the system and generally equals zero. $\sigma B(t_k)$ is a BM following a normal distribution $N(0,\sigma^2 t_k)$, which represents the temporal variability of the stochastic process. θ_k is a random variable following a normal distribution $N(\mu_\theta,\sigma_\theta^2)$, which describes the unit-to-unit variability. The nonlinear structure $\theta_k t_k^b$ denotes the severity of the degradation. In a linear model, b equals one and the structure is reduced to $\theta_k t_k$ which only describes the linear degradation process. However, in real cases, most mechanical systems generally present nonlinear degradation processes. In this model, b is adjusted to the degradation processes adaptively. Therefore, $\theta_k t_k^b$ is supposed to describe the nonlinear variability.

   By calculating the derivative of x_k with respect to the time, the degradation speed of the system at t_k is acquired.

$$dx_k = \theta_k b t_k^{b-1} dt + \sigma dB(t_k). \tag{6.56}$$

The first term $\theta_k b t_k^{b-1} dt$ denotes the deterministic part of the degradation speed related to a certain fatigue stress and mechanical material. The second term $\sigma dB(t_k)$ represents the stochastic part of the degradation speed, which describes the random errors caused by the fluctuation of the stress or the inhomogeneity of the material. According to the characteristics of the BM, $\sigma dB(t_k)$ is time-independent and follows a normal distribution of $N(0,\sigma^2)$. That is to say, the random errors at different time points are i.i.d. This is reasonable because the random errors are generated stochastically and they just influence the degradation speed at the current time point.

The health states of the mechanical systems are difficult or even impossible to be measured directly. Thus, they should be estimated through some indirect measurements, such as the health indicators extracted from vibration signals. The relationship between the measurements and the states is formulated as

$$z_k = x_k + v \tag{6.57}$$

where z_k is the measurement at t_k, and v is the random measurement noise representing the measurement variability of the stochastic process model, which is assumed to be i.i.d. following $v \sim N(0,\gamma^2)$.

2. *Parameter initialization with MLE*

There are five unknown parameters, that is, σ^2, μ_θ, σ_θ^2, b, and γ^2, which need to be initialized in the polynomial model. The maximum likelihood estimation (MLE) algorithm is utilized to initialize the unknown parameters according to available measurements. The unknown parameters are denoted by $\Theta = \left[\sigma^2, \mu_\theta, \sigma_\theta^2, b, \gamma^2\right]^T$, where $[\cdot]^T$ represents the vector transposition. It is assumed that $B(t_k)$, θ_k, b and v are mutually statistically independent. The available measurements at ordered times $t_1, \ldots, t_M$ are denoted as $\boldsymbol{z} = \left[z_1, \ldots, z_M\right]^T$, where M is the number of the measurements. According to Eqs. (6.55) and (6.57), z_k is expressed as follows:

$$z_k = x_0 + \theta_k t_k^b + \sigma B(t_k) + v. \tag{6.58}$$

Let $\boldsymbol{t} = \left[t_1^b, \ldots, t_M^b\right]^T$. According to the statistically independent increments property of BM, it is acquired that $\boldsymbol{z} = \left[z_1, \ldots, z_M\right]^T$ is multivariate normally distributed which is denoted as:

$$\boldsymbol{z} \sim N\left(\mu_\theta \boldsymbol{t}, \sigma_\theta^2 \boldsymbol{t}\boldsymbol{t}' + \Omega\right) \tag{6.59}$$

with

$$\Omega=\sigma^2 Q+\gamma^2 I_M \quad (6.60)$$

where $Q=\left[\min\{t_i,t_j\}\right]_{1\le i,j\le M}$ and I_M is an identity matrix of order M.

The log-likelihood function of the initial parameter vector $\Theta_0=\left[\sigma^2,\mu_{\theta,0},\sigma^2_{\theta,0},b,\gamma^2\right]^T$ based on the measurements is expressed as the following equation:

$$\begin{aligned}\ell(\Theta_0\,|\,z)=&-\frac{M}{2}\ln(2\pi)-\frac{1}{2}\ln\left|\sigma^2_{\theta,0}tt'+\Omega\right| \\ &-\frac{1}{2}\left(z-\mu_{\theta,0}t\right)'\left(\sigma^2_{\theta,0}tt'+\Omega\right)^{-1}\left(z-\mu_{\theta,0}t\right)\end{aligned}. \quad (6.61)$$

To simplify the calculation process, the unknown parameters σ^2 and γ^2 are transformed into $\tilde{\sigma}^2=\sigma^2/\sigma^2_{\theta,0}$ and $\tilde{\gamma}^2=\gamma^2/\sigma^2_{\theta,0}$. And Ω is correspondingly denoted as $\tilde{\Omega}=\Omega/\sigma^2_{\theta,0}$. Then Eq. (6.61) is transformed into

$$\begin{aligned}\ell(\Theta_0\,|\,z)=&-\frac{M}{2}\ln(2\pi)-\frac{M}{2}\ln\sigma^2_{\theta,0}-\frac{1}{2}\ln\left|tt'+\tilde{\Omega}\right| \\ &-\frac{1}{2\sigma^2_{\theta,0}}\left(z-\mu_{\theta,0}t\right)'\left(tt'+\tilde{\Omega}\right)^{-1}\left(z-\mu_{\theta,0}t\right)\end{aligned}. \quad (6.62)$$

Then the first partial derivatives of $\ell(\Theta_0\,|\,z)$ with respect to $\mu_{\theta,0}$ and $\sigma^2_{\theta,0}$ are expressed as

$$\begin{cases}\dfrac{\partial\ell(\Theta\,|\,z)}{\partial\mu_{\theta,0}}=\dfrac{t'\left(tt'+\tilde{\Omega}\right)^{-1}\left(z-\mu_{\theta,0}t\right)}{\sigma^2_{\theta,0}} \\ \dfrac{\partial\ell(\Theta\,|\,z)}{\partial\sigma^2_{\theta,0}}=-\dfrac{M}{2\sigma^2_{\theta,0}}+\dfrac{1}{2\sigma^4_{\theta,0}}\left(z-\mu_{\theta,0}t\right)'\left(tt'+\tilde{\Omega}\right)^{-1}\left(z-\mu_{\theta,0}t\right)\end{cases}. \quad (6.63)$$

Let $\partial\ell(\Theta_0\,|\,z)/\partial\mu_{\theta,0}=0$ and $\partial\ell(\Theta_0\,|\,z)/\partial\sigma^2_{\theta,0}=0$. The results of the MLE for $\mu_{\theta,0}$ and $\sigma^2_{\theta,0}$ are represented as follows:

$$\begin{cases}\mu_{\theta,0}=\dfrac{t'\left(tt'+\tilde{\Omega}\right)^{-1}z}{t'\left(tt'+\tilde{\Omega}\right)^{-1}t} \\ \sigma^2_{\theta,0}=\dfrac{\left(z-\mu_{\theta,0}t\right)'\left(tt'+\tilde{\Omega}\right)^{-1}\left(z-\mu_{\theta,0}t\right)}{M}\end{cases} \quad (6.64)$$

which are functions of the residual three parameters $\tilde{\sigma}^2$, b and $\tilde{\gamma}^2$. The log-likelihood function is reduced into a three-variable function by substituting Eq. (6.64) into Eq. (6.62).

$$\ell(\Theta_0 \mid z) = -\frac{M}{2}\ln(2\pi) - \frac{M}{2} - \frac{M}{2}\ln\sigma^2_{\theta,0} - \frac{1}{2}\ln\left|tt' + \tilde{\Omega}\right|. \quad (6.65)$$

The MLE values of $\tilde{\sigma}^2$, b, and $\tilde{\gamma}^2$ are obtained by maximizing the log-likelihood function in Eq. (6.65) through a three-dimensional optimizing. By substituting the MLE of $\tilde{\sigma}^2$, b and $\tilde{\gamma}^2$ into Eq. (6.64), we can acquire the values of $\mu_{\theta,0}$ and $\sigma^2_{\theta,0}$. Then σ^2 and γ^2 are calculated with $\tilde{\sigma}^2$ and $\tilde{\gamma}^2$ multiplied by $\sigma^2_{\theta,0}$.

3. *State estimation with KPF*

After the model parameter initialization using the MLE, the KPF algorithm is utilized to estimate the system states and the model parameters adaptively according to new measurements. In this process, the time-varying parameter θ_k is treated as a hidden state of the system, and the polynomial model is transformed into the following state space model (Si et al., 2014):

$$\begin{cases} x_k = x_{k-1} + \theta_{k-1}\left(t_k^b - t_{k-1}^b\right) + \omega_{k-1} \\ \theta_k = \theta_{k-1} \\ z_k = x_k + v \end{cases} \quad (6.66)$$

where $\omega_{k-1} = \sigma\left(B(t_k) - B(t_{k-1})\right)$ following a normal distribution $N(0, \sigma^2 \Delta t_k)$ with $\Delta t_k = t_k - t_{k-1}$. The state space of the system is extended from x_k to $y_k = \left[x_k, \theta_k\right]^T$ with θ_k being deemed as a hidden state and the state space model is further organized as

$$\begin{cases} y_k = A_k y_{k-1} + \eta_{k-1} \\ z_k = C y_k + v \end{cases} \quad (6.67)$$

where

$$\eta_{k-1} = \begin{bmatrix} \omega_{k-1} \\ 0 \end{bmatrix}, \quad A_k = \begin{bmatrix} 1 & t_k^b - t_{k-1}^b \\ 0 & 1 \end{bmatrix} \quad \text{and} \quad C = \begin{bmatrix} 1 \\ 0 \end{bmatrix}^T. \quad (6.68)$$

Based on the parameters initialized using the MLE algorithm, a series of initial particles $\left\{y_0^n\right\}_{n=1:Np}$ are sampled from $p\left(y_0^n \mid \Theta_0\right) \sim N\left(y_0, H_0\right)$ with corresponding particle weights $\left\{w_0^n\right\}_{n=1:Np}$, where

$$y_0 = \begin{bmatrix} x_0 \\ \mu_{\theta,0} \end{bmatrix}, \quad H_0 = \begin{bmatrix} \sigma^2 \Delta t_1 & 0 \\ 0 & \sigma^2_{\theta,0} \end{bmatrix}, \quad \text{and} \quad w_0^n = \frac{1}{Np}, \quad (n = 1:Np). \quad (6.69)$$

At time t_k, a series of new particles should be sampled from a predefined importance PDF $q(y_k \mid y_{k-1}^n, z_k)$. In the traditional PF algorithm (Arulampalam et al., 2002), the prior PDF of the state is chosen as the importance PDF.

$$q(y_k \mid y_{k-1}^n, z_k) = p(y_k \mid y_{k-1}^n). \tag{6.70}$$

However, the prior PDF does not contain the real-time information of the new measurement z_k. The particles sampled from it are subject to the influence of the stochasticity and possibly deviate from the actual state distribution. After a few iterations, most particle weights become negligible and the problem of particle degeneracy appears inevitably. To deal with this problem, the KPF algorithm is applied to estimate the states of the system. In this algorithm, the prior PDF of the state is first updated into the posterior PDF using the Kalman filtering (KF) algorithm. Then, the posterior PDF is employed as the importance PDF for the particle sampling.

The KF updating process of the prior PDF contains two steps.

State updating:

$$\begin{cases} \boldsymbol{H}_{k|k-1} = \boldsymbol{A}_k \boldsymbol{H}_{k-1|k-1} \boldsymbol{A}_k^T + \boldsymbol{P}_k \\ \boldsymbol{K}_k = \boldsymbol{H}_{k|k-1} \boldsymbol{C}^T \left[\boldsymbol{C} \boldsymbol{H}_{k|k-1} \boldsymbol{C}^T + \gamma^2 \right]^{-1} \\ \boldsymbol{y}_{k|k-1}^n = \boldsymbol{A}_k \boldsymbol{y}_{k-1}^n \\ \boldsymbol{y}_{k|k}^n = \boldsymbol{y}_{k|k-1}^n + \boldsymbol{K}_k \left(z_k - \boldsymbol{C} \boldsymbol{y}_{k|k-1}^n \right) \end{cases}, \tag{6.71}$$

Variance updating:

$$\boldsymbol{H}_{k|k} = \boldsymbol{H}_{k|k-1} - \boldsymbol{K}_k \boldsymbol{C} \boldsymbol{H}_{k|k-1}, \tag{6.72}$$

with

$$\boldsymbol{P}_k = \begin{bmatrix} \sigma^2 \Delta t_k & 0 \\ 0 & 0 \end{bmatrix} \text{ and } \boldsymbol{H}_{k|k} = \begin{bmatrix} \sigma_{x,k}^2 & \sigma_{x\theta,k}^2 \\ \sigma_{x\theta,k}^2 & \sigma_{\theta,k}^2 \end{bmatrix} = \operatorname{cov}(\boldsymbol{y}_k \mid z_{1:k}) \tag{6.73}$$

where $\sigma_{x,k}^2 = \operatorname{var}(x_k \mid z_{1:k})$, $\sigma_{\theta,k}^2 = \operatorname{var}(\theta_k \mid z_{1:k})$ and $\sigma_{x\theta,k}^2 = \operatorname{cov}(x_k \theta_k \mid z_{1:k})$.

The posterior PDF of the nth particle follows a normal distribution $N(\boldsymbol{y}_{k|k}^n, \boldsymbol{H}_{k|k})$, where $\boldsymbol{y}_{k|k}^n$ is the updated state value and $\boldsymbol{H}_{k|k}$ is the updated variance. It is seen from the KF updating process that, the posterior PDF incorporates the information of the new measurement, thus avoiding the severe deviation from the actual states. It is employed as the importance PDF and used to generate a series of new particles $\{y_k^n\}_{n=1:Np}$. Then, each particle weight w_{k-1}^n is updated by (Doucet et al., 2001b)

$$w_k^n = \frac{w_{k-1}^n p(z_k \mid y_k^n) p(y_k^n \mid y_{k-1}^n)}{q(y_k^n \mid y_{k-1}^n, z_k)} \tag{6.74}$$

where $p(z_k \mid y_k^n) \sim N(Cy_k^n, \gamma^2)$, $p(y_k^n \mid y_{k-1}^n) \sim N(y_{k|k-1}^n, H_{k|k-1})$ and $q(y_k^n \mid y_{k-1}^n, z_k) \sim N(y_{k|k}^n, H_{k|k})$. After that, each particle weight is normalized using

$$w_k^n = \frac{w_k^n}{\sum_{n=1}^{Np} w_k^n}. \tag{6.75}$$

Finally, the state of the system is estimated with the particles and their corresponding weights by

$$\hat{y}_k = \sum_{n=1}^{Np} (w_k^n y_k^n) = \begin{bmatrix} \hat{x}_k \\ \hat{\theta}_k \end{bmatrix}. \tag{6.76}$$

4. *RUL prediction*
The RUL L_k at t_k is defined as:

$$L_k = \inf\{l_k : x(l_k + t_k) \geq \lambda \mid z_{1:k}\} \tag{6.77}$$

where λ is a prespecified failure threshold, and

$$x(l_k + t_k) = x_k + \theta_k \varphi(l_k) + \sigma W(l_k), \quad l_k \geq 0 \tag{6.78}$$

with

$$\varphi(l_k) = (l_k + t_k)^b - t_k^b \tag{6.79}$$

and

$$W(l_k) = B(l_k + t_k) - B(t_k). \tag{6.80}$$

Let $\lambda_k = \lambda - x_k$ and $\psi(l_k) = b l_k (l_k + t_k)^{b-1}$. If x_k and θ_k are determined values, the PDF of the RUL based on x_k and θ_k is denoted as (Si, 2015)

$$f_{L_k \mid x_k, \theta_k}(l_k \mid x_k, \theta_k) \cong \frac{\lambda_k - \theta(\varphi(l_k) - \psi(l_k))}{\sqrt{2\pi\sigma^2 l_k^3}} \exp\left[-\frac{(\lambda_k - \theta\varphi(l_k))^2}{2\sigma^2 l_k}\right]. \tag{6.81}$$

However, x_k and θ_k are uncertain and their PDF based on the measurements $\boldsymbol{y}_{1:k}$ is acquired from the estimated system state $\hat{\boldsymbol{z}}_k$ in Eq. (6.76) and the variance $\boldsymbol{H}_{k|k}$ in Eq. (6.72).

$$\begin{cases} p(x_k \mid \boldsymbol{z}_{1:k}) \sim N(\hat{x}_k, \sigma_{x,k}^2) \\ p(\theta_k \mid \boldsymbol{z}_{1:k}) \sim N(\hat{\theta}_k, \sigma_{\theta,k}^2) \\ p(x_k \mid \theta_k, \boldsymbol{z}_{1:k}) \sim N(\mu_{x|\theta,k}, \sigma_{x|\theta,k}^2) \end{cases} \tag{6.82}$$

with

$$\begin{cases} \mu_{x|\theta,k} = \hat{x}_k + \rho_k \frac{\sigma_{x,k}}{\sigma_{\theta,k}} (\theta_k - \hat{\theta}_k) \\ \sigma_{x|\theta,k}^2 = \sigma_{x,k}^2 (1 - \rho_k^2) \end{cases} \tag{6.83}$$

where $\rho_k = \dfrac{\sigma_{x\theta,k}^2}{(\sigma_{x,k}\sigma_{\theta,k})}$.

The goal is to derive the PDF of the RUL based on the measurements $\boldsymbol{z}_{1:k}$, that is, $f_{L_k|\boldsymbol{z}_{1:k}}(l_k \mid \boldsymbol{z}_{1:k})$. The PDF of the RUL is derived using the total probability formula shown as

$$\begin{aligned} & f_{L_k|\boldsymbol{z}_{1:k}}(l_k \mid \boldsymbol{z}_{1:k}) \\ & = \int_{-\infty}^{+\infty} p(\theta_k \mid \boldsymbol{z}_{1:k}) \int_{-\infty}^{+\infty} p(x_k \mid \theta_k, \boldsymbol{z}_{1:k}) f_{L_k|x_k,\theta_k}(l_k \mid x_k, \theta_k) dx_k d\theta_k \\ & = E_{\theta_k|\boldsymbol{z}_{1:k}}\left[E_{x_k|\theta_k,\boldsymbol{z}_{1:k}}\left[f_{L_k|x_k,\theta_k}(l_k \mid x_k, \theta_k) \right] \right]. \end{aligned} \tag{6.84}$$

The double integral in Eq. (6.84) is separated into two one-dimension integrals. First, the PDF of the RUL based on θ_k and $\boldsymbol{z}_{1:k}$ is calculated with the PDF $p(x_k \mid \theta_k, \boldsymbol{z}_{1:k})$ in Eq. (6.82) taken into account.

$$\begin{aligned} & f_{L_k|\theta_k,\boldsymbol{z}_{1:k}}(l_k \mid \theta_k, \boldsymbol{z}_{1:k}) \\ & = \int_{-\infty}^{+\infty} p(x_k \mid \theta_k, \boldsymbol{z}_{1:k}) f_{L_k|x_k,\theta_k}(l_k \mid x_k, \theta_k) dx_k \\ & = E_{x_k|\theta_k,\boldsymbol{z}_{1:k}}\left[f_{L_k|x_k,\theta_k}(l_k \mid x_k, \theta_k) \right]. \end{aligned} \tag{6.85}$$

Then, the PDF of the RUL based on $\boldsymbol{z}_{1:k}$ is calculated by considering the PDF $p(\theta_k \mid \boldsymbol{z}_{1:k})$ in Eq. (6.82).

$$
\begin{aligned}
& f_{L_k|z_{1:k}}\left(l_k \mid z_{1:k}\right) \\
& = \int_{-\infty}^{+\infty} p\left(\theta_k \mid z_{1:k}\right) f_{L_k|\theta_k, y_{1:k}}\left(l_k \mid \theta_k, z_{1:k}\right) d\theta_k \\
& = E_{\theta_k|z_{1:k}}\left[f_{L_k|\theta_k, z_{1:k}}\left(l_k \mid \theta_k, z_{1:k}\right)\right].
\end{aligned}
\tag{6.86}
$$

To calculate the integrals of Eqs. (6.85) and (6.86), a lemma is provided as follows (Si, 2015).
Lemma: If $x \sim N(\mu, \sigma^2)$, and λ_1, λ_2, $A, B \in R$, $C \in R^+$, then the following equation is acquired.

$$
\begin{aligned}
E_x\left[\left(\lambda_1 - Ax\right)\exp\left[-\frac{(\lambda_2 - Bx)^2}{2C}\right]\right] = \sqrt{\frac{C}{B^2\sigma^2 + C}}\left(\lambda_1 - A\frac{B\lambda_2\sigma^2 + \mu C}{B^2\sigma^2 + C}\right) \\
\exp\left[-\frac{\left(\lambda_2 - B\mu\right)^2}{2(B^2\sigma^2 + C)}\right].
\end{aligned}
\tag{6.87}
$$

Let

$$
\begin{cases}
\lambda_1 = \lambda - \theta_k\left(\varphi\left(l_k\right) - \psi\left(l_k\right)\right), \ \lambda_2 = \lambda - \theta_k \varphi\left(l_k\right), \\
A = 1, \quad B = 1, \quad C = \sigma^2 l_k.
\end{cases}
\tag{6.88}
$$

Eq. (6.86) is calculated as

$$
\begin{aligned}
& f_{L_k|\theta_k, z_{1:k}}\left(l_k \mid \theta_k, z_{1:k}\right) \\
& \cong E_{x_k|\theta_k, z_{1:k}}\left[\frac{\lambda_k - \theta_k\left(\varphi\left(l_k\right) - \psi\left(l_k\right)\right)}{\sqrt{2\pi\sigma^2 l_k^3}}\exp\left[-\frac{\left(\lambda_k - \theta_k \varphi\left(l_k\right)\right)^2}{2\sigma^2 l_k}\right]\right] \\
& \frac{\left(\lambda - \hat{x}_k + \rho_k \frac{\sigma_{x,k}}{\sigma_{\theta,k}}\hat{\theta}_k\right)\sigma^2 l_k - \left(\left(\rho_k \frac{\sigma_{x,k}}{\sigma_{\theta,k}} + \varphi\left(l_k\right) - \psi\left(l_k\right)\right)\sigma^2 l_k - \psi\left(l_k\right)\sigma^2_{x|\theta,k}\right)\theta_k}{\sqrt{2\pi l_k^2\left(\sigma^2_{x|\theta,k} + \sigma^2 l_k\right)^3}} \\
& \times \exp\left[-\frac{\left(\lambda - \hat{x}_k + \rho_k \frac{\sigma_{x,k}}{\sigma_{\theta,k}}\hat{\theta}_k - \left(\rho_k \frac{\sigma_{x,k}}{\sigma_{\theta,k}} + \varphi\left(l_k\right)\right)\theta_k\right)^2}{2\left(\sigma^2_{x|\theta,k} + \sigma^2 l_k\right)}\right].
\end{aligned}
\tag{6.89}
$$

Let

$$
\begin{cases}
\lambda_{1,k} = \left(\lambda - \hat{x}_k + \rho_k \frac{\sigma_{x,k}}{\sigma_{\theta,k}}\hat{\theta}_k\right)\sigma^2 l_k, \ \lambda_{2,k} = \lambda - \hat{x}_k + \rho_k \frac{\sigma_{x,k}}{\sigma_{\theta,k}}\hat{\theta}_k, \\
A_k = \left(\rho_k \frac{\sigma_{x,k}}{\sigma_{\theta,k}} + \varphi\left(l_k\right) - \psi\left(l_k\right)\right)\sigma^2 l_k - \psi\left(l_k\right)\sigma^2_{x|\theta,k}, \\
B_k = \rho_k \frac{\sigma_{x,k}}{\sigma_{\theta,k}} + \varphi\left(l_k\right), \ C_k = \sigma^2_{x|\theta,k} + \sigma^2 l_k.
\end{cases}
\tag{6.90}
$$

Then, Eq. (6.89) is further transformed as

$$f_{L_k|z_{1:k}}(l_k \mid z_{1:k}) \cong \frac{\lambda_{1,k}\left(B_k^2\sigma_{\theta,k}^2 + C_k\right) - A_k B_k \lambda_{2,k}\sigma_{\theta,k}^2 - A_k\hat{\theta}_k C_k}{\sqrt{2\pi l_k^2 C_k^2\left(B_k^2\sigma_{\theta,k}^2 + C_k\right)^3}} \times \exp\left[-\frac{\left(\lambda - \hat{x}_k - \varphi(l_k)\hat{\theta}_k\right)^2}{2\left(B_k^2\sigma_{\theta,k}^2 + C_k\right)}\right]. \tag{6.91}$$

The expectation of the RUL is calculated using

$$E_{L_k|z_{1:k}}(l_k \mid z_{1:k}) = \frac{\int_{0^+}^{+\infty} l_k f_{L_k|z_{1:k}}(l_k \mid z_{1:k})\,dl_k}{\int_{0^+}^{+\infty} f_{L_k|z_{1:k}}(l_k \mid z_{1:k})\,dl_k}. \tag{6.92}$$

6.3.3.3 Simulation Verification

In this subsection, a simulated degradation process is utilized to evaluate the effectiveness of the proposed method based on the polynomial model. The degradation process is simulated using the state space model presented in (6.66). The model parameters are prespecified as shown in Table 6.5. The initial state value of the process is set to be zero. The time interval is set to be 10 s and the time length is 500 s. Measurements of the simulated degradation process are shown in Fig. 6.23.

The measurements of the simulated degradation process are input into the polynomial model in (6.58) and the model parameters are initialized using the MLE algorithm. For comparison, the measurements are also input into a linear model in Si et al. (2014). The initialized model parameters of the two models are shown in Table 6.5. To quantitatively compare the fitness of the models for

Table 6.5 Model Parameters of the Simulated Degradation Process

Model Parameter	b	μ_θ	σ_θ^2	σ^2	γ^2	Max ℓ	AIC
Prespecified value	1.5000	0.1000	0	3.0000	3.0000	—	—
Nonlinear model	1.5338	0.1432	5.3280e-06	1.8582	4.0349	−122.2750	252.5501
Linear model	1.0000	1.1791	5.5978e-06	2.0066	4.1845	−123.5099	255.0198

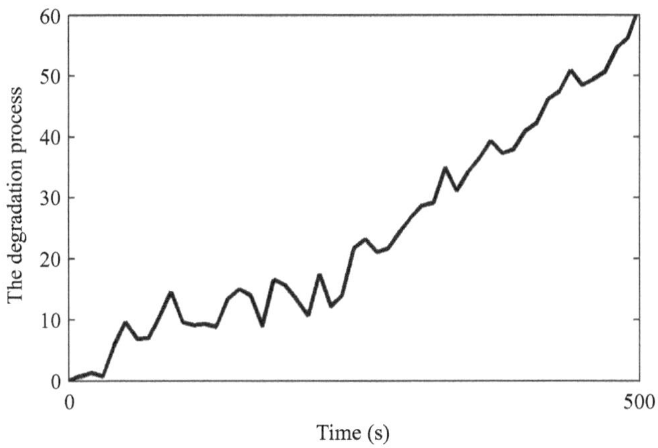

Figure 6.23. Simulated degradation process.

the simulated dataset, two criteria, that is, the maximum of the log-likelihood function (max ℓ) and the Akaike information criterion (AIC) (Akaike, 1974) are calculated. A larger max ℓ value and a smaller AIC value represent a higher fitting accuracy. The values of the max ℓ and AIC are also displayed in Table 6.5.

It is seen from Table 6.5 that most initialized model parameters of the nonlinear model coincide with the prespecified values. The results of the linear model, however, have large deviations with the prespecified values. It is also observed that the nonlinear model has a larger max ℓ value and a smaller AIC value than the linear model, identifying the superiority of the nonlinear model in fitting the simulated dataset. The reason is analyzed as follows. The degradation severity of the models are described by $\theta_k t_k^b$, where the parameter b in the linear model identically equals one, smaller than the prespecified value. In order to coincide with the actual severity of the degradation, the parameter θ_k needs to be overestimated, leading to a larger μ_θ than the prespecified value. In addition, the linear model is incapable of handling the nonlinear variability of the stochastic process. The temporal variability and the measurement variability are incorrectly considered to be responsible for the nonlinear variability. As a result, the linear model has larger σ^2 and γ^2 values than the nonlinear model.

After the parameter initialization, the next step is to estimate the states and to further predict the RUL for the simulated degradation process. To access the performance of the proposed method, four methods are employed for comparisons.

M1 The linear model combined with the KF algorithm.
M2 The nonlinear model combined with the KF algorithm.

M3 The nonlinear model combined with the PF algorithm.

M4 The nonlinear model combined with the KPF algorithm, that is, the proposed method.

M1 and M2 share the same state estimation algorithm but different types of models. They are used to compare the effectiveness of the two types of models in the RUL prediction. M2, M3, and M4 share the same model type but different state estimation algorithms. Correspondingly, they are used to compare the performance of the three state estimation algorithms in the RUL prediction.

Fig. 6.24 gives the state estimation results of the four methods, where the measurement curve shows random fluctuations combined with a gradual degradation trend. A good estimation method is supposed to remove the random fluctuation and reveal the gradual degradation trend. The results of M1 and M2 almost coincide with the measurements thus still including the random measurement noise. While the results of M3 and M4 present smoother degradation trends and succeed in restraining the measurement noise. Therefore, M3 and M4 perform better than M1 and M2 in the state estimation of the degradation process. To demonstrate the superiority of the KPF algorithm compared with the PF in handling the particle degeneracy problem, a criterion named effective sample size (Arulampalam et al., 2002) is used to measure the severity of the particle degeneracy in the state estimation processes

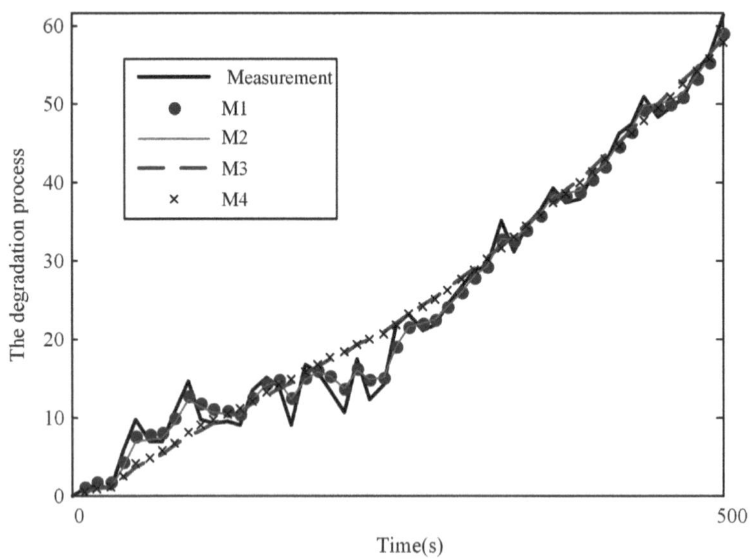

Figure 6.24. Comparison of the state estimation results for the simulated degradation process.

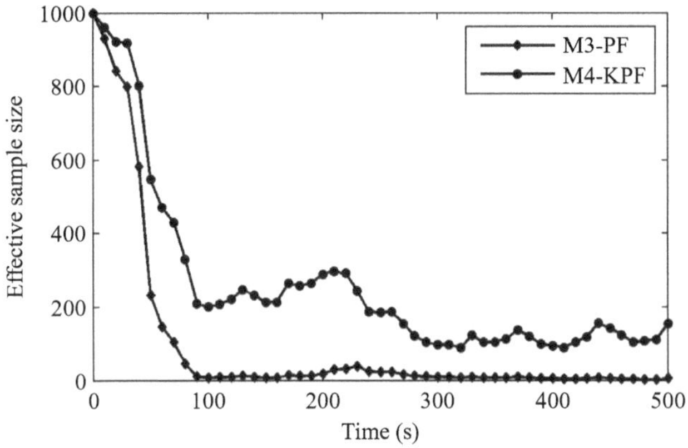

Figure 6.25. Variation of the effective sample sizes of M3 and M4.

of M3 and M4. The particle numbers are 1000 for both of them. The variation of the effective sample size of M3 and M4 during the whole lifetime is shown in Fig. 6.25. It is observed that the effective sample size of M3 abruptly reduces to a small value after a few iterations, indicating the occurrence of the severe particle degeneracy. While M4 has larger effective sample sizes than M3 and reduces the problem of particle degeneracy.

The PDF of the RUL is predicted using the four methods and the prediction results are shown in Fig. 6.26. It is seen from the results that there is obvious discrepancy between the result of M1 and the results of M2 – M4, while the results of M2 – M4 almost coincide with each other. In addition, since the degradation process has reached the final failure, all results of the four methods equal zero at the end of the lifetime $t = 500$ s. To further compare the prediction accuracy of the four methods, the expectations of

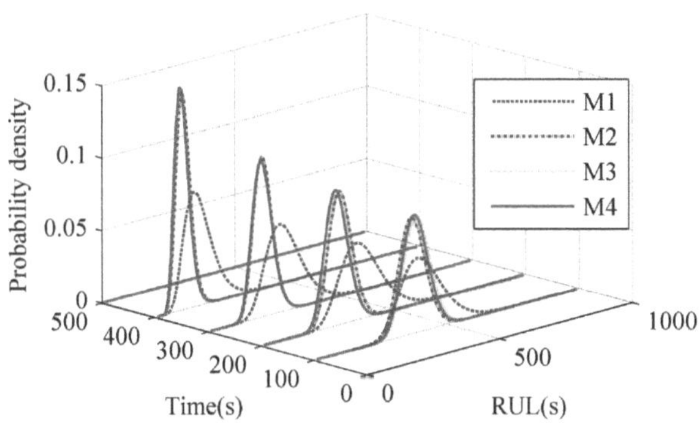

Figure 6.26. PDF of the predicted RUL for the simulated degradation process.

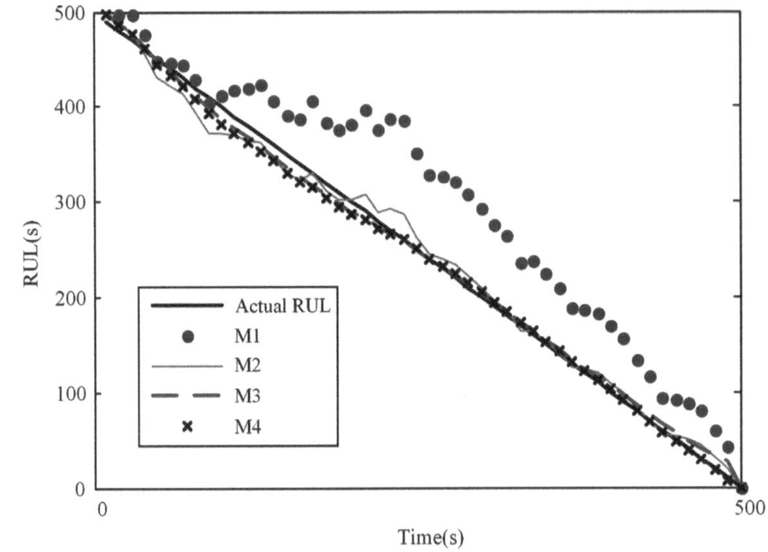

Figure 6.27. Comparison of the RUL prediction results for the simulated degradation process.

the predicted RUL are calculated and displayed in Fig. 6.27. It is observed that the predicted RUL of M1 deviates severely from the actual RUL. The other three methods perform better than M1. This verifies the superiority of the methods based on the nonlinear model in the RUL prediction. For quantitative comparison, a commonly used performance criterion named CRA (Saxena et al., 2010) is calculated. CRA quantifies how accurately a prediction method performs at specific time indexes and the perfect value is 1. The CRA values for M1 – M4 are 0.5341, 0.9132, 0.9153, and 0.9645, respectively. It is seen that the CRA value of M1 is much smaller than other methods, and M4 obtains the largest CRA value among them. The superiority of M4 is explained as follows. In the KPF algorithm, each particle is updated using KF, thus avoiding the severe deviation from the actual states. Then, these particles are weighted using PF according to their fitting degree with the measurement curve. With the integration of KF and PF, KPF performs better than both of them in RUL prediction.

6.3.3.4 RUL Prediction of Gearboxes: a Case Study

To demonstrate the effectiveness of the proposed polynomial model-based prediction method, accelerated degradation tests of gears are conducted on an experimental system in Fig. 6.5. More details about the tests have been described in Section 6.2.2.4.

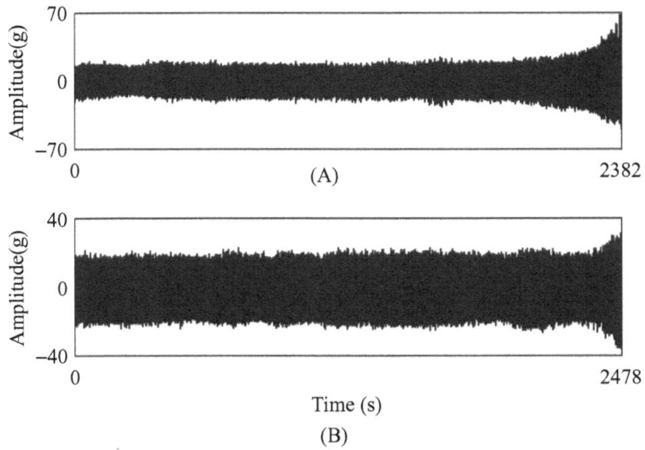

Figure 6.28. Vibration signals of the tested gears. (A) Gear 1 and (B) gear 2.

The vibration signals of two tested gears during the whole lifetime are shown in Fig. 6.28. After the tests, different kinds of faults, that is, the wear, the pitting and the crack, are observed on several teeth of gear 1, and a serious crack is found on a tooth of gear 2, respectively. Fig. 6.28 shows that the vibration amplitudes increase gradually with the degeneracy of the gears. The RMS is used as the health indicator of the tested gears and shown in Fig. 6.29. In the normal operation stage, the RMS presents little variation and it is hard to find any information about the degradation process of the gears, while the RMS in the accelerated degradation stage contains more information about the gear degradation. What's more, predicting the RUL in this stage is more practical than in the normal operation stage. Therefore, we start to predict the RUL of the

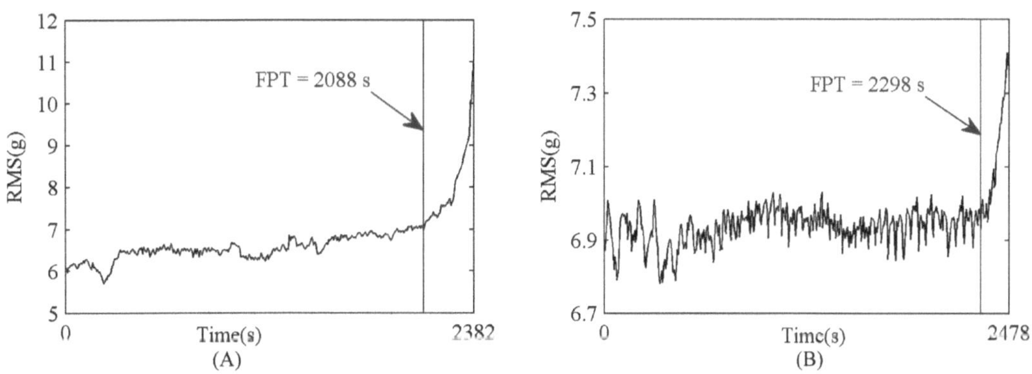

Figure 6.29. RMS and FPT of the tested gears. (A) Gear 1 and (B) gear 2.

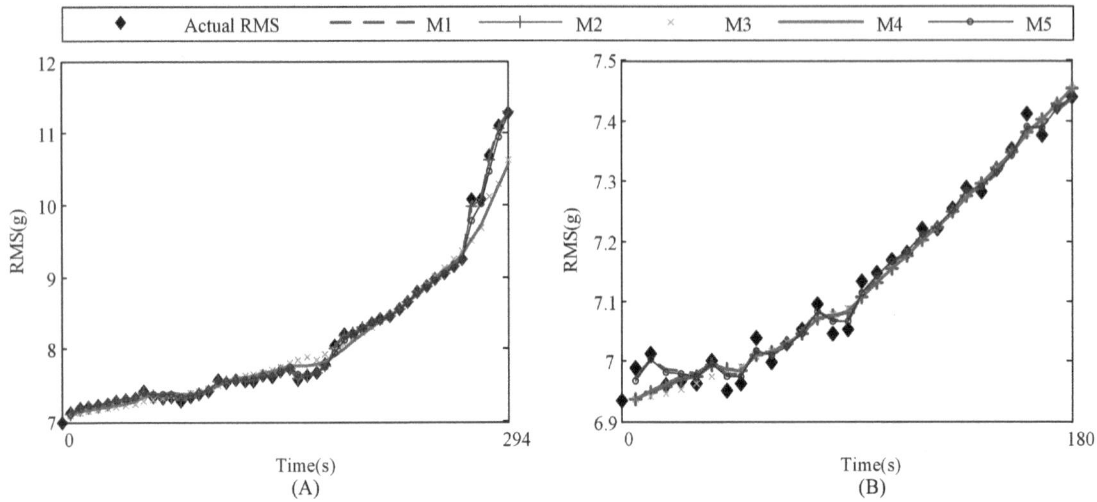

Figure 6.30. Health condition estimation results of the tested gears. (A) Gear 1 and (B) gear 2.

gears from the accelerated degradation stage. The FPT of the gears is selected using the adaptive FPT selection method discribed in Section 6.3.2.3 and the results are shown in Fig. 6.29.

The RMS is input into the stochastic process models and the RUL is predicted using the four methods M1 − M4. To further evaluate the effectiveness of the proposed method against the method based on a physics model, a commonly used method based on the Paris–Erdogan model (An et al., 2013) is employed to predict the RUL of the tested gears. This method is defined as M5 following the above four methods M1 − M4. To ensure a fair comparison, the model parameters are initialized via the MLE algorithm and the particle number is set to be 1000 as well. The health condition estimation results of the gears are shown in Fig. 6.30 and the RUL prediction results are given in Fig. 6.31. The CRA values of the RUL prediction results are displayed in Table 6.6. Fig. 6.30 shows that M3 and M4 present smoother curves than M1, M2, and M5, which implies that M3 and M4 may exclude the measurement noises and reveal the health conditions of the gears. In Fig. 6.31, M4 provides more accurate prediction results than others as time goes on. This corresponds to our requirements for RUL prediction that more accurate prediction results are expected when the machine is closer to the final failure. In Table 6.6, M4 acquires the highest CRA values for both tested gears. These results demonstrate the superiority of the proposed method in the RUL prediction of the tested gears.

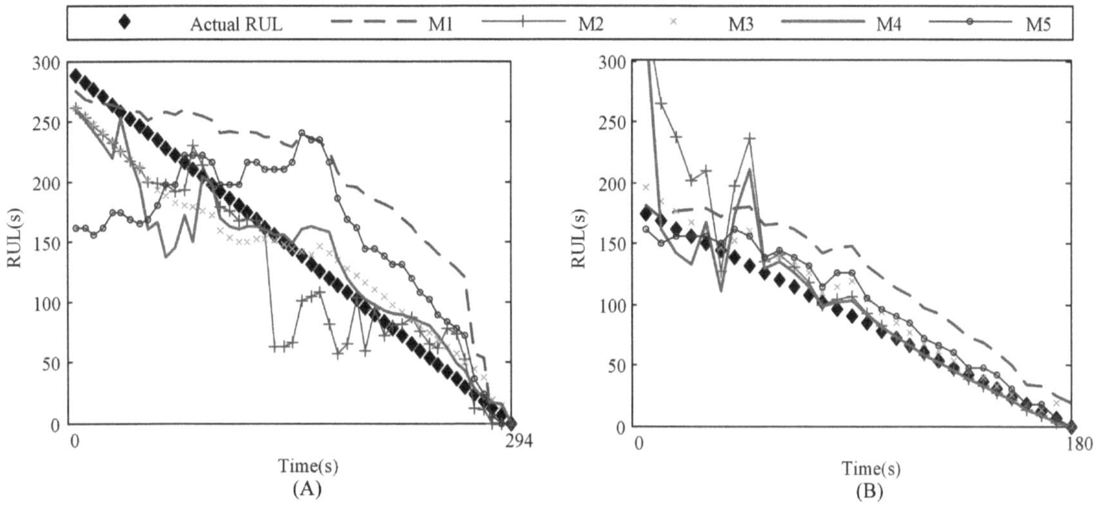

Figure 6.31. RUL prediction results of the tested gears. (A) Gear 1 and (B) gear 2.

6.3.4 Paris–Erdogan Model-Based Methods

6.3.4.1 Motivation

The above two subsections give examples of RUL prediction methods based on the exponential model and the polynomial model, both belonging to the stochastic process models. Another commonly used model type is physical models. In this subsection, a prediction method based on a typical physical model, that is, the Paris–Erdogan model (Paris and Erdogan, 1963) is presented as an example. Paris–Erdogan model was first proposed by Paris and Erdogan (1963) to explain the propagation of micro fatigue cracks in the mechanical material. Since most degradation processes of machinery are directly or indirectly caused by the micro fatigue cracks, Paris–Erdogan model is effective in describing a general

Table 6.6 CRA of the RUL Prediction Results for the Tested Gears

Case	M1	M2	M3	M4	M5
Gear 1	0.2309	0.7310	0.7394	0.7732	0.5037
Gear 2	0.3704	0.7757	0.8031	0.8411	0.8100

degradation process of machinery. Until now, many variants have been developed from the Paris–Erdogan model and utilized to the RUL prediction of machinery (An et al., 2013). The RUL prediction methods based on Paris–Erdogan model could incorporate both expert knowledge and real-time information from machinery. Consequently, they may work well in RUL prediction of machinery. However, there still exist some problems in the Paris–Erdogan model–based methods.

The first problem is how to develop a health indicator which is suitable for the RUL prediction. The degradation process of rotating machinery is generally irreversible. Therefore a suitable indicator is supposed to have obvious monotonic trends which is able to simplify the modeling process and produce accurate prediction results (Camci et al., 2013). Many researchers have attempted to construct such health indicators. Camci et al. (2013) extracted time–domain indicators from vibration signals of rolling element bearings and evaluated the effectiveness of the indicators using a monotonicity metric. Li and He (2012) extracted acoustic emission indicators based on the EMD technique and used these indicators to monitor the health condition of gearboxes. Qian et al. (2013) used recurrence quantification analysis to extract recurrence plot entropy indicators and applied them into the bearing RUL prediction. The above research attempts to use original features directly captured from measured data as the health indicators. In fact, each original feature is always sensitive to a certain fault in a certain degradation stage (Qiu et al., 2003). A suitable health indicator should incorporate mutual information from multiple features for the health state estimation and RUL prediction. Consequently, many researchers have tried to construct a fusion health indicator by fusing multiple original features. For example, Qiu et al. (2003) used a self organizing map (SOM) to fuse multiple time–domain features into a health indicator which is named as minimum quantization error (MQE). Then, Huang et al. (2007) applied the MQE indicator into the RUL prediction of rolling element bearings. Yu (2012) used the dynamic principal component analysis to extract principal components from multiple features and constructed a health indicator based on the Mahalanobis distance of the hidden Markov model. These fusion indicators may incorporate mutual information related to the degradation process of machinery from multiple features. However, some original features may not be sensitive enough to the degradation processes or even have a negative influence on the fusion result. Therefore, it is essential to estimate the value of original features for the RUL prediction and select typical ones before fusion.

The second problem in the RUL prediction method based on the Paris–Erdogan model is how to estimate model parameters according to the real-time information from measurements. The PF algorithm is suitable to deal with this problem because of its capability in combining information from real-time data and expert knowledge from empirical models. The basic theory of PF has been introduced in Section 6.2.2.2. PF is particularly effective in estimating states and model parameters of degradation models according to real-time information. However, a set of initialized model parameters should be provided before the model parameter updating. The existing research either sets the initial parameters based on empirical knowledge or predetermines the initial parameters arbitrarily, which may influence the performance of the PF-based prediction algorithm.

To deal with the aforementioned two problems in the RUL prediction method based on the Paris–Erdogan model, an enhanced Paris–Erdogan model–based RUL prediction method Lei et al. (2016). Two modules are contained in the enhanced method, that is, indicator construction and RUL prediction. The first module focuses on the health indicator construction, where a new health indicator named weighted minimum quantization error (WMQE) is constructed. This indicator is developed from the MQE proposed by Qiu et al. (2003). Different from the MQE indicator, original features are selected and weighted before fusion in the construction of the WMQE indicator, which further enhances the superiority of the fusion health indicator for RUL prediction. In the second module, the degradation process of the rotating machinery is described using the Paris–Erdogan model (Paris and Erdogan, 1963), and the RUL is predicted using a PF-based algorithm (An et al., 2013). In this module, the model parameters are initialized using the MLE algorithm instead of prespecified artificially. The initialization process is conducted according to the available measurements from a certain degradation process of rotating machinery. Therefore, it is able to adjust the model parameters adaptively according to the degradation trend of the specified machinery.

The flowchart of the enhanced Paris–Erdogan model-based method is shown in Fig. 6.32. In the indicator construction module, original features are extracted from vibration signals of rotating machinery. Then the degradation trends of these features are estimated and some features which have obvious degradation trends are selected from them. The selected features are further clustered using a correlation clustering algorithm and typical features of the clusters are fused using the SOM to construct the health indicator WMQE. In the RUL prediction module, the WMQE is input into the Paris–Erdogan model and model parameters are initialized using the MLE algorithm. Then a PF-based

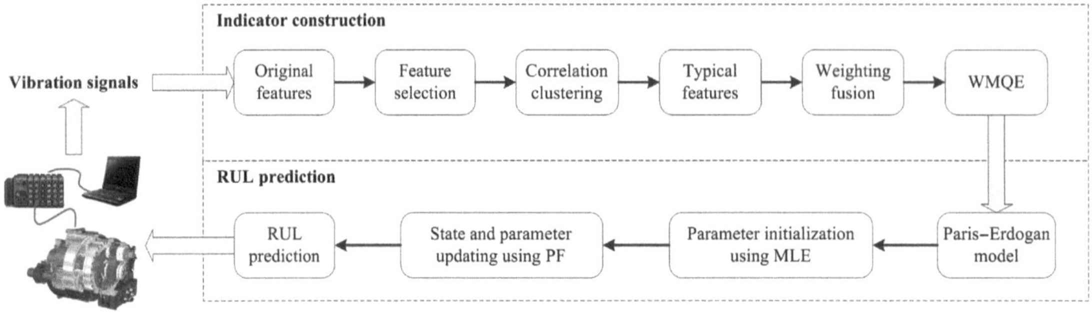

Figure 6.32. Flow chart of the Paris–Erdogan model–based method.

algorithm is utilized to update model parameters and predict RUL of the rotating machinery. More details about these two modules are presented as follows.

6.3.4.2 Health Indicator Construction

In this module, original features are extracted from vibration signals of rotating machinery. Since rotating machinery always suffers from an irreversible degradation process, an original feature which is sensitive to the degradation process should have a monotonic degradation trend. To evaluate the monotonicity of the original features, a trendability measure of each feature is calculated, which is defined as the feature's Spearman coefficient with time (Spearman, 1904).

$$\rho^i = \frac{\left|\sum_{k=1}^{K}\left(T_k - \bar{T}\right)\left(F_k^i - \bar{F}^i\right)\right|}{\sqrt{\sum_{k=1}^{K}\left(T_k - \bar{T}\right)^2 \sum_{k=1}^{K}\left(F_k^i - \bar{F}^i\right)^2}}, \tag{6.93}$$

where $\left\{T_k\right\}_{k=1:K}$ and $\left\{F_k^i\right\}_{k=1:K}$ are the ranks of the time $\left\{t_k\right\}_{k=1:K}$ and the ith feature $\left\{f_k^i\right\}_{k=1:K}$, respectively, K is the length of the time sequence, and $\bar{T}$ and $\bar{F}^i$ are the means of $\left\{T_k\right\}_{1:K}$ and $\left\{F_k^i\right\}_{1:K}$, respectively. When a feature monotonously increases or decreases with time, the trendability value is one. On the contrary, if a feature is stable or changes randomly with time, the trendability value equals zero. So a higher trendability value means a better monotonicity.

The Spearman coefficient is widely used in assessing how well the relationship between two variables can be described using a monotonic function. It is developed from the Pearson coefficient

Table 6.7 Process of the Correlation Clustering Algorithm

Step 1:	Calculate the absolute correlation matrix of the selected features and let $H = 1$.
Step 2:	Find the first center and the second center which have the lowest correlation coefficient among these features.
Step 3:	Find the next center which has the lowest mean correlation coefficient with the determined centers.
Step 4:	Repeat Step 3 until all of the H centers are determined.
Step 5:	Classify each of the remaining features into one cluster whose center has the highest correlation coefficient with this feature.
Step 6:	Set the membership degree $u_{hc,k}$ according to the clustering result, that is, $u_{hc,k} = 1$ if the kth feature belongs to the hth cluster, otherwise $u_{hc,k} = 0$. Calculate the value of $PBM(H)$ using (6.94).
Step 7:	Let $H = H + 1$ and repeat Step 3 to Step 6 until H equals the number of the selected features.
Step 8:	Report the cluster number H which has the maximum PHM value and output the clustering result.

which has been applied into the trendability measurement of features by Javed et al. (2015). But the Pearson coefficient is only effective in measuring the linear relationship between a feature and the time, whereas rotating machinery generally performs nonlinear degradation trends in real cases. The major difference of the Spearman coefficient with the Pearson coefficient is that, the Spearman coefficient evaluates the trendability of a feature using the ranks of the feature instead of the feature itself. Therefore, it is suitable to evaluate the nonlinear relationship and is more effective in measuring the monotonicity of the features.

The trendability value calculated by Eq. (6.93) belongs to the interval [0,1]. The original features which have the trendability values higher than 1/2 are selected to insure the high monotonic trendability of the selected features. Some selected original features may have redundant information which represents the similar degradation trends of the rotating machinery. The original features containing the similar degradation trend information generally have high correlation coefficients. To identify the redundant information, a correlation clustering algorithm is developed to classify these features. This algorithm evaluates the similarity of these selected features and classifies them into different clusters according to their correlation matrix. The process of the correlation clustering algorithm is shown in Table 6.7. A PBM index is applied to determine an appropriate number of clusters in this algorithm, which is commonly used for verifying the cluster validity (Pakhira et al., 2004). The formulation of the PBM index is as follows:

$$PBM(H) = \left(\frac{1}{H} \cdot \frac{E_H}{E_1} \cdot D_H \right)^2, \tag{6.94}$$

where H is the number of clusters, $E_H = \sum_{h=1}^{H} \left(\sum_{k=1}^{K} u_{hc,k} \cdot R_{hc,k} \right)$, $D_H = \max_{i,j=1}^{H} R_{ic,jc}$, $u_{hc,k}$ is the membership degree of the kth feature to the hth cluster, $R_{hc,k}$ is the correlation coefficient of the hth center and the kth feature, and $R_{ic,jc}$ is correlation coefficient of the ith center and the jth center.

After the correlation clustering process, all of the selected features are classified into H different clusters. Features in the same cluster are highly correlated and assumed to have the similar information for the degradation process. To reduce the redundancy of the features, one typical feature which has the highest trendability in each cluster is selected. Thus, a total of H typical features are selected from the original features.

After the feature selection, the H typical features are fused via a SOM based method (Qiu et al., 2003). The SOM is first trained by the feature vectors during the normal operation stage. Then a feature vector extracted under the unidentified condition is compared with the weight vector of its best matching unit (BMU) in the SOM. An indicator named MQE has been proposed in Qiu et al. (2003) depending on the distance between the BMU and the input feature vector. From the point of view of the degradation monitoring, the distance between the BMU and input feature vector indicates how far the unidentified condition deviates from the normal operation stage. The MQE indicator ignores the phenomenon that different features generally present distinct sensitivities to the degradation process of machinery. The features which are more sensitive to the degradation process are supposed to make more contribution to the fusion process. Therefore, an enhanced indicator of WME, that is, weighted minimum quantization error (WMQE), is proposed and denoted as follows:

$$s_k = \sum_{h=1}^{H} \left(\tilde{\rho}^h \left| f_k^h - m_{\text{BMU}}^h \right| \right), \tag{6.95}$$

where s_k is the WMQE value at t_k, $\tilde{\rho}^h$ is the normalized trendability of the hth typical feature with $\tilde{\rho}^h = \rho^h \Big/ \sum_{h=1}^{H} \rho^h$, which measures the sensitivity of the feature to the degradation process, f_k^h is the value of the hth typical feature at t_k, and m_{BMU}^h is the hth weight of the BMU.

6.3.4.3 RUL Prediction

The constructed indicator WMQE is further input into the RUL prediction module. In this module, a PF-based prediction algorithm is utilized to predict RUL of the rotating machinery whose degradation processes are described using a variant of Paris–Erdogan model. The Paris–Erdogan model is formulated as

$$\frac{dx}{dt} = c(\Delta\delta)^{\gamma}, \quad \Delta\delta = m\sqrt{x} \tag{6.96}$$

where x represents the semicrack length, t is the number of stress cycles (i.e., the fatigue life), c, γ, and m are material constants which are determined by tests, and $\Delta\delta$ is amplitude of stress intensity factor roughly proportional to the square root of x.

It is seen from Eq. (6.96) that there are several model parameters in the Paris–Erdogan model, that is, c, γ, and m, which are difficult to measure during the operation process of the rotating machinery. For convenient application, the Paris–Erdogan model is transformed into the following format with $\alpha = cm^{\gamma}$ and $\beta = \gamma/2$.

$$\frac{\mathrm{d}x}{\mathrm{d}t} = \alpha x^{\beta}. \tag{6.97}$$

Then, the above function is rewritten into the following state space model.

$$\begin{cases} x_k = x_{k-1} + \alpha_{k-1} x_{k-1}^{\beta} \Delta t_k \\ \alpha_k = \alpha_{k-1} \\ z_k = x_k + v_k \end{cases}, \tag{6.98}$$

where α_{k-1} is a random variable following a normal distribution of $N\left(\mu_\alpha, \sigma_\alpha^2\right)$, β is a constant parameter, $\Delta t_k = t_k - t_{k-1}$, z_k is the measured WMQE value at t_k and v_k is the measurement noise following the normal distribution of $N\left(0, \sigma_v^2\right)$. With the transformation of the Paris–Erdogan model, the model parameters are more convenient to estimate according to the measurements. In addition, the state space model inherits the superiority of the Paris–Erdogan model in describing the general degradation processes. Therefore, it is supposed to be a good model for a general degradation process.

After the transformation, the unknown model parameters are changed to be $\Theta = \left(\mu_\alpha, \sigma_\alpha^2, \beta, \sigma_v^2\right)'$, where $(\cdot)'$ denotes the vector transposition. Then, the measured WMQE values constructed from vibration signals are input into the model, and the model parameters are initialized using MLE. It is assumed that there are a series of measurements $z_{0:M} = \left(z_0, \ldots, z_M\right)'$ at ordered times $t_0, \ldots, t_M$. According to Eq. (6.98), z_k is formulated as follows:

$$z_k = x_{k-1} + \alpha_k x_{k-1}^{\beta} \Delta t_k + v_k. \tag{6.99}$$

The degradation state x_{k-1} has the following relationship with the measurement z_{k-1}.

$$x_{k-1} = z_{k-1} - v_{k-1}. \tag{6.100}$$

The degradation state x_{k-1} is hard to be acquired in real applications. If the measurement noise v_{k-1} is small enough compared with the measurement itself, it is negligible and x_{k-1} is approximated by z_{k-1}. Let $\boldsymbol{T} = \left(z_0^{\beta}\Delta t_1, \ldots, z_{M-1}^{\beta}\Delta t_M\right)$. $\boldsymbol{z}_{1:M} = \left(z_1, \ldots, z_M\right)'$ is multivariate normally distributed, which is denoted as follows:

$$\boldsymbol{z}_{1:M} \sim N\left(\boldsymbol{z}_{0:M-1} + \mu_{\alpha}\boldsymbol{T}, \sigma_{\alpha}^2 \boldsymbol{T}\boldsymbol{T}' + \sigma_v^2 \boldsymbol{I}_M\right), \tag{6.101}$$

where I_M is an identity matrix of order M.

Let $\Delta\boldsymbol{z}_{1:M} = \left(z_1 - z_0, \ldots, z_M - z_{M-1}\right)'$, and the log-likelihood function of the unknown parameters based on the measurements is expressed as

$$\begin{aligned} \ell\left(\Theta \mid \boldsymbol{z}_{0:M}\right) &= -\frac{M}{2}\ln(2\pi) - \frac{1}{2}\ln\left|\sigma_{\alpha}^2 \boldsymbol{T}\boldsymbol{T}' + \sigma_v^2 \boldsymbol{I}_M\right| \\ &\quad -\frac{1}{2}\left(\Delta\boldsymbol{z}_{1:M} - \mu_{\alpha}\boldsymbol{T}\right)'\left(\sigma_{\alpha}^2 \boldsymbol{T}\boldsymbol{T}' + \sigma_v^2 \boldsymbol{I}_M\right)^{-1}\left(\Delta\boldsymbol{z}_{1:M} - \mu_{\alpha}\boldsymbol{T}\right) \\ &= -\frac{M}{2}\ln(2\pi) - \frac{M}{2}\ln\sigma_{\alpha}^2 - \frac{1}{2}\ln\left|\boldsymbol{T}\boldsymbol{T}' + \tilde{\sigma}_v^2 \boldsymbol{I}_M\right| \\ &\quad -\frac{1}{2\sigma_{\alpha}^2}\left(\Delta\boldsymbol{z}_{1:M} - \mu_{\alpha}\boldsymbol{T}\right)'\left(\sigma_{\alpha}^2 \boldsymbol{T}\boldsymbol{T}' + \sigma_v^2 \boldsymbol{I}_M\right)^{-1}\left(\Delta\boldsymbol{z}_{1:M} - \mu_{\alpha}\boldsymbol{T}\right) \end{aligned}, \tag{6.102}$$

with $\tilde{\sigma}_v^2 = \sigma_v^2/\sigma_{\alpha}^2$. The first partial derivatives of $\ell\left(\Theta \mid \boldsymbol{z}_{0:M}\right)$ with respect to μ_{α} and σ_{α}^2 are calculated and formulated with

$$\frac{\partial \ell\left(\Theta \mid \boldsymbol{z}_{0:M}\right)}{\partial \mu_{\alpha}} = \frac{1}{\sigma_{\alpha}^2}\boldsymbol{T}'\left(\boldsymbol{T}\boldsymbol{T}' + \tilde{\sigma}_v^2 \boldsymbol{I}_M\right)^{-1}\left(\Delta\boldsymbol{z}_{1:M} - \mu_{\alpha}\boldsymbol{T}\right), \tag{6.103}$$

$$\begin{aligned} \frac{\partial \ell\left(\Theta \mid \boldsymbol{z}_{0:M}\right)}{\partial \sigma_{\alpha}^2} &= -\frac{M}{2\sigma_{\alpha}^2} + \frac{1}{2\sigma_{\alpha}^4}\left(\Delta\boldsymbol{z}_{1:M} - \mu_{\alpha}\boldsymbol{T}\right)' \\ &\quad \left(\boldsymbol{T}\boldsymbol{T}' + \tilde{\sigma}_v^2 \boldsymbol{I}_M\right)^{-1}\left(\Delta\boldsymbol{z}_{1:M} - \mu_{\alpha}\boldsymbol{T}\right). \end{aligned} \tag{6.104}$$

Let $\partial \ell\left(\Theta \mid \boldsymbol{z}_{0:M}\right)/\partial \mu_{\alpha} = 0$ and $\partial \ell\left(\Theta \mid \boldsymbol{z}_{0:M}\right)/\partial \sigma_{\alpha}^2 = 0$. The MLE results of μ_{α} and σ_{α}^2 are

$$\mu_\alpha = \frac{T'\left(TT' + \tilde{\sigma}_v^2 I_M\right)^{-1} \Delta z_{1:M}}{T'\left(TT' + \tilde{\sigma}_v^2 I_M\right)^{-1} T}, \tag{6.105}$$

$$\sigma_\alpha^2 = \frac{\left(\Delta z_{1:M} - \mu_\alpha T\right)' \left(TT' + \tilde{\sigma}_v^2 I_M\right)^{-1} \left(\Delta z_{1:M} - \mu_\alpha T\right)}{M}. \tag{6.106}$$

With Eqs. (6.105) and (6.106) substituted into Eq. (6.102), the log-likelihood function is reduced into a two-variable function about β and $\tilde{\sigma}_v^2$, which is denoted by

$$\ell\left(\Theta \mid z_{0:M}\right) = -\frac{M}{2}\ln(2\pi) - \frac{M}{2}\ln\sigma_\alpha^2 - \frac{1}{2}\ln\left|TT' + \tilde{\sigma}_v^2 I_M\right| - \frac{M}{2}. \tag{6.107}$$

The MLE values of β and $\tilde{\sigma}_v^2$ are obtained by maximizing the log-likelihood function (6.107) through two-dimensional optimizing. Then the MLE values of β and $\tilde{\sigma}_v^2$ are substituted into Eqs. (6.105) and (6.106), and the MLE values of μ_α and σ_α^2 are acquired. The value of σ_v^2 is calculated with $\tilde{\sigma}_v^2$ multiplied by σ_α^2. Finally, all of the unknown parameters $\Theta = \left(\mu_\alpha, \sigma_\alpha^2, \beta, \sigma_v^2\right)'$ are initialized.

After parameter initialization, the model parameters are further updated and the RUL is predicted using a PF-based prediction algorithm. Based on the initialized parameters, a series of initial particles $\left\{y_0^n\right\}_{n=1:Ns}$ are sampled from the initial PDF of the system state $p\left(y_0^n \mid \Theta_0\right) \sim N\left(y_0, Q_0\right)$ with

$$y_0 = \begin{bmatrix} x_0 \\ \mu_\alpha \end{bmatrix} \text{ and } Q_0 = \begin{bmatrix} 0 & 0 \\ 0 & \sigma_\alpha^2 \end{bmatrix}. \tag{6.108}$$

Ns is the number of particles and the weight of each particle is set to be $1/Ns$. Then new particles $\left\{y_k^n\right\}_{i=1:Ns}$ are obtained following

$$y_k^n = \begin{bmatrix} x_k^n \\ \mu_\alpha^n \end{bmatrix} = \begin{bmatrix} x_{k-1}^n + \mu_\alpha^n \left(x_{k-1}^n\right)^\beta \Delta t_k \\ \mu_\alpha^n \end{bmatrix}. \tag{6.109}$$

When the new measurement z_k at t_k is available, each particle weight is updated and normalized by

$$w_k^n = w_{k-1}^n p\left(z_k \mid y_k^n\right), \quad \tilde{w}_k^n = \frac{w_k^n}{\sum_{n=1}^{N_s} w_k^n}, \tag{6.110}$$

where

$$p\left(z_k \mid y_k^n\right)=\frac{1}{\sqrt{2\pi}\sigma_v}\exp\left[-\frac{1}{2}\left(\frac{z_k-x_k^n}{\sigma_v}\right)^2\right]. \tag{6.111}$$

The particles are resampled according to the particle weights and their weights are reset to be $1/Ns$. After that, the RUL is predicted based on the resampled particles. The RUL l_k at t_k is defined as

$$l_k=\inf\left\{l_k : x\left(l_k+t_k\right)\geq\lambda \mid x_{1:k}\right\}, \tag{6.112}$$

where λ is a prespecified failure threshold. Each particle is transmitted following the transition function of Eq. (6.98) from current state until the state value exceeds the failure threshold, and the RUL $\left\{l_k^n\right\}_{n=1:Ns}$ predicted using each particle is acquired. Then the PDF of the RUL is approximated by

$$p\left(l_k \mid z_{0:k}\right)=\sum_{n=1}^{Ns}\tilde{\omega}_k^n\delta\left(l_k-l_k^n\right). \tag{6.113}$$

6.3.4.4 RUL Prediction of Rolling Element Bearings: a Case Study

In this subsection, the proposed method is evaluated by predicting RUL of rolling element bearings. Accelerated degradation tests of rolling element bearings are conducted and vibration signals are acquired during the tests to verify the effectiveness of the method. The accelerated degradation tests are conducted on the experimental system named PRONOSTIA. This experimental system is able to conduct accelerated degradation tests on bearing in a few hours. Seventeen bearings are tested under three different operation conditions. The operation conditions are displayed in Table 6.8. Two bearings under each operation condition are used for training, and the others are used for testing. Accelerometers are fixed on the outer race of the bearings and vibration signals are captured. The sampling frequency is 25.6 kHz. Each sample contains 2560 points, that is, 0.1 s, and sampling is repeated every 10 s. Vibration signals of a tested bearing are shown in Fig. 6.33.

Original features listed in Table 6.9 are extracted from vibration signals of the tested bearings, including 10 time–domain features (Lei and Zuo, 2009), 16 time–frequency–domain features and 2 features based on trigonometric functions (Javed et al., 2015). The 16 time–frequency–domain features are energies and energy ratios of 8 frequency bands generated by performing 3-level wavelet packet decomposition on vibration signals. The 2 features based on trigonometric functions are calculated by Eq. (6.114), which

Table 6.8 Operation Conditions of Tested Bearings

Operation Conditions	Load (N)	Speed (rpm)	Training Dataset	Testing Dataset
Condition 1	4000	1800	Bearing 1_1 Bearing 1_2	Bearing 1_3 Bearing 1_4 Bearing 1_5 Bearing 1_6 Bearing 1_7
Condition 2	4200	1650	Bearing 2_1 Bearing 2_2	Bearing 2_3 Bearing 2_4 Bearing 2_5 Bearing 2_6 Bearing 2_7
Condition 3	5000	1500	Bearing 3_1 Bearing 3_2	Bearing 3_3

are the standard deviation of inverse hyperbolic cosine (SD of IHC) and the standard deviation of inverse hyperbolic sine (SD of IHS), respectively.

$$\begin{cases} \text{SD of IHC} = \sigma\left[\log\left(x_i + \sqrt{x_i^2 - 1}\right)\right] \\ \text{SD of IHS} = \sigma\left[\log\left(x_i + \sqrt{x_i^2 + 1}\right)\right] \end{cases}. \quad (6.114)$$

These 28 original features are utilized to construct the fusion health indicator WMQE. We take bearing 1_1, for example, and give the detailed information of the WMQE construction process. First, the trendability values of the 28 original features are calculated and 23 features whose trendability values are higher than 1/2 are selected from the original features. The trendability values of the 28 original features are shown in Fig. 6.34. Then, the 23 selected features are input into the correlation clustering algorithm presented in Table 6.7. The 23 selected features are classified into

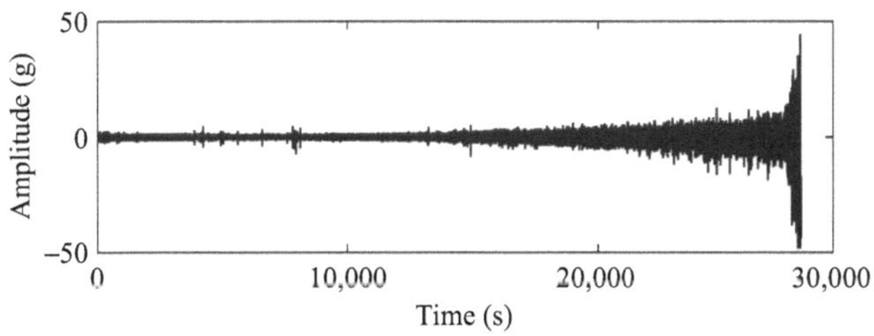

Figure 6.33. Vibration signals of a tested rolling element bearing.

Table 6.9 Original Features Extracted From Vibration Signals

Type	Feature	
Time–domain features	F1: Standard deviation	F6: Crest factor
	F2: Peak-Peak	F7: Impulse factor
	F3: Mean-absolute	F8: Skewness
	F4: Root mean square	F9: Kurtosis
	F5: Shape factor	F10: Entropy
Time–frequency–domain features	F11–F18: Energies of eight bands	
	F19–F26: Energy ratios of eight bands	
Features based on trigonometric functions	F27: SD of IHC	
	F28: SD of IHS	

four different clusters according to the variation of the PBM index with the number of clusters shown in Fig. 6.35. One typical feature is selected from each cluster and totally four typical features are selected and shown in Fig. 6.36. It is seen that these four typical features present different types of degradation trends during the whole lifetime and reveal distinct information about the degradation process. F11 does not show any degradation information until the tested bearing approaches the end of the lifetime. F18 shows abrupt fluctuation from the middle of the lifetime. F21

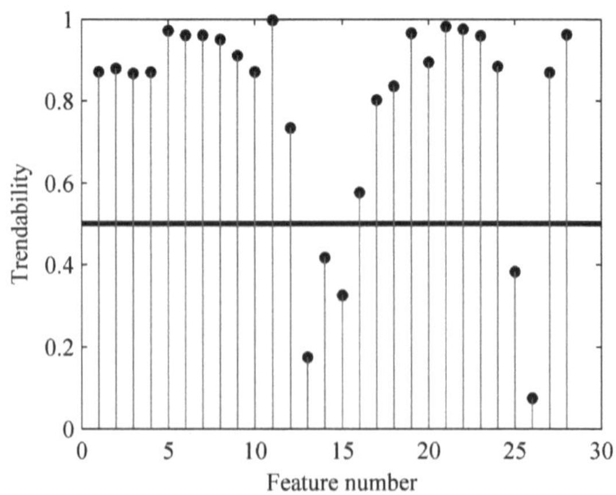

Figure 6.34. Feature selection result of Bearing 1_1.

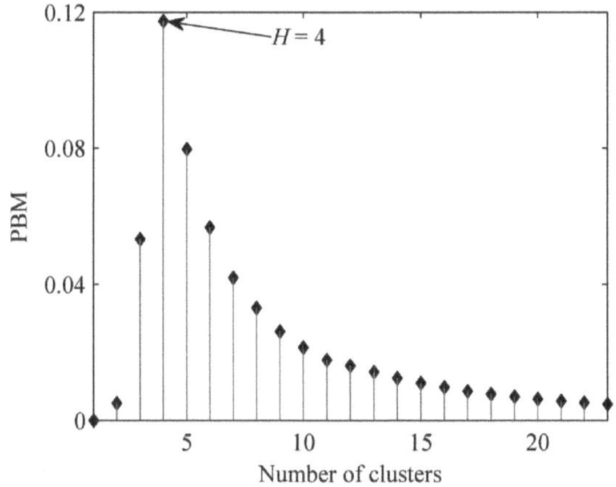

Figure 6.35. Variation of the PBM index with the number of clusters for Bearing 1_1.

decreases abruptly at the beginning but gradually changes to be stable. F22 has a more obvious degradation trend than other typical features, but it is not informative for the final failure. To incorporate the mutual information from these typical features, they are fused into WMQE. With the fusion of these typical features, WMQE shows more obvious degradation trends than each of the typical features and comprehensively describes the degradation process of bearing 1_1 during the whole lifetime, which can be seen from Fig. 6.36.

The WMQE values of the 17 tested bearings are constructed using the same method. Fig. 6.37 (A) shows the WMQE of the six

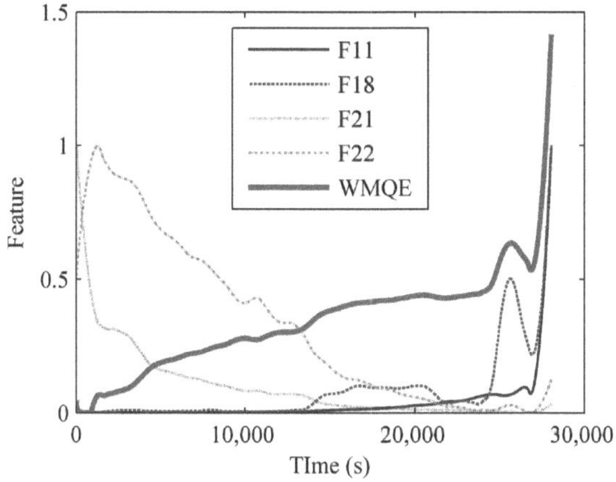

Figure 6.36. Typical features and the WMQE of Bearing 1_1.

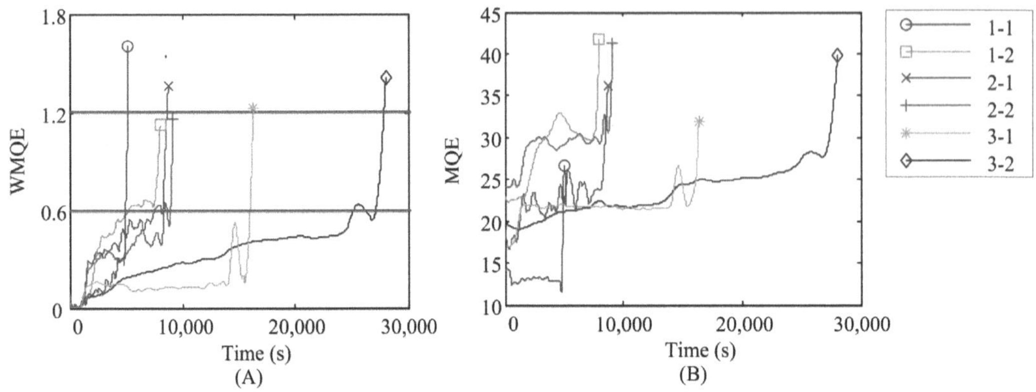

Figure 6.37. WMQE and MQE of the six training datasets. (A) WMQE and (B) MQE.

training datasets. In order to identify the benefits of the WMQE for RUL prediction of the bearings, the MQE (Qiu et al., 2003) indicators of the bearings are also constructed using the 28 original features displayed in Table 6.9, and the MQE values of the six training datasets are shown in Fig. 6.37B. It is seen that the WMQE has more obvious degradation trends than the MQE. To compare the degradation trends of the WMQE with MQE quantitatively, the trendability values of them are calculated and displayed in Table 6.10. It is seen that the trendability values of the WMQE are larger than those of the MQE for most bearings. In addition, the

Table 6.10 Trendability of WMQE and MQE for the Tested Bearings

Bearing	WMQE	MQE	Bearing	WMQE	MQE	Bearing	WMQE	MQE
1_1	0.9960	0.9863	2_1	0.9314	0.7146	3_1	0.9764	0.0577
1_2	0.9797	0.7770	2_2	0.9852	0.7466	3_2	0.5000	0.1613
1_3	0.9818	0.9976	2_3	0.1368	0.1359	3_3	0.9706	0.8250
1_4	0.9868	0.9790	2_4	0.9854	0.8651			
1_5	0.9905	0.9403	2_5	0.6764	0.8932			
1_6	0.9533	0.9190	2_6	0.9909	0.9774			
1_7	0.9895	0.8921	2_7	0.8434	0.3363			

WMQE values have similar amplitude ranges. Therefore, it is easy to determine a uniform threshold value for all of these bearings. On the contrary, the MQE values have distinct amplitude ranges among different bearings. In Fig. 6.37A the degradation process is roughly classified into two distinct stages. From 0 to 0.6, the WMQE values increase gradually corresponding to the long-time fault development process. From 0.6 to 1.2, the WMQE values show rapid increasing trends, which implies that the bearings run to failure quickly once the faults get severe. To prevent the happening of accidents, the bearings should be shut down before the faults get severe. Therefore, the threshold value of the bearings is set to be 0.6 for RUL prediction.

The WMQE values are input into the Paris–Erdogan model in Eq. (6.99) and the model parameters are initialized using the MLE algorithm. Then the RUL values of the 11 testing datasets are predicted using the PF-based prediction algorithm. The particle number is set to be 1000. Taking bearing 1_3, for example, the parameter initialization results are μ_α = 8.2325e-4, σ_α^2 = 5.2877e-7 β = 0.7840 and σ_v^2 = 3.6655e-6. The updating process of the parameter α is shown in Fig. 6.38. As time goes on, the distribution of the parameter α gradually converges to a narrow range. From about t_k = 7000 s, α converges to a constant value 1.0116e-3.

The degradation trend of bearing 1_3 is predicted and displayed in Fig. 6.39A. It is seen that, the prediction result is close to the actual degradation trend of the WMQE curve. Fig. 6.39B shows the distribution of the predicted RUL of 1000 particles. Each particle represents a possible degradation process in the future. It is seen that the predicted RUL values of the particles are

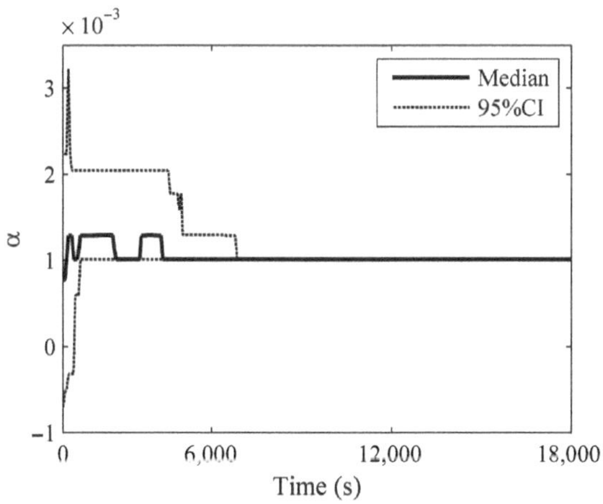

Figure 6.38. Updating process of parameter α for bearing 1_3.

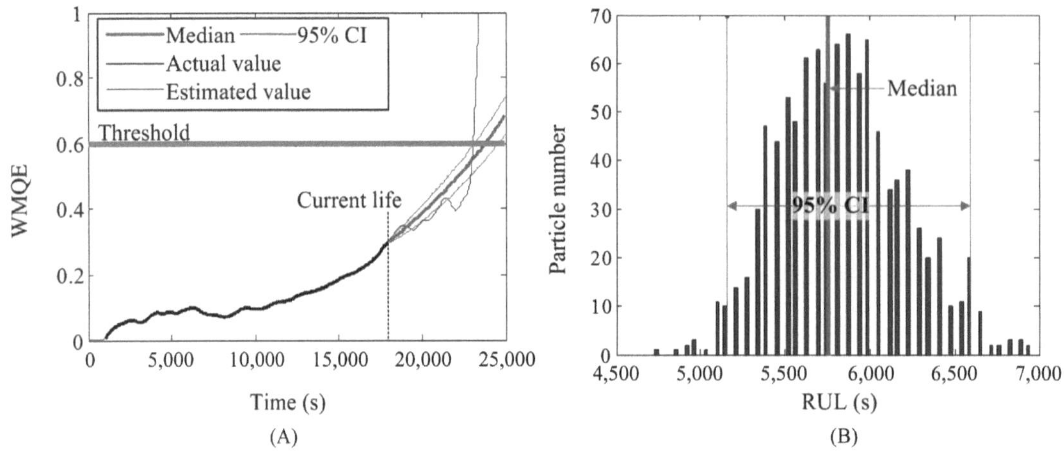

Figure 6.39. RUL prediction result of bearing 1_3. (A) Degradation trend prediction and (B) distribution of the predicted RUL.

approximately normally distributed. To evaluate the uncertainty associated with the estimates of the RUL, the median and the 95% confidence interval (CI) of the predicted RUL are calculated and given in Fig. 6.39B. The median and 95% CI are 5750 s and [5150 s, 6590 s], respectively. For convenient comparison with the results of other methods, the median of the distribution is calculated as the predicted RUL of the bearing. The predicted RUL values of the 11 testing datasets are displayed in Table 6.11. Based on the predicted RUL, the percent errors of prediction results are calculated by (Nectoux et al., 2012):

$$Er_i = \frac{ActRUL_i - RUL_i}{ActRUL_i} \times 100\%, \tag{6.115}$$

where $ActRUL_i$ and RUL_i are the actual RUL and the predicted RUL of the ith testing dataset, respectively.

The percent errors of the proposed method are displayed in the column of Er 1 in Table 6.11. To identify the effectiveness of the proposed method in the RUL prediction of rolling element bearings, the prediction results are compared with those of two similar studies which use the same bearing datasets (Nectoux et al., 2012). The first study was presented by Hong et al. (2014). The second one is the winner of the IEEE PHM 2012 Prognostic challenge of the academic teams (Sutrisno et al., 2012). The percent errors of the prediction results in the two studies are displayed in the column of Er 2 and Er 3, respectively. The score in the IEEE PHM 2012

Table 6.11 RUL Prediction Results of the Testing Datasets

Bearing Case	Current Life (s)	Actual RUL (s)	Predicted RUL (s)	Er 1 (%)	Er 2 (%)	Er 3 (%)
1_3	18010	5730	5750	−0.35	−1.04	37
1_4	11380	339	320	5.60	−20.94	80
1_5	23010	1610	0	100	−278.26	9
1_6	23010	1460	1050	28.08	19.18	−5
1_7	15010	7570	9050	−19.55	−7.13	−2
2_3	12010	7530	9050	−20.19	10.49	64
2_4	6110	1390	1270	8.63	51.80	10
2_5	20010	3090	2370	23.30	28.80	−440
2_6	5710	1290	530	58.91	−20.93	49
2_7	1710	580	550	5.17	44.83	−317
3_3	3510	820	490	40.24	−3.66	90
Score				0.4302	0.3550	0.3066

Prognostic challenge is used to measure the performance of the three methods, which is formulated as (Nectoux et al., 2012):

$$Score = \frac{1}{11}\sum_{i=1}^{11} A_i, \tag{6.116}$$

where

$$A_i = \begin{cases} \exp\left(-\ln(0.5)\cdot\left(Er_i/5\right)\right) & \text{if } Er_i \le 0 \\ \exp\left(+\ln(0.5)\cdot\left(Er_i/20\right)\right) & \text{if } Er_i > 0 \end{cases}. \tag{6.117}$$

The scores of the three methods are given in the last row of Table 6.11. We can see that the score of the proposed method is the highest among them, which means that the proposed method performs better than the other two methods for the RUL prediction of the tested bearings.

6.3.5 Epilog

This section addresses the model-based RUL prediction methods. At first, a comprehensive introduction of the model-based methods is given including the basic theories and the categories of the commonly used models. Then two stochastic process model-based methods and a physical model-based method are described in the following three subsections, respectively. In these subsections, we first introduce the existing methods and discuss some shortcomings of them. To deal with the shortcomings of these methods, some improved methods are developed from them. Finally, accelerated degradation tests of rotating machinery are employed to demonstrate the benefits of the improved methods compared with the original ones. The results verify the effectiveness of the improvement of these methods in the RUL prediction of rotating machinery.

6.4 Conclusions

In this chapter, we focus on the major process in PHM of rotating machinery, that is, prognostics, and summarize recent research and development in RUL prediction of rotating machinery. Basic concepts, various techniques and algorithms are introduced and some advanced data-driven methods and model-based methods are illustrated. However, it should be noticed that most of the existing prediction studies are conducted in research laboratories, where certain practical considerations are easily neglected and the prediction process is simplified. The degeneracy of rotating machinery in real situation is generally complicated and influenced by complex operation environment. The ultimate goal of the RUL prediction is to establish reliable CBM/PHM systems and to offer economic and social benefits. To fulfill the goal, there are still many challenges and problems in the field of the RUL prediction of rotating machinery. In other words, it also provides opportunities for researchers to further their research study. We provide the prospects of the RUL prediction of rotating machinery as follows.

1. *Data capture and utilization*
   Capturing degradation data is the first step and the fundament in CBM/PHM strategies. The degradation data is classified into the event data and real-time data. Both of them are hard to be captured due to the limitation of the data acquisition equipment and complex operating environment in industrial applications. The lack of effective capture and management of degradation data places a huge constraint on the research and development of the RUL prediction of rotating machinery. To collect accurate information from the operating rotating machinery, sensors are supposed to be embedded into the structure of rotating machinery. Advanced sensor techniques, such as wireless sensor networks, should be developed for robust on-line data

acquisition. The degradation data information would be valuable for model building and model validation as well.

Common datasets consisting of relevant degradation data, maintenance event records and cost data may be available to all researchers in the field. In this way, the technical and economic feasibility of each RUL prediction method and maintenance strategy can be evaluated and compared consistently. A uniform set of benchmark criteria are also needed for convenient methods comparison. This practice will significantly accelerate the advancement of the RUL prediction research.

Since the degradation data is difficult to be acquired, it appears to be more significant to make full use of the limited available data information. Researchers are suggested to pay more attention to the following aspects. (1) How to properly utilize the incomplete trending data? (2) How to integrate the event data and real-time data? (3) How to fuse the multidimensional degradation data? Typically we can see in practice that multichannel CM data may be available, which may be correlated and not on the same scale and measurement units. This will also impose a severe challenge to threshold-based models which are mostly established under a single threshold level. In this case how to effectively use all available information to estimate the underlying RUL is a question not being well answered.

2. *Health indicator extraction*

Failure interaction between components and the probabilistic natures of fault symptoms and growth, as well as varying operating conditions, add to the complexity of the collected degradation data. There exists a nonlinear relationship between the degradation data and the health condition of rotating machinery. The direct method to reveal the relationship between them is to extract a health indicator from the degradation data.

A suitable indicator which may simplify modeling and produce accurate prediction results generally have obvious monotonic trends and properly correlate to the degradation of machinery. Some researchers try to use one-dimensional features which are directly extracted from the measured data as the health indicators. But each one-dimensional feature is only sensitive to a certain fault in a certain degradation stage. A suitable health indicator should take advantage of mutual information from multidimensional features for degradation assessment and RUL prediction. Therefore, it is necessary to fuse multidimensional features and construct a fusion health indicator. Although such a health indicator is constructed, it is also difficult to evaluate its effectiveness for the RUL prediction, and whether it really reveals the degradation trend of the rotating machinery.

3. *Prediction model development*
   It is important to develop a RUL prediction model which is supposed to describe the degradation trend of rotating machinery. However, this task appears to be difficult in the situation where very few or even no data is available. This is typical for newly designed equipment where no observed degradation data exist. In this case, the quantity and completeness of data is insufficient to construct a statistic model or stochastic process model. In this situation, a physical model is necessary, which is developed based on the subjective expert knowledge from design and manufacturing. However, how to establish a physical model and how to take advantage of subjective judgmental information in RUL estimation remain a challenge. Up to now, little research considers the effect of subjective information in RUL estimation.
   Another challenge in the development of prediction models is how to deal with multiple failure modes for a single component, which is a common scenario observed in real practice. Obviously this will further complicate the issue since for some RUL prediction techniques, the measured degradation data is heavily influenced by the underlying failure modes, and therefore the RUL is dependent on failure mode. In addition, a component may suffer from various failure modes during the whole lifetime resulting in the change degradation trends. Besides, a single prediction model is only efficient for a certain degradation trend. To deal with this phenomenon, the degradation process during the whole lifetime should be divided into different stages according to the variation of failure modes. A specified model is supposed to be selected for each stage, or a multistage model is necessary with model parameters adjusted adaptively in different degradation stages.
4. *Influence from external environmental variables*
   The forth challenge is how to model the influence of external environmental variables, such as the speed, the loading and the effects from maintenance actions on RUL prediction. Actually, most prediction models are based on the assumption that the external operation condition of the rotating machinery is stable. However, in real applications, external environment variables may change the degradation speed or failure modes and therefore influence the effectiveness of the prediction model. It is a complicated issue to develop a prediction model considering the influence from external environment variables, since those variables will impact on the observed degradation data and the observations will in turn influence the RUL prediction. If it is not done properly, overfitting will occur, which may reduce the robustness of the developed model.

5. *From theoretical formulations to real applications*
It should also be noted that many prediction models in literature are simply theoretical formulations or just demonstrated using accelerated degradation data captured in experimental systems, which is different from real applications. Therefore, much more work should be done to speed up the progress of engineering-oriented studies.
With the rapid development of the microelectro-mechanical systems technology, a future trend of RUL prediction research and developments would be the design of intelligent device which has the capability of continuously monitoring its own health state using on-line data acquisition, signal processing, diagnostic and prognostic tools, and finally making an optimal maintenance strategy as suggestion for operators. Many facts should be considered in designing such a sophisticated CBM/PHM system. (1) We should investigate the requirements from consumers and the practicability of the work. To realize the greatest economic and social benefits, it is important to design every aspect of a system by considering the asset manager's perspective. (2) The designed system should be easily updated to coincide with the frequent change of design, technologies, business policies and management executives. (3) A uniform performance measurement framework and efficient validation module should be incorporated into the system. (4) One of the most overlooked measures in the development of the CBM/PHM system is the degree of user-friendliness. The users often do not have in-depth knowledge about the underling system physics and model algorithms, but these applications should be convenient to be operated by the users. (5) Collaborations between research groups from different areas of expertise should be encouraged.

References

Akaike, H., 1974. A new look at the statistical model identification. IEEE Trans. Automatic Control 19, 716–723.
An, D., Choi, J.H., Kim, N.H., 2013. Prognostics 101: A tutorial for particle filter-based prognostics algorithm using Matlab. Reliab. Eng. Syst. Saf. 115, 161–169.
Arulampalam, M.S., Maskell, S., Gordon, N., Clapp, T., 2002. A tutorial on particle filters for online nonlinear/non-Gaussian Bayesian tracking. IEEE Trans. Signal Process. 50, 174–188.
Baraldi, P., Cadini, F., Mangili, F., Zio, E., 2013. Model-based and data-driven prognostics under different available information. Prob. Eng. Mech. 32, 66–79.
Camci, F., Medjaher, K., Zerhounib, N., Nectouxb, P., 2013. Feature evaluation for effective bearing prognostics. Qual. Reliab. Eng. Int. 29, 477–486.
Chen, C., Zhang, B., Vachtsevanos, G., Orchard, M., 2011. Machine condition prediction based on adaptive neuro-fuzzy and high-order particle filtering. IEEE Trans. Ind. Electron. 58, 4353–4364.

Chen, C., Vachtsevanos, G., Orchard, M.E., 2012. Machine remaining useful life prediction: An integrated adaptive neuro-fuzzy and high-order particle filtering approach. Mech.l Syst. Signal Process. 28, 597–607.

Cox, D.R., 1972. Regression models and life-tables. J. Royal Stat. Soc. 34, 187–220.

Doucet, A., De Freitas, N., Gorsdon, N., 2001a. An introduction to sequential Monte Carlo methods. Sequential Monte Carlo Methods in Practice. Springer, New York.

Doucet, A., Freitas, N.D., Gordon, N., 2001b. An introduction to sequential Monte Carlo methods. Sequential Monte Carlo Methods in Practice. Springer, New York.

Elwany, A.H., Gebraeel, N.Z., 2008. Sensor-driven prognostic models for equipment replacement and spare parts inventory. IIE Trans. 40, 629–639.

Gebraeel, N., 2006. Sensory-updated residual life distributions for components with exponential degradation patterns. IEEE Transactions on Automation Science and Engineering 3, 382–393.

Gebraeel, N., Lawley, M., Liu, R., Parmeshwaran, V., 2004. Residual life predictions from vibration-based degradation signals: a neural network approach. IEEE Trans. Ind. Electron. 51, 694–700.

Gebraeel, N.Z., Lawley, M.A., Li, R., Ryan, J.K., 2005. Residual-life distributions from component degradation signals: a Bayesian approach. IIE Trans. 37, 543–557.

Ginart, A., Barlas, I., Goldin, J., Dorrity, J.L., 2006. Automated feature selection for embeddable prognostic and health monitoring (PHM) architectures. IEEE Autotestcon, 18–21. Anaheim, CA. IEEE, 195–201.

Heng, A., Zhang, S., Tan, A.C.C., Mathew, J., 2009. Rotating machinery prognostics: state of the art, challenges and opportunities. Mech. Syst. Signal Process. 23, 724–739.

Hong, S., Zhou, Z., Zio, E., Hong, K., 2014. Condition assessment for the performance degradation of bearing based on a combinatorial feature extraction method. Digital Signal Process. 27, 159–166.

Hosmer, D.W., Lemeshow, S., 1989. Applied Logistic Regression. John Wiley & Sons, New York.

Huang, R., Xi, L., Li, X., Richard Liu, C., Qiu, H., Lee, J., 2007. Residual life predictions for ball bearings based on self-organizing map and back propagation neural network methods. Mech. Syst. Signal Process. 21, 193–207.

Condition Monitoring and Diagnostics of Machines-Prognostics-Part1: General Guidelines: International Standards Organization, 2004. Directed by ISO13381-1.

Jardine, A.K.S., Lin, D.M., Banjevic, D., 2006. A review on machinery diagnostics and prognostics implementing condition-based maintenance. Mech. Syst. Signal Process. 20, 1483–1510.

Javed, K., Gouriveau, R., Zerhouni, N., Nectoux, P., 2015. Enabling health monitoring approach based on vibration data for accurate prognostics. IEEE Trans. Ind. Electron., 62 (1), 647–656.

Lei, Y., Zuo, M.J., 2009. Gear crack level identification based on weighted K nearest neighbor classification algorithm. Mech. Syst. Signal Process. 23, 1535–1547.

Lei, Y., He, Z., Zi, Y., Hu, Q., 2007. Fault diagnosis of rotating machinery based on multiple ANFIS combination with GAS. Mech. Syst. Signal Process. 21, 2280–2294.

Lei, Y., Li, N., Gontarz, S., Lin, J., Radkowski, S., Dybala, J., 2016. A model-based method for remaining useful life prediction of machinery. IEEE Trans. Ind. Electron., 1314–1326.

Li, R., He, H., 2012. Rotational machine health monitoring and fault detection using EMD-based acoustic emission feature quantification. IEEE Trans. Instrum. Meas. 61, 990–1001.

Li, N.P., Lei, Y.G., Liu, Z.Y., Lin, J., 2014. A particle filtering-based approach for remaining useful life predication of rolling element bearings. International

Conference on Prognostics and Health Management, June 22–25. Cheney, WA, United states. IEEE, 1–8.
Li, N.P., Lei, Y.G., Lin, J., Ding, S.X., 2015. An improved exponential model for predicting remaining useful life of rolling element bearings. IEEE Trans. Ind. Electron. 62, 7762–7773.
Liu, J.S., Chen, R., 1998. Sequential monte carlo methods for dynamic systems. J. Am. Stat. Assoc. 93, 1032–1044.
Liu, J., Wang, W., Golnaraghi, F., 2009. A multi-step predictor with a variable input pattern for system state forecasting. Mech. Syst. Signal Process. 23, 1586–1599.
Liu, J., Wang, W., Ma, F., Yang, Y.B., Yang, C.S., 2012. A data-model-fusion prognostic framework for dynamic system state forecasting. Eng. Appl. Artif. Intell. 25, 814–823.
Liu, J., Seraoui, R., Vitelli, V., Zio, E., 2013. Nuclear power plant components condition monitoring by probabilistic support vector machine. Ann. Nucl. Energy 56, 23–33.
Nectoux, P., Gouriveau, R., Medjaher, K., Ramasso, E., Chebel-Morello, B., Zerhouni, N., Varnier, C., 2012. PRONOSTIA: an experimental platform for bearings accelerated degradation tests. IEEE International Conference on Prognostics and Health Management, June 18–21. Colorado, United States. IEEE, 1–8.
Oppenheimer, C.H., Loparo, K.A., 2002. Physically based diagnosis and prognosis of cracked rotor shafts. Component and Systems Diagnostics, Prognostics, and Health Management II, July 15. Bellingham. SPIE, 122–132.
Orchard, M.E., Hevia-Koch, P., Zhang, B., Tang, L., 2013. Risk measures for particle-filtering-based state-of-charge prognosis in lithium-ion batteries. IEEE Trans. Ind. Electron. 60, 5260–5269.
Orsagh, R., Roemer, M., Sheldon, J., Klenke, C.J., 2004 A comprehensive prognostics approach for predicting gas turbine enging bearing life. Proceedings of ASME Turbo Expo 2004, June 14–17. Vienna, Austria. ASME, 777–785.
Pakhira, M.K., Bandyopadhyay, S., Maulik, U., 2004. Validity index for crisp and fuzzy clusters. Pattern Recognit. 37, 487–501.
Pandey, M.D., Yuan, X.X., Noortwijk, J.M.V., 2009. The influence of temporal uncertainty of deterioration on life-cycle management of structures. Structure Infrastructure Eng. 5, 145–156.
Pang, C.K., Zhou, J.H., Yan, H.C., 2014. PDF and breakdown time prediction for unobservable wear using enhanced particle filters in precognitive maintenance. IEEE Trans. Instrum. Meas. 64, 649–659.
Paris, P.C., Erdogan, F., 1963. A critical analysis of crack propagation laws. J. Fluids Eng. 85, 528–533.
Park, J.I., Bae, S.J., 2010. Direct prediction methods on lifetime distribution of organic light-emitting diodes from accelerated degradation tests. IEEE Trans. Reliab. 59, 74–90.
Park, C., Padgett, W.J., 2005. Accelerated degradation models for failure based on geometric Brownian motion and gamma processes. Lifetime Data Anal. 11, 511–527.
Peng, C.-Y., Tseng, S.-T., 2009. Mis-specification analysis of linear degradation models. IEEE Trans. Reliab. 58, 444–455.
Qian, Y.N., Yan, R.Q., 2015. Remaining useful life prediction of rolling bearings using an enhanced particle filter. IEEE Trans. Instrum. Meas. 64, 2696–2707.
Qian, Y., Hu, S., Yan, R., 2013. Bearing performance degradation evaluation using recurrence quantification analysis and auto-regression model. Instrumentation and Measurement Technology Conference (I2MTC), 2013 IEEE International, 6–9 May. Minneapolis, MN. IEEE, 1713–1716.

Qiu, H., Lee, J., Lin, J., Yu, G., 2003. Robust performance degradation assessment methods for enhanced rolling element bearing prognostics. Adv. Eng. Inform. 17, 127–140.

Rausand, M., Hoyland, A., 2004. System Reliability Theory: Models, Statistical Methods, and Applications. Wiley-Interscience, Hoboken, New Jersey.

Saxena, A., Celaya, J., Saha, B., Saha, S., Goebel, K., 2010. Metrics for offline evaluation of prognostic performance. Int. J. Progn. Health Manag. 1, 1–20.

Si, X.S., 2015. An adaptive prognostic approach via nonlinear degradation modeling: application to battery data. IEEE Trans. Ind. Electron. 62, 5082–5095.

Si, X.S., Wang, W., Hu, C.H., Zhou, D.H., 2011. Remaining useful life estimation—a review on the statistical data driven approaches. Eur. J. Oper. Res. 213, 1–14.

Si, X.S., Wang, W., Chen, M.Y., Hu, C.H., Zhou, D.H., 2013a. A degradation path-dependent approach for remaining useful life estimation with an exact and closed-form solution. Eur. J. Oper. Res. 226, 53–66.

Si, X.S., Wang, W.B., Hu, C.H., Chen, M.Y., Zhou, D.H., 2013b. A Wiener-process-based degradation model with a recursive filter algorithm for remaining useful life estimation. Mech. Syst. Signal Process. 35, 219–237.

Si, X.S., Wang, W.B., Hu, C.H., Zhou, D.H., 2014. Estimating remaining useful life with three-source variability in degradation modeling. IEEE Trans. Reliab. 63, 167–190.

Sikorska, J.Z., Hodkiewicz, M., Ma, L., 2011. Prognostic modelling options for remaining useful life estimation by industry. Mech. Syst. Signal Process. 25, 1803–1836.

Smits, G.F., Jordaan, E.M., 2002. Improved SVM regression using mixtures of kernels. Proceedings of the 2002 International Joint Conference on Neural Networks. IEEE.

Spearman, C., 1904. The Proof and Measurement of Association between Two Things. Am. J. Psychol. 15, 72–101.

Sutrisno, E., Oh, H., Vasan, A.S. S., Pecht, M., 2012. Estimation of remaining useful life of ball bearings using data driven methodologies. Prognostics and Health Management (PHM), IEEE Conference, 2012. IEEE, 1–7.

Tran, V.T., Yang, B.-S., 2012. An intelligent condition-based maintenance platform for rotating machinery. Expert Syst. Appl. 39, 2977–2988.

Ye, Z.S., 2013. On the conditional increments of degradation processes. Stat. Probab. Lett. 83, 2531–2536.

Ye, Z.S., Wang, Y., Tsui, K.-L., Pecht, M., 2013. Degradation data analysis using Wiener processes with measurement errors. IEEE Trans. Reliab. 62, 772–780.

You, M.-Y., Liu, F., Wang, W., Meng, G., 2010. Statistically planned and individually improved predictive maintenance management for continuously monitored degrading systems. IEEE Trans. Reliab. 59, 744–753.

Yu, J.B., 2012. Health condition monitoring of machines based on hidden markov model and contribution analysis. IEEE Trans. Instrum. Meas. 61, 2200–2211.

Zio, E., Di Maio, F., 2012. Fatigue crack growth estimation by relevance vector machine. Expert Syst. Appl. 39, 10681–10692.

GLOSSARY

AACO	Accumulative amplitudes of carrier orders
AIC	Akaike information criterion
ANFIS	Adaptive neuro-fuzzy inference systems
ANN	Artificial neural network
APE	Absolute percentage errors
AR	Autoregressive
ARMA	Autoregressive moving average
BM	Brownian motion
BMU	Best matching unit
BP	Back propagation algorithm
BPNN	Back propagation neural network
CBM	Condition-based maintenance
CDET	Compensation distance evaluation technique
CI	Confidence interval
CRA	Cumulative relative accuracy
DAQ	Data acquisition
DNN	Deep neural network
EDM	Electrical discharge machining
EEMD	Ensemble empirical mode decomposition
EM	Expectation maximization
EMD	Empirical mode decomposition
EOP	Energy operator
ER	Energy ratio
ERDS	Energy ratio based on difference spectra
ERM	Empirical risk minimization
FC	Frequency center
FCM	Fuzzy c-means
FFT	Fast Fourier transform
FIR	Finite impulse response
FPT	First predicting time
FT	Fourier transform
GA	Genetic algorithms
HHT	Hilbert–Huang transform
HL	Hidden layer

HP	High-pass
HT	Hilbert transform
IIR	Infinite impulse response
IMFs	Intrinsic mode functions
KF	Kalman filtering
***K*NN**	*K* nearest neighbor
KPF	Kalman particle filtering
MAPE	Mean of absolute percentage errors
MF	Mean frequency
MLE	Maximum likelihood estimation
MLP	Multilayer perceptron
MPE	Multiscale permutation entropy
MQE	Minimum quantization error
MRVM	Multiclass relevance vector machine
NN	Neural network
OAA	One-against-all
OAO	One-against-one
OM	Operation and maintenance
PCA	Principal component analysis
PCs	Principal components
PDF	Probability density function
PE	Permutation entropy
PF	Particle filtering
PHM	Prognostics and health management
PWD	Pseudo Wigner–Ville distribution
RBF	Radial basis function
RMS	Root mean square
RMSE	Root mean square of errors
RMSF	Root mean square frequency
RUL	Remaining useful life
RVM	Relevance vector machine
SD of IHC	Standard deviation of inverse hyperbolic cosine
SD of IHS	Standard deviation of inverse hyperbolic sine
SIS	Sequential importance sampling
SLT	Statistical learning theory
SNR	Signal-to-noise ratio
SOM	Self organizing map
SPRI	Spectrum peak ratio of bearing inner race
SPRO	Spectrum peak ratio of bearing outer race
SPRR	Spectrum peak ratio of bearing roller
SRM	Structural risk minimization
STFT	Short-time Fourier transform
STDF	Standard deviation frequency
SVM	Support vector machine
TFSWT	Two-stage feature selection and weighting technique

TSA	Time synchronous average
VC	Vapnik–Chervonenkis
WFCM	Weighted fuzzy c-means
W*K*NN	Weighted *K* nearest neighbor classifier
WMQE	Weighted minimum quantization error
WNN	Wavelet neural network
WPT	Wavelet packet transform
WPTLS	Wavelet packet transform based on lifting scheme
WT	Wavelet transform
WVD	Wigner–Ville distribution
WWEA	World wind energy association

INDEX

Kolibri 360 GmbH
Ferdinand-Jühlke-Straße 7
99095 Erfurt, Deutschland
produktsicherheit@kolibri360.de